1 YEAR UPGRADE

BUYER PROTECTION PLAN

Bluetooth

APPLICATION DEVELOPER'S GUIDE:

The Short Range Interconnect Solution

David Kammer

Gordon McNutt

Brian Senese

Jennifer Bray Technical Editor

KEY	SERIAL NUMBER
001	D8LDE945T5
002	AKLRTGY7M4
003	2XW4L3N54N
004	SGBBT639UN
005	8LU8CA2H7H
006	7KG4RN5TM4
007	BW2QV7R46T
008	JPF5R565MR
009	83N5M77UBS
010	GT6YH2XZ52

PUBLISHED BY
Syngress Publishing, Inc.
800 Hingham Street
Rockland, MA 02370

Bluetooth Application Developer's Guide: The Short Range Interconnect Solution

Printed in the United States of America

1 2 3 4 5 6 7 8 9 0

ISBN: 1-928994-42-3

Technical Editor: Jennifer Bray
Co-Publisher: Richard Kristof
Acquisitions Editor: Catherine B. Nolan
Developmental Editor: Kate Glennon

Cover Designer: Michael Kavish
Page Layout and Art by: Reuben Kantor
Copy Editor: Michael McGee
Indexer: Robert Saigh

Distributed by Publishers Group West in the United States and Jaguar Book Group in Canada.

Acknowledgments

We would like to acknowledge the following people for their kindness and support in making this book possible.

Richard Kristof and Duncan Anderson of Global Knowledge, for their generous access to the IT industry's best courses, instructors, and training facilities.

Ralph Troupe, Rhonda St. John, and the team at Callisma for their invaluable insight into the challenges of designing, deploying and supporting world-class enterprise networks.

Karen Cross, Lance Tilford, Meaghan Cunningham, Kim Wylie, Harry Kirchner, Kevin Votel, Kent Anderson, and Frida Yara of Publishers Group West for sharing their incredible marketing experience and expertise.

Mary Ging, Caroline Hird, Simon Beale, Caroline Wheeler, Victoria Fuller, Jonathan Bunkell, and Klaus Beran of Harcourt International for making certain that our vision remains worldwide in scope.

Annabel Dent of Harcourt Australia for all her help.

David Buckland, Wendi Wong, Marie Chieng, Lucy Chong, Leslie Lim, Audrey Gan, and Joseph Chan of Transquest Publishers for the enthusiasm with which they receive our books.

Kwon Sung June at Acorn Publishing for his support.

Ethan Atkin at Cranbury International for his help in expanding the Syngress program.

Jackie Gross, Gayle Vocey, Alexia Penny, Anik Robitaille, Craig Siddall, Darlene Morrow, Iolanda Miller, Jane Mackay, and Marie Skelly at Jackie Gross & Associates for all their help and enthusiasm representing our product in Canada.

Lois Fraser, Connie McMenemy, and the rest of the great folks at Jaguar Book Group for their help with distribution of Syngress books in Canada.

Contributors

David Kammer has been involved with the handheld industry since 1997. David is currently the Technical Lead for Bluetooth technologies at Palm Inc., and is one of the authors of the original Bluetooth specification. Before working on Bluetooth, David worked on IR technology, and on the Palm VII. In addition to his work at Palm, he also consults for several companies, including In2M and Microsoft, in the field of wireless communications and PalmOS programming. David has spoken at a number of events, including The Bluetooth Developers Conference, The Bluetooth World Congress, and PalmSource, and has been interviewed about Bluetooth for the New York Times. David holds a B.A. from Oberlin College in Computer Science, and currently lives in Seattle. David would like to thank his folks for the education, Meredith Krieble and Sebastian for a nice space to work in, the excellent folks of the Palm Bluetooth Team, and Vanessa Pepoy for her understanding and patience.

Tracy Hopkins is an Applications Engineering Manager at Cambridge Silicon Radio (CSR). She and her group offer consultancy application services on all aspects of integrating Bluetooth into customer's products from initial conception through to production. She has a 2:1 BSc degree with honors in Electronic Engineering and after completing a 6-year apprenticeship with Phillips Telecommunications has worked in numerous engineering disciplines designing hardware for Satellite communications, production engineering at Studio Audio and Video (SADiE) and managed the international post-production technical support for broadcast giant Snell and Wilcox. She has written and presented many technical papers for both the communications and broadcast TV industries including the SMPTE technical conference and designs all of CSR's technical training seminars.

Brian P. Senese has directly participated in the development of state of the art wireless communications networks and associated components for

15 years. He has worked for Nortel, Uniden, ADC Telecommunications, and other aggressive technology companies and has held positions from designer to senior engineering manager. Currently, as an Applications Engineer for Extended Systems Inc., he gives seminars, is a regular speaker at conferences, and has published several articles on Bluetooth technology and its practical application in realizing products. He has spoken extensively on a wide variety of technical topics, is internationally published, and has another book entitled *Successful High Tech Product Introduction*. He holds an M.E.Sc. and B.E.Sc. in Electrical Engineering from the University of Western Ontario, London, Ontario, Canada.

Radina (Jiny) Bradshaw graduated with a first in Computer Science from Kings College, Cambridge University. She received her Ph.D. in the Laboratory for Communications Engineering, also in Cambridge, with Professor Andy Hopper, investigating power efficient routing in radio peer networks. She is currently a Software Engineer at Cambridge Silicon Radio (CSR).

David McCall graduated from Edinburgh University with an MEng in Electronics. He worked for Visteon, designing circuitry for car stereos, before joining Cambridge Silicon Radio (CSR) in July of 2000. As a Senior Applications Engineer he is responsible for helping CSR's customers with all aspects of their Bluetooth product design RF, hardware and software, from concept through production.

Wajih A. Elsallal received his B.S. degree in Electrical Engineering from the King Fahd University of Petroleum and Minerals in 1998 and continued his education at Georgia Institute of Technology where he received the M.S. degree in Electrical and Computer Engineering in early 2000. Currently, he is pursuing a Ph.D. in Electrical and Computer Engineering from Georgia Institute of Technology with a minor in Public Policy. His fields of expertise include development of antenna and phased array antenna design, electromagnetic computational methods, Bluetooth wireless LAN for handheld devices, Inter-Satellite-Link networking, microstrip and packaging technologies and sidelobe cancellor algorithms for radar applications. He has held internships at Lucent Technology and 3Com Palm Computings, Inc. and is currently a co-op staff member at the Antennas and Passives Section within the Advanced Technology Center of Rockwell Collins, Inc., a graduate teaching assistant at Georgia Tech, and a research assistant for Georgia Tech Research Institute (GTRI/SEAL).

Patrick Connolly was educated at Trinity College, Dublin, where he received a Bachelors and Masters degree in Computer Science. He has been involved with the design and development of leading edge systems for over fifteen years, using such technologies as DCE, CORBA, and J2EE. Patrick is the Chief Architect at Rococo Software, where he plays a leading technical role in setting and driving product direction. His chapter in this book was co-authored by Patrick and two of his Rococo colleagues: Karl McCabe, Rococo's CTO, and Sean O'Sullivan, Rococo's CEO.

Gordon McNutt is a Kernel Developer for RidgeRun, Inc, responsible for porting Linux to embedded devices containing multiple processors. After receiving his B.S. in Computer Science from Boise State University in 1999, he spent one year at Hewlett Packard developing I/O firmware to support USB, IR, and 1284.4 for LaserJet printers.

Bill Munday is one of the founders of blueAid, which started as an organization to help those companies who could not afford the high consultancy rates for Bluetooth technology. He graduated from UMIST (Manchester, UK) in 1991 with a double degree of BSc(Hons) and MEng in Microelectronics Systems Engineering. He was sponsored by NORTEL and joined them upon graduation as a Systems Designer. He worked on first and second generation SDH and SONET transmission systems, then pioneered new time-to-market concepts while working on an innovative next-generation Voice over ATM distributed switching product. In 1997 he moved to Tality (nee Cadence, Symbionics) to start a career in wireless communications. His first project was implementing the HiperLAN 2 standard before moving on to Bluetooth. He was the first person in the

United Kingdom to have access to Bluetooth technology as he managed and created the Ericsson Bluetooth Development Kit. He quickly became an expert and continued to work on dozens of prototype Bluetooth products including Tality's own Bluetooth IP. He presented and attended all the Bluetooth seminars and Unplugfest sessions around the world. In 2001 he moved on to start blueAid and working on 3G mobile phones for a start-up company 3GLabs.

Robin Heydon is a Section Owner of HCI as a member of the Bluetooth Special Interest Group (SIG). He obtained his degree in Computer Science and worked for nine years in the computer gaming industry on multiplayer flight simulator games. Robert began working with Bluetooth technology in February 2000, specifically working on the baseband, inquiry, sniff, and hold development, and writing the USB device driver. Robin lives in Cambridge, UK.

Technical Editor and Contributor

Jennifer Bray is a consultant at Cambridge Silicon Radio (CSR), the single-chip Bluetooth company. She is currently working in the group developing software for their BlueCore family of integrated circuits (ICs). Jennifer currently holds the positions of Associate Councillor and Errata Program Manager on the Bluetooth Architectural Review Board (BARB). She has a bachelor's degree in Physics with Microcomputer Electronics, a master's degree in Satellite Communications Engineering, and a doctorate in the field of wireless communications. More recently, she gained a distinction in the Open University's Management of Technology course. Her decade of experience in communications product development includes working on Nortel and 3Com's first ATM systems, as well as wireless ATM, the first secure Ethernet repeater, ADSL to ATM gateways, FDDI, CDMA, CDMA, and Bluetooth. In addition to her communications development experience, she has worked on cutting-edge control and monitoring systems for Formula One and Indy cars, and acted as an ISO 9001 and CMM auditor advising blue-chip companies on how to improve their development and support processes. Jennifer has written and delivered technology training courses (naturally including Bluetooth), and is a frequent speaker at conferences. She co-authored with Charles Sturman *Bluetooth: Connect without Cables*.

Contents

Connecting Devices

The page scanning device's Bluetooth Device Address can be obtained in several ways:

- From an inquiry response via FHS
- From user input
- By preprogramming at manufacture

Chapter 2
Exploring the Foundations of Bluetooth 69

Relationship between SP Mode and Mandatory Page Scan Period

Scan Period Mode	$T_{mandatory\_pscan}$
P0	>20 seconds
P1	>40 seconds
P2	>60 seconds

**Using Power
Management: When and
Why Is It Necessary?**

- Consider whether your application is suitable for power-managed operation.
- Consider the constraints imposed by the application (e.g., maximum response times, characteristics of the data traffic, and so on).

Chapter 5
Service Discovery

Answers to Your Frequently Asked Questions

Q: How are services represented in SDP?

A: A service on a Bluetooth device is described in an SDP service record, which is stored in the device's "Service Discovery Database." A service record consists of service attributes, each of which describes some information about the available service.

Chapter 6
Linux Bluetooth Development   211

Security Alert

Never remove the
Bluetooth driver while the
sdp_server daemon is
using /proc/sdp_srv. If you
do so in the current
release version of the
stack (0.0.2 at the time of
this writing), you will get
a kernel panic when you
stop the daemon. Future
versions of the stack will
probably not allow you to
remove the driver while
the sdp_server daemon is
using it.

Chapter 7
Embedding Bluetooth Applications   265

Chapter 8
Using the Palm OS for Bluetooth
Applications   **317**

The Casira Development Kit

The Casira development kit provides a variety of useful interfaces:

- **SPI interface** Connects to a PC parallel port, and allows you to reconfigure the Casira using the PSTool utility.
- **Serial interface** Connects to a PC serial port.
- **USB port** Connects to a PC USB port, and supports the Bluetooth Specification's USB protocol (H2).
- **Audio I/O** An audio jack which connects to the headsets supplied with the Casira.
- **LEDs** These can be used to monitor applications running on the BlueCore chip.
- **PIO lines** Parallel Input-Output lines; useful for connecting custom hardware.

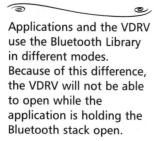

Warning

Applications and the VDRV
use the Bluetooth Library
in different modes.
Because of this difference,
the VDRV will not be able
to open while the
application is holding the
Bluetooth stack open.

Chapter 9
Designing an Audio Application   379

Chapter 10
Personal Information Base Case Study   419

Choosing a Codec

The Bluetooth specification supports three different audio coding schemes on the air interface:

- Continuous Variable Slope Delta Modulation (CVSD)
- Log Pulse Code Modulation (PCM) coding using A-law compression
- Log PCM with μ-law compression

Foreword

Every so often, a new technology comes along that, by its very nature, will change the world. The automobile, the television, and the Internet are obvious examples of technologies whose impact upon the entire population has been so far-reaching that it is truly beyond measure. Bluetooth is not one of these technologies. Despite the massive amount of media hype that has surrounded it, the effect of Bluetooth on the average person will be more like the invention of the automatic transmission than the invention of the car itself: it will make things easier for the user, but not fundamentally change the nature of the way we live and work. Simply put, for the average person, Bluetooth will probably merit a "Cool!" or a "What will they think of next?" response, but probably won't leave them stunned or slack-jawed. This is not to say that Bluetooth will be unimportant. I've invested several years working on Bluetooth, and I think it will be a valuable technology that millions of people will use, but I also think it's important to be realistic about it.

There is, however, a small group of people for whom I think Bluetooth could fundamentally change the way things are perceived, and if you are reading this introduction, in all probability you are one of those people—a software developer. Traditionally, software developers have tended to look at the communication between two devices in terms of big and small, primary or secondary (terminal and mainframe, client and server, apparatus and accessory). While these terms are certainly still relevant in some situations, Bluetooth definitely presents us with scenarios in which the lines become blurry. If two people exchange business cards between PDAs, which one is the client and which one is the server? Traditionally, both a cell phone and a printer might be considered *accessories*, but when you use Bluetooth to print an SMS message from your phone, which one is the accessory? We may still use the terms *client* and *server* to refer to certain aspects of an interaction (like who initiates the connection), but it is easy to see that many of the other ideas and assumptions associated with these terms are no longer relevant.

In the world of the Internet, the term *peer-to-peer* has come to describe applications that are decentralized—a relationship between equals. I believe this is a good way to think of the relationship between devices using Bluetooth. In the Bluetooth peer-to-peer paradigm, devices are more or less equal, dealing with data in ways that are appropriate to their nature; sending vCard data to a phone or PDA might cause the device to store the information in its address book, while sending the same vCard to a printer may cause the printer to render the data and then print it. Certainly, not all categories of Bluetooth applications will fall under the peer-to-peer paradigm. There are many good applications out there that will retain a server-client approach, but I think the realm of peer-to-peer applications that Bluetooth opens to developers will prove to be exciting and extensive.

At this point, you are hopefully saying to yourself "Great, so let's get down to the nitty-gritty; how does it work and how do I get started?" This book will take you through the most important aspects of Bluetooth technology, and offer guidance on writing Bluetooth applications for some of today's most popular operating systems. Bluetooth is still a very young technology, but the authors of these chapters are among those who have helped to see it through its infancy, and the experience they have gained should prove valuable to everyone interested in creating Bluetooth applications.

Who Should Read This Book

In general, this book is aimed at software application developers who are interested in creating Bluetooth-aware applications. Its principle goal is to provide information and examples that are pertinent to application developers. This does not mean, however, that only application developers will find benefit in reading this book. As someone who worked at integrating a Bluetooth protocol stack into an OS, I know that I would have found many of the insights in this book valuable. It is important that an OS developer understand what the world looks like from an application developer's point of view, and the insights that other OS developers have gained should certainly prove useful. In addition to developers, anyone who is evaluating a Bluetooth application for review, corporate use, or bundling may find the information in this book valuable in making an informed evaluation. For example, I know that if I were evaluating an application for enterprise use, I would want to have a good understanding of how security is handled in Bluetooth, so I could decide whether a given application met my company's security requirements.

What This Book Will Teach You

Simply put, this book will teach you what Bluetooth technology is all about, and how to write Bluetooth applications for several popular operating systems. This is a technical book, and it assumes that the reader has a solid background in application development and has a reasonable understanding of the issues involved in creating communications applications. The book is roughly divided into three sections: Bluetooth technology in general, Bluetooth applications on various operating systems, and a Bluetooth usage case study. The flow of the book is designed to introduce things to you in the most helpful order—first, supplementing your general knowledge with information about ideas and situations unique to Bluetooth, then showing you how these situations are handled in various operating systems, and finally by stimulating your imagination from looking at several real-world scenarios in which Bluetooth might be used.

It is probably worth noting a few things that this book does not cover. It is not designed to serve as a detailed investigation of the low-level particulars of the Bluetooth specification. The specification itself is publicly available, and there already exist books that do a good job providing a detailed, blow-by-blow, examination of the specification specifics. Although this is probably already clear, you should be aware that this is not a general applications programming book. If you don't already know how to write applications for Windows, this book is not going to teach you.

Further Information

By the time you finish this book, you should have all the information you need to get started writing your Bluetooth application. In fact, I wouldn't be surprised if 98 percent of all developers discover that this book will be the only Bluetooth reference they ever need. Of course, no author can anticipate every situation, so for the other 2 percent of you out there, here are some other Bluetooth references that I think are worthwhile:

- **www.bluetooth.com** Home of the Bluetooth specification. In general, I think most people will find reading the specification itself is not terribly helpful. In a good OS implementation, most of the protocols and procedures defined in the specification should be nicely abstracted. Still, sometimes you have to go straight to the source.

- **Bluetooth: Connect Without Cables** (by Jennifer Bray and Charles F. Sturman, published by Prentice Hall, 2000). If you choose to look at the Bluetooth Specification, I think you will find that this book is an excellent companion. It goes into detailed explanation, and does a good job explaining many of the oddities, ambiguities, and occasional paradoxes of the Bluetooth specification.

- **www.syngress.com** The Syngress Publishing Web site. Bluetooth technology will unquestionably evolve over time. As it does, Syngress will help you keep up by releasing updates and new publications.

I hope you enjoy the book, and have a great time creating new and exciting applications.

—David Kammer

Chapter 1

Introducing Bluetooth Applications

Solutions in this chapter:

- Why Throw Away Wires?

- Considering Product Design

- Investigating Product Performance

- Assessing Required Features

- Deciding How to Implement

☑ Summary

☑ Solutions Fast Track

☑ Frequently Asked Questions

Introduction

As human beings, we accept without question that we have the ability to communicate, that if we speak or write according to a pre-defined set of linguistic rules that we will succeed in conveying information to one another. The tools of human communication, producing sounds that are perceived as speech or creating words on a page, once learnt are used without thought. The limitation on these physical processes that we take for granted is the actual translation of thoughts into effective and meaningful statements. When it comes to electronic communication, however, there is very little that can be assumed or taken for granted. Communication between electronic devices can only be achieved when they also abide by a set of predetermined rules and standards—the Open Systems Interconnect (OSI) model for communications systems protocol stacks being the primary example, and the basis from which many others have evolved.

These standards need to be applied to every aspect of the communication process, from the manipulation of data at the highest level to the utilization of physical transmission media at the lowest. Electronic communication has evolved significantly over the last decade from the earliest packet switched data networks (PSDNs) and the Xerox, Ethernet, and IBM Token Ring local area network (LAN) technologies, to the now common-place mobile telephony and dedicated high-speed data communication. (How would we survive without e-mail and the WWW?)

New technologies are now emerging that allow wireless communication. The IEEE 802.11b or Wi-Fi standard is becoming accepted as the choice for the networking community as it supports features that enable it to perform handovers between access points, and it can effectively become a transparent wireless network, expanding the static wired network. IEEE 802.11b has a data throughput of up to 11 Mbps, which gives it viability against wired networks. This is evolving further with the advent of IEEE 802.11a and its competitor HyperLAN2 with even greater data rates. This technology is expensive and therefore not compatible with price-conscious consumer products, but we have now been provided with the means to create wireless, low-power, cost-effective, unconscious and ad-hoc connectivity between our devices. Its name: Bluetooth. If we believe all of the hype surrounding Bluetooth technology, we can expect our fridge to use our mobile phone to order groceries over the Internet, and, of course, end up ordering an extremely expensive new car instead of a steak! Yes, we have all seen the jokes, but in reality we can utilize this technology *now* to develop products that will allow us to throw away all the wires—and communicate without cables.

Excellent, we all think, and our imagination races into the realms of Science Fiction, removing the wires from everything! Musing on using our mobile phone to communicate and control everything the same way we use the TV remote to operate our entertainment systems.

This is a book for engineers in the real world, so let's take a long hard look at what Bluetooth technology really does offer. For some applications, Bluetooth technology delivers the dream of convenient wireless connectivity. For other applications, however, it just isn't the right answer. You do not want to spend a lot of time and effort learning about Bluetooth technology only to realize it isn't for you, so we are going to start out by analyzing what the features of a really good Bluetooth product are. If your application does not fit into the Bluetooth scheme of things, you can put the book down after this chapter and go and look elsewhere.

If you make it past this chapter, you can be confident Bluetooth technology is right for you. There will still be quite a few make or break pitfalls before you have a killer application, but they are minor issues compared to choosing the wrong technology.

What you need to know before reading this chapter:

- There are no pre-requisites for this chapter, though a broad familiarity with communications products will be useful.

Why Throw Away Wires?

Wired or *wireless*? Let's examine just why we'd want to connect without wires, and what it might offer us in tangible terms; we can use the paradigm of our own personal area network (PAN). We have a PC with its ubiquitous mouse and keyboard, a laptop, a personal digital assistant (PDA), a mobile phone with a "hands free" kit and a printer. How do we currently communicate between these devices? The answer is: with a rather unwieldy network of cables, hubs, and connectors—plugging, unplugging, and synchronizing often with the compulsory intervention of the overworked and often less-than-friendly IT department!

In the wired solution scenario that we are all accustomed to, all of the mobile devices are used in the *singular*—the interaction between them is always user-initiated. We generally keep our contacts' addresses in our PCs or laptops, while their phone numbers also need to be entered into our mobile phone's directory. We are effectively forced to become database managers simply in order to maintain an up-to-date record of our contact's details. We connect to our company

LAN via user-initiated password entry and connect to a printer only if we have already installed the driver or have administrator rights on our PC's—nothing is unconscious.

Figure 1.1 illustrates the alternative scenario—to Bluetooth-enable *all* of these devices. The simple act of utilizing Bluetooth technology as cable replacement removes the problem of the actual physical connections and the unconscious and ad-hoc connection capability of the technology can allow communication between the devices with *no* user intervention at all (OK, after *some* software configuration and initial device setup!).

Figure 1.1 A Bluetooth PAN (Doesn't Include Power Cables to PC and Printer)

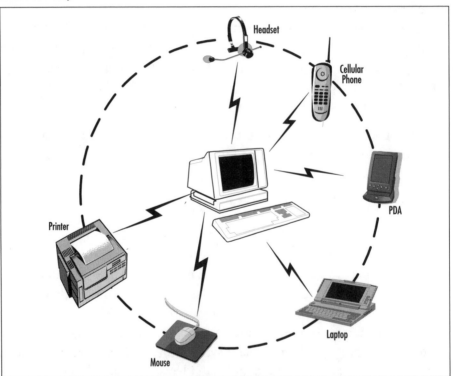

This fully wireless scenario can be achieved because of the master/slave nature of the Bluetooth technology. All devices are peers, identified by their own unique 48-bit address, and can be assigned as a *master* either by function or user intervention. A master can connect to up to seven *slaves* at the same time, forming a *piconet*—this "point-to-multipoint" feature is what sets Bluetooth apart from other wireless technologies. Figure 1.2 illustrates several connection scenarios.

Figure 1.2 Bluetooth Technology Connection Scenarios

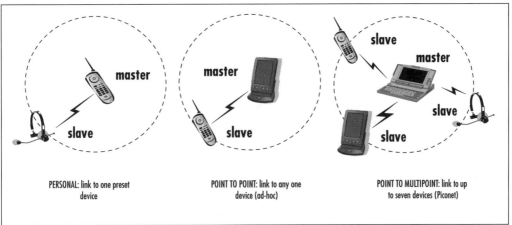

In the ultimate scenario, a member of one piconet can also belong to another piconet. Figure 1.3 illustrates the *scatternet*, wherein a slave in one piconet is also the master of a second piconet—thus extending the networking between devices. A device in my PAN can communicate with one in yours!

Figure 1.3 A Bluetooth Scatternet

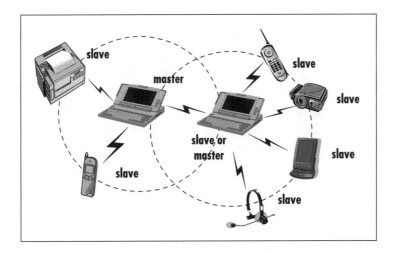

Let us put this into context by interpreting exactly what "unconscious and ad-hoc connections" can mean to us in real life, and how the fundamental components of the Bluetooth PAN in Figure 1.1 can be integrated into a wireless infrastructure to enhance our lives and even reduce the need to queue!

Adding Usability to Products

Mr. I.M. Wireless is embarking on a business trip. At the airport, as he gets within range of the airline's counter, his reservation is confirmed and a message is sent to his mobile phone detailing flight confirmation, personal boarding reference, seat information and departure gate number, which he listens to via a headset being that his phone is actually in his briefcase. While in the departure lounge, he connects to the Internet and accesses his e-mail via his mobile phone or the wireless LAN Access Point fitted in the lounge. He boards his flight and during the journey composes e-mails which will be sent as he enters the range of a LAN in the arrivals lounge or again via his mobile phone. He walks to the rental car company's counter to pick up his keys—as with the airline, all booking, payment, and car location details would have been transmitted between his PDA/mobile telephone and the rental company's computer. He starts to drive the rental car and his PDA downloads his hotel information into the car's on-board systems, which allows the navigation system to smoothly direct him to its location. On arrival, his room booking reservation is already confirmed. At his meeting, the normal 15-minute exchange of business cards is removed as all of the personal information is exchanged automatically via his PDA. He then uses his PDA to run his presentation from his laptop, which all attendees at the meeting are viewing simultaneously on their own laptops. Back in his hotel room after the meeting, his PDA synchronizes with both his laptop and mobile phone—now the telephone details of all the new contacts he met are stored in his mobile phone directory and the address and e-mail information in his laptop. Later, while relaxing, he listens to MP3 files stored on his laptop with the same headset that he answers his phone with. He also uses his digital camera to send "an instant postcard" via his mobile phone and the Internet to his wife's PC at home (obviously, it won't be a picture from the Karaoke evening arranged by his clients!)

If we extract some conclusions from this slightly excessive example, we find that wireless connectivity offers us immense freedom and convenience. It allows us to perform tedious tasks with a minimum of intervention, allows some of our devices to have dual functionality, and makes the vast array of cables we inevitably always leave in the office redundant. Bluetooth technology "will" change the assumptions we all have about our electronic devices. With the cables gone, the idea of having a particular gadget for a specific job will no longer be relevant. With many of the devices already available to consumers, this scenario grows closer to reality every day.

As for networking our homes, there are two ideologies. The first predicts a "master device" that will control everything from the video recorder to the security system, and which will replace the PC as the technological hub of the home. The other suggests the PC will remain at the centre of a networked home. Figure 1.4 illustrates how the PAN can be extended in our homes and combined with our wired infrastructure to provide a home area network (HAN) that utilizes wireless technologies for audiovisual (AV) control and distribution. The British mobile telephone company Orange is currently promoting a wireless house that will demonstrate various technologies in a "real-world" environment. More information can be found on the Orange Web site at www.orange.co.uk.

Figure 1.4 A Wireless HAN for AV Control and Distribution

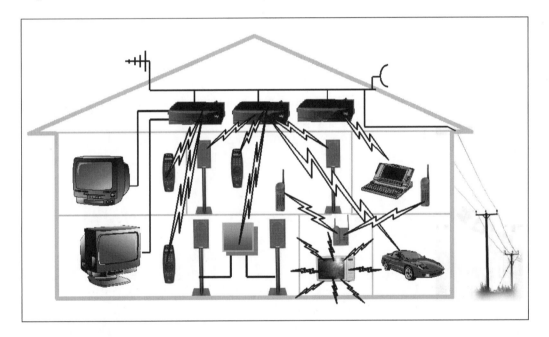

Allowing for Interference

Wireless means a radio link—and radio links are subject to interference. Interference can impact both the quality of an audio (Synchronous Connection Oriented [SCO]) connection or the throughput of a data (Asynchronous Connectionless [ACL]) connection. High levels of interference can interrupt communications for long enough to cause the protocol stack to timeout and abandon the link altogether. Although this is addressed in the Bluetooth

Specification with a frequency-hopping scheme which does provide robustness, it is still a serious consideration for some applications.

Bluetooth technology should not be used for safety-critical applications where data absolutely *must* get through, because there is always a possibility of a burst of interference stopping the link. Interference can come from a variety of sources: microwave ovens, thunderstorms, other communications systems (such as IEEE 802.11b), even other Bluetooth devices in the area (although these will not have a great effect as they are designed to cope with interference from one another in normal use).

It is possible to overcome the problem of link failure. For example, if you are relying on a Bluetooth link to monitor your baby and you know the environment is such that the link will only fail approximately once a week, then you might be happy to have the receiver alert you when the link fails. Once a week you may be out of touch, but an alert will let you know that the link has failed, so you have the option of returning within earshot of the infant. Since the Bluetooth links only operate up to around 100 meters, it shouldn't take you too long to get there!

There are other safety-critical applications where an unreliable link may be acceptable. An example is a system developed for Nokian tires, which allows tire pressure to be automatically monitored and sent to the car dashboard display. A wireless link will be subject to frequent failures in the harsh automotive environment, but the link can be re-established. Even if it only works a tenth of the time, it is still checking tire pressures far more often than will the average motorist! Here again, the system could be set to alert the driver if the tire pressures have not been reported recently. This way the driver knows that a manual check is needed.

So far, we have looked at effects of the Bluetooth link receiving interference, but, of course, it can also interfere with other devices. Bluetooth devices are obviously completely unsuitable for use in an environment where the Bluetooth link would interfere with sensitive control equipment—an aircraft being the primary example. Interference issues are explained in more depth later in this chapter.

Considering Connection Times

With a radio link, although the connections can be unconscious, connection times can be lengthy as transmitters and receivers all need to synchronize before communication can commence. These limitations could have serious consequences if the wireless link was of a critical nature—for example, a "panic button," a life-dependant medical monitor, or an engine management system.

There are two delays in setting up a Bluetooth link. First, it takes time to discover devices in the neighborhood. In device discovery, a device sends out inquiry packets, and receives responses from devices in the area, then reports these to the user. It can take ten seconds to find all the devices in an area, and even then you will only find those devices which are willing to report their presence. Some devices may not be set to scan for inquiries, in which case you will never find them!

A second delay occurs when you set up the connection itself. Again, this can take up to ten seconds. This lengthy connection time means that Bluetooth devices are unsuitable for systems where a fast response is needed, such as automatic toll collection on busy roads.

Coping with Limited Bandwidth

Wireless can also mean "slower." An Internet connection via a Bluetooth LAN is limited to the maximum data rate (723.2 Kbps) over the air interface. After allowing for management traffic and the capacity taken up by headers for the various protocol layers, even less is available to applications at the top of the stack. This will not compete with a high-speed wired link. Thus, for sending or downloading vast amounts of data, a Bluetooth wireless connection would not be the optimum method.

This also impacts on audio quality: Bluetooth technology simply does not have the bandwidth for raw CD quality sound (1411.2 Kbps). However, if a suitable compression technique is employed (using MP3 to compress an audio stream down to 128 Kbps, for example), it is feasible to use an ACL link for high-quality audio. The quality of a Bluetooth SCO link is certainly not high quality—it is approximately equivalent to a GSM telephone audio link (64 Kbps).

Compression can be useful for data devices. If large amounts of data are to be sent, using a compressed format will obviously speed up transfer time.

Considering Power and Range

Power is a critical consideration for wireless devices. If a product is to be made wireless, unleashed from its wired connection, where will its power come from? Often the communication cable also acts as a power cable. With the cable gone, the subject of batteries is brought into focus, and the inevitable questions arise concerning battery life, standby time, and physical dimensions.

Some devices, such as headsets, have no need for power when they are connected with wires. Audio signals come down a wire and drive speakers directly; a very simple system with no need of extra power connections. When the wires are replaced with a Bluetooth link, suddenly we need power to drive the link, power to drive the microprocessor that runs the Bluetooth protocol stack, and power to amplify the audio signal to a level the user can hear. With small mobile devices you obviously do not want to install huge batteries, so keeping the power consumption low is an important consideration.

Deciding on Acceptable Range

The Bluetooth specification defines three power classes for radio transmitters with an output power of 1 mW, 2.5 mW and 100 mW. The output power defines the range that the device is able to cover and thus the functionality of your product must be considered when deciding which power class to use. The user would not want to have to get up from his desk to connect to the LAN and therefore requires a higher power radio. Conversely, a cellular phone headset is likely to be kept close to the phone, making a lower range acceptable, which allows smaller batteries and a more compact design. Table 1.1 details the respective maximum output power versus range.

Table 1.1 Bluetooth Radio Power Classes

Power Class	Max Output Power	Range
Class 1	100 mW	100 meters+
Class 2	2.5 mW	10 meters
Class 3	1 mW	1 meter

It is important to realize that the range figures are for typical use. In the middle of the Cambridgeshire fens, where the land is flat and there is not much interference, a Class 1 device has been successfully tested at over a mile. But in a crowded office with many metal desks and a lot of people, the Bluetooth signal will be blocked and absorbed, so propagation conditions are far worse and ranges will be reduced.

Recognizing Candidate Bluetooth Products

Taking into account the preceding sections, we can see that for a product to be a candidate for Bluetooth technology, it needs to adhere to the six loosely defined conditions that follow:

- Adds usability (that is, convenience and ease-of-use—the Bluetooth Dream!)

- Interference or latency will not affect its primary function

- Is tolerant to the connection time overhead

- Can afford the limited Bluetooth bandwidth

- Battery life or power supply requirements are compatible

- The range is adequate

The remainder of this chapter will explore these issues in depth to attempt to provide an insight into what actually "does" make a good candidate for the Bluetooth technology. It will also present a case for the various implementation techniques available to the developer with their inherent advantages and limitations.

Considering Product Design

Your product may look like a candidate Bluetooth product, but there are practical considerations to take into account. It costs money to add a Bluetooth link, and for some products, that cost may be more than the customer is willing to pay.

You must look long and hard at the design of your product, how Bluetooth technology will affect the design, and whether in the final analysis that cost will be worth it. This section covers some of the issues you will have to take into account when moving from a wired product to a wireless one.

Are You Adding End User Value?

Having your product's packaging be anointed with the Bluetooth logo to announce you are part of the new technology revolution may persuade the consumer to purchase your product over a competitor's wired product. Your product may even command a premium price that will pay back your development efforts. But will the customer be satisfied when he gets it home? Will it give him the added value he has paid his extra dollars for? Will the "out-of-the-box" experience fulfill his notion of the promised ad-hoc wireless connectivity?

With mobile products that are not constrained by mains power cables, the added value of being wireless is easy for us to see. Who rushed out to try IrDA in their PDAs? Horrendous file transfer times and the "line-of-sight" constraint notwithstanding, the added value from simply being wireless convinced consumers to try it and *use* it! However, for products that are inherently static, the added value may just be initial "desire" and not really a viable investment in both resources and dollars.

Consider the static devices in our wired PAN (Figure 1.1)—for example, the ubiquitous mouse and keyboard. Both are dependant for their power supply requirements upon their host PC, so if made wireless, the subject of batteries becomes crucial. This added value of wireless connectivity can only be enjoyed if the user does not have to change or re-charge the batteries every week! Our static devices—desktop PCs with the obligatory mains power cable—would be perhaps better served by a wired Ethernet link rather than a Bluetooth LAN point (both cables embedded under the floor in your office as standard). Electric lights are another facet to consider—just think of the reduced installation costs in an office building of no wiring loom. Here, however, we do require power. So is wireless really adding value? It could be valuable if added as a control extra. The user could then connect via a handheld device or static panel to whichever light they wished to control. At the other end of the scale, the end user value of a Bluetooth PCMCIA card is easily visible, and will provide complete wireless connectivity.

Ensure that your product will really give the user added value by being wireless, not just offer a gimmick. If the consumer has to connect a power cable, then consider what other functionality can be offered. The desktop PC, although best served by a wired Ethernet connection, will still need to connect to our laptop and PDA, and thus requires both wired and wireless connectivity.

An intriguing application would be a wireless pen—consider its use for signature authentication provided by the credit company, bank, or reception desk, a super method to try and eliminate fraud. If a wireless implementation could be designed for the stringent size constraint, how would we stop users from walking off with it? Why are the ordinary pens always attached to the counters? Would being wireless really add value to this application?

Investigating Convenience

Added user value is a "big plus" for the consumer but wireless communications may not necessarily make the product more convenient to use. We assume that consumers are all comfortable with gadgets and electronic devices, but can your friends all program their VCRs yet?

Let's examine the traditional headset and mobile phone and decide if Bluetooth technology makes this more convenient for the user. With current hands-free technology, you have to decide in advance if you require the hands-free option. This involves fitting your car with a hands-free kit—a microphone or headset plugged in, with the wire trailing from it to your phone which is either in your pocket, clipped to your jacket/belt, in a cradle on your dashboard, or like most of us, fallen down between the seat and the handbrake!

When you receive a call, you answer by pressing a button on the cable; volume control is available via a button on the cable. The limitation is that you always have to have your telephone with you; it can only be as far away as the cable is long. Thus, it is always a conscious decision to use the headset, and to decide to plug it in! With a Bluetooth headset and phone, the phone can be inside your briefcase, in the boot of the car, in your jacket on the hook in the office, in fact, absolutely anywhere—as long as it's within the range of the headset. In much the same way as the conventional technology, you press a button on the headset to receive a call or to adjust the volume. The connection between the two devices is extremely different, however, and although virtually invisible to the user, it will incur a connection time overhead. First, the headset must "pair" with the Audio Gateway (AG), the Bluetooth part of the phone. This allows Bluetooth addresses to be swapped, and link keys to be established. The headset will then be able to make a connection to the AG or the AG will be able to connect to the headset—the exact operation is a software application issue. If the headset connects to the phone, then the phone needs to know why, either to set up voice dialing, action voice dialing, or some other function. If the phone connects to the headset, it patches a SCO link across and the headset can be used to take the incoming call.

The connection time could be a problem if you must connect every time a call comes in. After ten seconds of trying to make a connection, the caller has probably decided you are not going to answer and given up! A low power park mode allows headset and phone to stay constantly connected without draining their batteries; this overcomes the slow connection problem. So you must beware—if connection time is an issue for your product, make absolutely sure your system supports park mode—although it's becoming increasingly common, it's still possible to buy devices that do not support it.

My conclusion would be that Bluetooth technology would make answering my phone far more convenient, although extremely expensive at the moment! I do not have to worry where my phone is, per-equip my car, or have to endure a cable running from my ear. If the complex connection issues are invisible to me and I look as cool as Lara Croft (she wore the original Ericsson Bluetooth headset in the Tomb Raider movie), who really cares! However if it turns into a software setup nightmare and I have to read through vast user guides, I would not be so sure.

The medical sector offers many opportunities for Bluetooth technology to add convenience. In hospitals, patient medical data could be stored on PDA type devices that would update a central database when brought within range of an access point (small scale trials for this application in the neurology department at the University Hospital in Mainz, Germany, have already begun). Wireless foot controls for medical equipment, respiratory monitors that transmit data to a PDA rather than a body-worn data collection system, ambulatory monitoring equipment for easier patient access in emergency situations… the list goes on. The questions of interference and security will need to be addressed in some of these applications, but if they are not "life-dependant" these issues could be overcome.

Regarding the LAN access points, we need to consider the issue of range. If the consumer has to get up and walk to be within range, there is no added convenience—in fact, it would become very *in*convenient. A Class 1 Bluetooth device has a range of approximately 100 meters. In reality, this could be much further, which would be viable in an office, home, or a hotel/airport lounge scenario, thus making possible the unconscious convenience of the airport check-in and car rental confirmation detailed at the beginning of this chapter.

With our own personal "toys" the added convenience is unequivocal. Our laptops will be able to play multiuser Quake with our colleagues in the airport or the office! Our PDAs and phones will synchronise with our laptops—gone are the days of database management. Our presentations can be shown at meetings directly on the laptops of the attendees without the need for a projector or any worries about forgetting your laptop's I/O expander.

Against this optimistic picture there are a few inconveniences envisaged that will affect the consumer. I wouldn't be happy if my new wireless product spends longer attached to a battery charger than it can be used without one, if the poor placement of an antenna within a handheld product means I had be a contortionist to be able to hold it and have it function, or if calls get dropped while waiting for my headset to connect to my phone. But the BIG one is inevitably the man-machine interface (MMI)—it must be simple to use, it must be simple to set up, it simply must be simple: "connect to Adam's PDA, Petra's phone, or the fridge?" Using the word "convenience" in the product marketing blurb is a hollow promise if the consumers requires a software degree to get their new PDA to connect to their laptop! If people still can't program their home AV equipment, how will they know what a windows "system tray" is, where to put a .dll file, or where to find the setup section in their mutlilayered phone menu system?

It is your challenge as an applications writer to make sure that the MMI is usable. Succeed and your products could be extremely popular—fail, and your products will likewise fail in the marketplace.

Enhancing Functionality

Convenience is one attribute that Bluetooth technology can bring to our products, but how else can it benefit us? It can also add enhanced functionality—features that would not be an implementation consideration in a wired product. Central heating control? A programmable thermostat and a Bluetooth radio integrated into the common light switch, this integration would allow the mains wiring to the light switch to power the controller. When the room is at the temperature programmed by the user, it connects wirelessly to the boiler in the utility room and can turn the entire system off. Alternatively, if each individual radiator is equipped with Bluetooth technology, the controller can connect to each individual radiator and shut the solenoid valve, turning only that specific radiator off! In this application, we can see the enhanced functionality; no additional wiring is required to achieve single room climate control and the humble light switch becomes multifunctional. The Set Top Box that sits anonymously in our TV stand and has been delivering cable channels and e-mail to the TV screen could be made capable of connecting to our laptops, offering us another option to the modem in our homes.

As mentioned earlier, the people who make Nokian tires are adding Bluetooth links to pressure monitors built into car wheel rims. This is a good application since the data could not easily be transferred by other methods: wire and optical wouldn't work, other radio technologies are too expensive, and being able to remotely read tire pressure is a real gain in functionality.

Bluetooth technology in our digital cameras and mobile phones will provide us with the ability to send the "instant postcard" shown in Figure 1.5. This could become almost as popular as Short Message Service (SMS) text messages. We take a picture with our camera, which instantly transmits the photo to our mobile phone that has a connection to the Internet via the Global System for Mobile

Communication (GSM) network. From there, the picture is sent over the Internet to our friend's PC. It's a simple process which adds a new dimension to both products.

Figure 1.5 The "Instant Postcard"

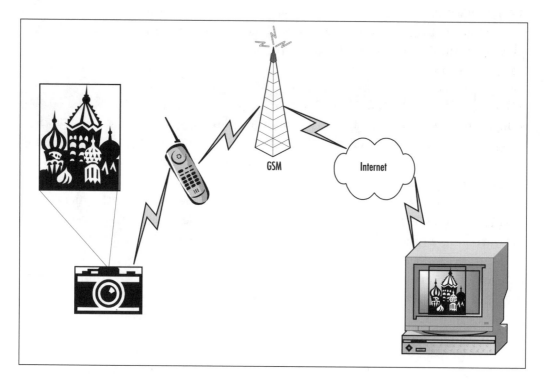

What if our gas and electricity meters could be read by the utility's serviceman simply by walking into the foyer of an apartment block and connecting to each apartment's meters individually to determine utility consumption? Not having to knock on each door would improve the efficiency of the job function but would inevitably mean that fewer personnel were required. With an application of this type, the cost implication and durability of Bluetooth technology comes to the fore. The ubiquitous gas and electricity meters have to last a long time, far longer than our favourite mobile phone or PDA which we change according to personal taste or consumer trends. The cost of replacing the meter infrastructure in our homes far exceeds the overhead of including Bluetooth technology, something which makes utility companies adverse to new technologies. Experiments have been conducted, but so far there has been no serious uptake.

With our children's toys, the possibilities become endless. Big soft toys are able to communicate with PC games allowing for communication and interaction external to the PC. Multiplayer handsets for our Playstations become possible without a mass huddle around the console and the constraint of the cable length. Action figures and robotic toys could be remotely controlled from a PC, or could transmit pictures from a camera accessory to the PC.

Far more serious is the added functionality that can be provided for the disabled consumer, a headset could provide a life enhancing benefit to the physically compromised user—voice control for their heating, lights, AV, and security systems—allowing control from anywhere in their home. Wireless Internet access can also be of benefit. For instance, the National Star College in Cheltenham, UK has just installed a Red-M Bluetooth network to allow their disabled students to wirelessly access online resources and submit their coursework directly from their laptops. Discrete intelligent proximity sensors communicating with a headset could help the visually compromised, or a vibrating dongle could indicate to a deaf consumer that the doorbell is ringing or could be programmed to vibrate on other sound recognitions. All of these applications simply extend the functionality of conventional products by being Bluetooth-enabled.

Do You Have Time?

Okay, so we've decided we want to be wireless. We "must have" Bluetooth technology in our next product. The consumer market is not quite sure why they want it yet, but they do, so the first and most difficult hurdle is over with. But what do we need to do? And how long will it take? Both of these are serious questions. After all, implementing any new technology often incurs risks that may outweigh the advantages of the technology itself.

First of all, the Bluetooth Specification by the Special Interest Group (SIG) is an extremely comprehensive document, which needs to be digested before any form of implementation can begin. Both the hardware and software implementation are required in order to adhere to this specification and be able to utilize the intellectual property (IP) contained within it. It is essential to stick with the specification to be able to interoperate with any other Bluetooth device irrespective of manufacturer or solution provider; interoperability is the "key" to consumer uptake of Bluetooth technology and the realization of the Bluetooth Dream. Going up the Bluetooth learning curve can take significant time. Courses are available which make it easier, but you must still allow significant learning time in your development cycle.

If you are late in the product implementation cycle, you may not have time to build in Bluetooth technology. Or you may not have enough market information to reassure yourself that it will add sufficient value to justify the cost of shipping Bluetooth components in every product. Many early adopters initially added Bluetooth technology to existing products as "add-ons," either as dongles or accessories to battery packs—mobile telephones being the principal example.

Using an "add-on" strategy allows you to decouple the Bluetooth development from your main product development. This means that you do not risk the Bluetooth development holding up your product launches. Since consumers can buy mobile phones, laptop computers and access points with Bluetooth technology fully integrated, this shows that the risks can be conquered successfully. Devices which implement Bluetooth technology as an "add-on" are likely to be less attractive to consumers when competing with built-in devices. So, when considering whether to build in or add on, you must survey the competition and decide whether your launch date means an "add-on" will not be as lucrative.

There is more to consider than the time to develop and manufacture your product. For any Bluetooth design to be able to display the Bluetooth logo, the design has to undergo a stringent qualification procedure and pass a vast array of tests on every aspect of the system from the radio, baseband, and software stack through to the supported profiles. This is achieved at a Bluetooth Qualification Test Facility (BQTF). Such test facilities can now be found globally, though they are becoming exceptionally busy and require booking many weeks in advance. In addition to the Bluetooth Qualification Program, product developers and manufacturers are required to meet all relevant national regulatory and radio emissions standards and requirements. This involves going through national type approval processes which vary from country to country. Qualification and type approval can significantly delay product launches, so they MUST be allowed for in your schedule.

Investigating Product Performance

In some of the applications previously mentioned, we can see that the many benefits of Bluetooth technology may outweigh the limitations, nevertheless we have only examined the subjective questions of added value and enhanced functionality. Now it's time to consider in depth some of the technical limitations that may actually influence our choice of adding Bluetooth technology to our products, despite the much desired benefits.

In this section, we shall look at connection times, quality of service in connections, voice communications, and the various sources of interference.

Evaluating Connection Times

As we have mentioned, Bluetooth devices can't connect instantly. It can take up to ten seconds to establish a Bluetooth link (although this is not a typical figure; tests with BlueCore chips show that 2.5 seconds is far more common). The connection time overhead is a limitation that could have serious consequences if you require an instant connection—a "panic button" would not be a viable application for Bluetooth technology. We will examine why and how this overhead can be reduced with a "known device" connection.

Wired networks are for the most part static. Components of the network are connected together with cables, and once connected, normally remain in the same position. A printer that was available on the network yesterday is expected to still be available tomorrow. However, you do have the initial overhead of configuring your PC to use it, the procedure being:

- Physically connect cables to new device.

- Type in address name on system that needs to use the new device.

- Install drivers and configure software on system which needs to use new device.

Bluetooth piconets are highly dynamic—they change rapidly, with devices appearing and disappearing. The members of a piconet may change, or the whole piconet may be dissolved in a moment. In such a dynamic network, it is not viable to spend significant time acquiring information about devices and configuring software to use them: this process must be automatic. The Bluetooth core specification provides this automatic discovery and configuration. For a Bluetooth device, the steps to using a new device are:

- Perform device discovery to find devices in the area.

- Perform service discovery to get information on how to connect to services on each device discovered.

- Choose a service to use, and use information obtained during service discovery to connect to it.

Potentially, the user could simply select the option to print, and the processes of device discovery, service discovery, and connection could happen automatically without further intervention from the user. The application software should present this to us transparently, but it is still a worthwhile exercise to understand the complete procedures; they are covered in the following sections.

Discovering Devices

Before any two devices can go through device discovery, they *must* be in *inquiry* and *inquiry scan* modes. The inquiring device must be trying to discover neighbouring devices, and the inquiry scanning device must be willing to be discovered (see Figure 1.6).

Figure 1.6 Bluetooth Device Discovery

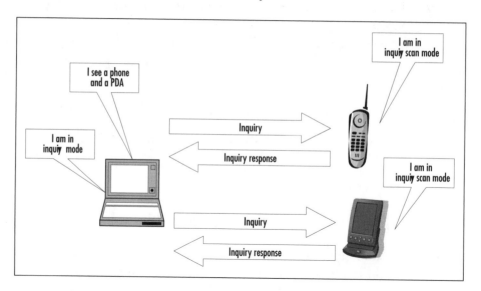

The inquiring device transmits a series of inquiry packets. These short packets are sent out rapidly in a sequence of different frequencies. The inquiring device changes frequencies 3200 times a second (twice the rate for a device in a normal connection). This fast frequency hopping allows the inquirer to cover a range of frequencies as rapidly as possible. These packets do not identify the inquiring device in any way; they are ID packets containing an inquiry access code which inquiry scanning devices will recognize.

The inquiry scanning device changes frequencies very slowly: just once every 1.28 seconds. Because the scanner changes very slowly while the inquirer changes rapidly, they will ultimately meet on the same frequency.

Scanning devices cannot stay on a fixed frequency, because any frequency chosen might be subject to interference, but hopping very slowly is the next best strategy for seeking the inquiring device. It responds to inquiries by sending a Frequency Hop Synchronisation (FHS) packet, which tells the inquiring device all the relevant information needed to be able to establish a connection.

NOTE

To guarantee that the inquiring device can locate all the devices in inquiry scan mode that are within range, the Bluetooth Specification defines an inquiry time of 10.24 seconds.

When a device that is scanning for inquiries receives an inquiry, it waits for a short random period, then if it receives a second inquiry, it transmits a response back. It does not transmit this response immediately, because this may lead to all devices in a single area responding to the first inquiry sent out, causing an undesirable high-power coordinated pulse of radiation in the ISM band. The random delay prevents this coordinated effect.

Connecting Devices

Before two devices can establish a connection, they *must* be in *page* and *page scan* mode; the paging device initiates the connection, while the page scanning device responds. In order to be able to page, the paging device must know the ID of the page scanning device; it can calculate the ID from the page scanning device's 48-bit Bluetooth device address. The page scanning device's Bluetooth device address can be obtained in several ways:

- From an inquiry response via FHS

- From user input

- By preprogramming at manufacture

NOTE

Each Bluetooth device has its own unique 48 bit IEEE MAC Bluetooth address (BD_ADDR), which identifies it to other devices; if the device is a master, the connection timing and the hopping sequence are also derived from this address. Addresses are obtainable from the SIG in blocks and need to be programmed into every Bluetooth product at manufacture—all silicon is shipped with the same default address that must be changed. A "friendly name" may also be programmed into your product either by the user or at manufacture to enable the MMI to connect to "CSR development module," "Daisy's phone," "Lara's headset," or "Amy's little black book," concealing the actual address. The address is concealed from the user because it is a string of numbers (typically expressed in hexadecimal) which is not a very user-friendly format. An example of a Bluetooth device address is 0x0002 5bff 1234.

By programming the device information that would normally be received in the FHS packet directly into the device, the inquiry and inquiry scanning can be avoided—devices move directly to paging, thus saving the 10.24 seconds required for inquiry. As previously noted, this could either be performed at manufacture, or carried out by the users. If we are manufacturing a mobile phone and a headset to be packaged together, the "out-of-the-box" experience will be one of disappointment if they do not communicate—they could be programmed such that they are both aware of each others' BD_ADDR. This way they become "known devices" to each other and can avoid the inquiry stage—what's called a *preset link*. We are also able to create a list of "known devices"—perhaps all the devices in our PAN.

Quantifying Connection Times

Now, we are aware of why connection times can be so long, but how long is long? What does this mean in minutes and seconds? The actual time is variable, depending upon the application software you are using, so you should look at what the Baseband Specifications specify. These, however, can be very confusing in giving definite minimum/maximum times used in inquiry and paging operations between devices, with the result that there may be a lot of speculation as to what these times actually are. Detailed in Table 1.2 are what the theory states *should* be the time taken to complete a typical successful Inquiry and Page operation, (that is, the typical time taken to set up an active Bluetooth link). To enable us to understand the basis of these figures, we will also briefly look at their origin.

Table 1.2 Connection Times to Set Up an Active Bluetooth Link

Operation	Minimum Time (sec)	Average Time (sec)	Maximum Time (sec)
Inquiry	0.00125	3 – 5	10.24 – 30.72
Paging	0.0025	1.28	2.56
Total	0.00375	4.28 – 6.28	12.8 – 33.28

Inquiry Times   An inquiry train must be repeated at least 256 times (2.56s duration), before the other train is used. Typically, in an error-free environment, three train switches must take place. This means that 10.24s could elapse unless the inquirer collects enough responses and determines to abort the procedure. However, during a 1.28s window, a slave on average responds four times, but on different frequencies and at different times.

Minimum Inquiry Time   A minimum time for an inquiry operation is two slots (1.25ms). The master transmits an inquiry message at the f(k) frequency in the first instant, and the slave scans the inquiry at the f(k) frequency at the same time. So, the slave receives the inquiry message in the first slot. The slave could respond with a FHS packet to the master's inquiry message in the next slot. So, in total two slots are needed. This is highly unlikely as the slave will not respond after receiving the first inquiry message but rather, wait a random number of slots. This random value varies between 0 and 1023.

Average Inquiry Time   As stated previously, 10.24s could elapse unless the inquirer receives enough responses and decides to abort the procedure. This value can vary considerably, depending on alignment of the device clocks and their respective states. This, however, is not sufficient to guarantee all the devices within range will be "found"!

Maximum Inquiry Time   10.24s is what the user would typically expect for a maximum inquiry time—the amount of time specified until the inquiry is halted. 30.72 seconds has been suggested as a maximum time, although specifications state this can be up to a minute.

Paging Times   Assuming you are employing the mandatory paging scheme (using page mode R1, where each train is repeated 128 times, before switching to the next one), then the average time for connection should be 1.28s. The maximum time for connection is 2.56s. During this, the A+ B train will have been repeated 128 times each, and a response returned.

Minimum Page Time   This is similar to the Minimum Inquiry Time. When the master transmits a page message at the f(k) frequency in the first instant, the slave scans the inquiry at the f(k) frequency at the same time. Thus, the slave receives the page message in the first slot. The slave responds with an ID packet for the master's page message in the next slot. Then in the third slot, the master transmits a FHS packet to the slave. Finally, in the next slot, the slave answers. Thus four slots (2.5ms) are needed for the minimum page duration.

Performing Service Discovery

When a Bluetooth-enabled device first enters an area there may be numerous other devices offering services it wishes to use. How does it tell which of these devices supports which service—in other words, which device will allow it to send an e-mail, print a fax, or exchange a business card? The Service Discovery Protocol (SDP) allows a device to retrieve information on services offered by a neighbouring device. (A service is any feature that another device can use.) A basic data connection must be set up before Service Discovery can be used. Then a special higher layer connection for use by Service Discovery is set up. Once the connection to service discovery is established, requests for information can be transmitted, and responses received back containing information on services. This information is known as the service's *attributes*. If a device is finding out information about many other devices in an area, then it makes sense to disconnect after finding information on any particular device. This relieves system resources (memory, processor power), which can be more effectively used establishing new connections to other devices to determine what they have to offer. Because SDP uses ACL, connection devices must use inquiry and paging before they can exchange SDP information. As a result, SDP can be slow. SDP is mandatory for all the profiles released with version 1.1 of the Bluetooth specification.

Quality of Service in Connections

In Bluetooth technology, the ACL link supports data traffic. The ACL link is based on a polling mechanism between master and up to seven active slaves in a piconet. It can provide both symmetric and asymmetric bandwidth, which is determined by the ACL packet type and the frequency with which the device is polled.

The ACL payload is protected by a CRC check, which may be used in a retransmission scheme. The delay involved with retransmissions on the ACL link is small, as an acknowledgement can be received within 1.25ms. Further, the number of unsuccessful retransmissions can be limited by a Flush Timeout setting, which flushes the transmission buffer after a specified period of unsuccessful retransmissions. This opens the possibility to perform retransmissions for delay-sensitive applications such as interactive real-time and streaming (IP-based) audio/video applications. In most implementations currently available, the ACL link only provides a best-effort type of service (i.e., there are no Quality of Service (QoS) guarantees associated with the transfer of packets). It especially does not provide any guarantees of bandwidth and delay.

The Bluetooth specification does provide mechanisms to balance traffic between slaves in a piconet, allowing a so-called "guaranteed" Quality of Service. However, because the quality of the underlying radio link can never be guaranteed, in practice all that Bluetooth technology *can* do is to make an attempt to support the QoS it *has* guaranteed.

The unpredictability of radio interference means that if a guaranteed bandwidth is absolutely necessary for your product, then a wired link is really your best choice.

However, it is worth considering whether guaranteed bandwidth is really necessary. By compressing data and buffering it on reception, it is possible to overcome glitches in transmission. This can make a radio link appear far more reliable at the application level than it really is down at the baseband level!

Data Rate

If a Bluetooth device transmitted constantly on only one frequency, the maximum raw data rate would be 1 Mbps. However, this is *not* the data rate we will obtain over the air interface. Bandwidth is required for a 72-bit access code to identify the piconet, and a 54-bit packet header to identify the slave—total slot time: 405µs. The radio requires a guard band of 220µs between packets to allow it to retune and stabilize on the next hopping frequency. This guard band consumes the rest of the slot.

Within a one slot packet these requirements leave only one-third of the bandwidth for the payload data—and this can only be transmitted every other slot, or every 1250μs. One way to mitigate this limitation is to transmit for a longer period of time: 3 or 5 slots. All of the extra bandwidth is used for payload data with a consequent improvement in efficiency (illustrated in Figure 1.7). While transmitting over more than one slot, the devices remain at the same frequency, moving to the next frequency in the hopping sequence at the end of the packet. Thus, in a five slot packet, the master will transmit on f(k), and after the five slots will transmit on f(k+5). (A 16-bit CRC is also included in every ACL packet, but this is not illustrated in Figure 1.7.)

Figure 1.7 The Payload in Bluetooth Packets

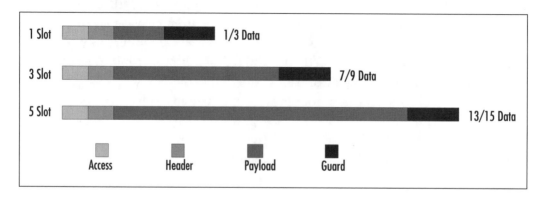

Bluetooth ACL packets can either be of Data Medium (DM) or Data High (DH) type. The DH packets achieve a higher data rate by using less error correction in the packet. A DH5 packet which utilizes five slots can carry the maximum amount of data: 339 bytes, or 2712 bits. So, if we take account of the packet overheads already discussed, 2858 bits are transmitted over the air interface for every 2712 bits of data payload. This gives us the maximum baseband data rate in a single direction of 723.2 Kbps – the single slot packets in this asymmetric link would carry 57.6 Kbps. If we chose to send five slot packets in both directions, the data rate would be reduced to 433.9 Kbps!

The choice of symmetric or asymmetric links allows our user scenarios to take account of the improvement in data rate in one direction of the asymmetric link (for example, our PDA browsing the Web via a server will require more bandwidth while downloading pages than it will require for us to specify the next link to browse.) Table 1.3 illustrates the maximum data rates with all of the packet types in both symmetric and asymmetric links.

Table 1.3 Bluetooth ACL Packet Maximum Data Rates

ACL Packet Type	Payload Header (Bytes)	User Payload (Bytes)	FEC	CRC	Symmetric Max Data Rate (Kbps)	Asymmetric Max Data Rate (Kbps)	
						Forward	Reverse
DM1	1	0 – 17	2/3	Yes	108.8	108.8	108.8
DH1	1	0 – 27	0	Yes	172.8	172.8	172.8
DM3	2	0 – 120	2/3	Yes	258.1	387.2	54.4
DH3	2	0 – 180	0	Yes	390.4	585.6	86.4
DM5	2	0 – 224	2/3	Yes	286.7	477.8	36.3
DH5	2	0 – 338	0	Yes	433.9	723.2	57.6

Latency

Bluetooth technology achieves reliability by retransmitting packets. Each packet carries a header with an acknowledgement bit in it. When a device sends a packet, it uses the acknowledgement bit to signal whether the last packet it received was good or corrupted. When a device receives a packet with the acknowledgement bit set to indicate that its last packet was corrupted in transmission, it simply retransmits the corrupted packet. This retransmission carries on until it receives an acknowledgement that the packet got through correctly.

This can add delays (latency), and sometimes these delays can be variable (a bursty link). This may cause problems for applications needing a constant feed of data (e.g., compressed video). The effects of bursty links can be smoothed out by writing data into buffers as it is received, and reading it out a short time afterwards. As the on air link speeds up and slows down, the amount of data in the buffers gets greater or less, but as long as data is read out at the same average rate as it arrives, buffers can be used to smooth out a bursty link.

Some applications do not care if data comes in bursts, but they do need low latency (fresh) data. An example might be a monitoring application. If data has to be retransmitted, the monitor might freeze momentarily, but it is more important to get the most recent data than to have a smooth flow of packets. In this case, flushing can be used: at the transmitting end, data from the monitor could back up in the device's buffers. A flush command tells a Bluetooth device to dump all stale data and start transmitting fresh data. It is possible to set up automatic flushing to avoid stale data accumulating.

Delivering Voice Communications

The voice quality on a Bluetooth SCO link is roughly what you'd get from a cell phone—in other words, it's not hi-fi quality.

The audio data is carried on SCO channels, and to establish a SCO channel, you must first set up an ACL (data) channel. This is because the ACL channel is used by the Link Manager to send control messages to set up and manage the SCO channel.

SCO channels use prereserved slots; reservation of slots ensures the integrity of the SCO packet. There are three different types of SCO packets, each of which requires a different pattern of reserved slots.

- An HV3 packet carries 30 bytes of encoded speech with no error correction. A SCO link using HV3 packets reserves every third pair of time slots available to a device.

- An HV2 packet carries 20 bytes of encoded speech plus 2/3 Forward Error Correction (for every 2 bits of data, 1 bit of error correction is added to give a total of 3 bits). A SCO link using HV2 packets reserves every second pair of time slots available to a device.

- An HV1 packet carries just 10 bytes of encoded speech protected with 1/3 Forward Error correction (for every bit of data, 2 bits of error correction is added to give a total of 3 bits). A SCO link using HV1 packets reserves every pair of time slots available to a device.

Because the SCO links reserve slot pairs for voice packets, they prevent the use of 3 or 5 slot packets for data transmission. The multislot packets can support higher data rates than single slot packets, this combines with the slots used by the voice link to reduce the maximum data throughput if SCO and ACL transmission occur concurrently.

The Bluetooth specification supports several coding schemes: Log PCM A-law, Log PCM μ-Law, and CVSD. Log PCM coding with either A-law or μ-law compression was adopted by the Bluetooth specification because it is popular in cellular phone systems. Continuous Variable Slope Delta (CVSD) modulation is supported in the Bluetooth specification because it can offer better voice quality in noisy environments. The Bluetooth audio quality is approximately the same as a GSM mobile phone—this translates to audio transmitted at a fixed data rate of 64 Kbps.

A master is capable of supporting up to three duplex audio channels simultaneously. These channels could be either to the same slave or to different slaves. Because voice transmissions are inherently time-dependant, SCO packets are never retransmitted, so any packets that are not received correctly are lost. In noisy environments, the errors introduced by lack of retransmission capabilities can have a serious impact on the quality or intelligibility of the received audio.

Bluetooth technology does not have the bandwidth for raw CD quality sound: 1411.2 Kbps. However, if a suitable compression technique is employed (for example, MP3 compressing an audio stream to 128 Kbps), it is feasible to use an ACL link for high-quality audio. An audio-visual workgroup is currently working within the Bluetooth SIG to provide a profile which will improve the maximum audio quality that can be delivered across Bluetooth links. As compressed audio incurs a delay in transmission, the existing SCO scheme will be retained for applications (such as cell phone headsets) where the bandwidth of the audio signal is already low.

Investigating Interference

The Bluetooth system operates in the 2.4GHz band. This band is known as the Industrial Scientific and Medical (ISM) band. In the majority of countries around the world, this band is available from 2.40–2.4835GHz and thus allows the Bluetooth system to be global. It is available for free unlicensed use in most of the world, although some countries have restrictions on which parts of the band may be used. However this freedom has a price—many other technologies also reside in the band:

- 802.11b
- Home RF
- Some Digital Enhanced Cordless Communications (DECT) variants
- Some handheld short-range two-way radio sets (walkie-talkies)

These are all intentional emitters—one way or another their function is to generate microwave radiation in the ISM band. In addition to the intentional emitters, Bluetooth technology is subject to interference from a variety of sources which emit accidentally:

- Microwave ovens
- High-power sodium lights
- Thunderstorms
- Overhead cables
- Communications channels in other bands—e.g., GSM, CDMA
- Spark generators such as poorly suppressed engines

There are also problems from signal fading due to distance or blockers such as walls, furniture, and human bodies. The more water content in the object, the more significant the effect of blocking. Old brick walls will have a higher water concentration than modern ones due to the nature of their constitution. This tends to cause fading in European houses where brick is a common construction material. In the USA, where timber frames are more popular, signals are much less affected by internal walls.

As with any radio technology, Bluetooth technology is prone to interference from its co-residents in the ISM band and will produce interference to them. To achieve a degree of robustness to interference, the Bluetooth system utilizes a frequency-hopping scheme: Frequency Hopping Spread Spectrum (FHSS). Constantly hopping around the different radio channels ensures that packets affected by interference can be retransmitted on a different frequency, which will hopefully be interference free. Bluetooth radios hop in pseudo random sequences around all the available channels. During a connection, they hop every 625 microseconds. When establishing a connection, they can hop every 312.5 microseconds.

The screenshot in Figure 1.8 is taken from a Sony/Tektronix WCA380 spectrum analyser and illustrates 30MHz of spectrum in the centre of the ISM band. The upper section shows a snapshot of output power against frequency at a single instant in time. The lower section shows time against frequency with the power level displayed by way of shading.

Figure 1.8 Bluetooth Packets in a Noisy Environment

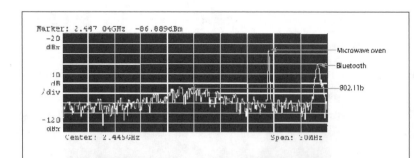

The screenshot clearly illustrates the spectral characteristics of microwave ovens with a strong but narrow spike of power, on the lower section of the screenshot. This wanders around the center of the ISM band as the oven operates, showing on the analyser screen as a curving red line. Our Bluetooth FHSS system can be seen to be hopping with 1MHz channel spacing with a strong central peak. The IEEE 802.11b or Wi-Fi DSSS system can be seen to have lower output power, indicated by the broad seep of power in the center, but the signal can spread across about 16MHz. (This is why co-located Wi-Fi networks cannot use adjacent channels.)

A Bluetooth FHSS system operating near an interfering signal can cope if a packet is hit by interference. The affected device simply retransmits the packet contents in the next slot when it has moved to a different frequency which is no longer affected by interference. This will impact on the throughput of an ACL link—the more interference, the more retransmissions. With a SCO link, it's a different matter. SCO data is not reliable, due to its inherent nature of being in real time, and retransmission is not tangible, so audio clarity becomes significantly worse with any interference. This can be overcome by sending SCO data via an ACL link.

Transfer of ACL information will still be reliable in a noisy environment. No information is lost as each dropped packet is retransmitted. The impact manifests itself in the data rate: the more noisy the environment, the more retransmissions will be required.

Figure 1.9 illustrates the effect of Bluetooth technology throughput in the presence of Wi-Fi interference. We can see that our Bluetooth device's throughput is degraded when a Wi-Fi device is very near. However, when the Wi-Fi device is relocated ten meters away, the throughput significantly improves. It is actually approximately 90 percent of the baseline throughput independent of range, thus illustrating that when Bluetooth and Wi-Fi devices are at a reasonable distance, the degradation in performance is tolerable.

Interfering with Other Technologies

Figure 1.10 illustrates the degradation our Bluetooth devices can have on Wi-Fi when they are extremely close to a Wi-Fi station. The impact on performance due to interference is significant. However, when our Bluetooth devices are relocated as little as ten meters away, the throughput is only minimally reduced compared to the baseline.

The last two figures indicate that the two wireless technologies can easily coexist as long as we are sensible in our expectations and attempt to combine

Figure 1.9 The Effect of Bluetooth Throughput with Wi-Fi Interference (Courtesy of Texas Instruments)

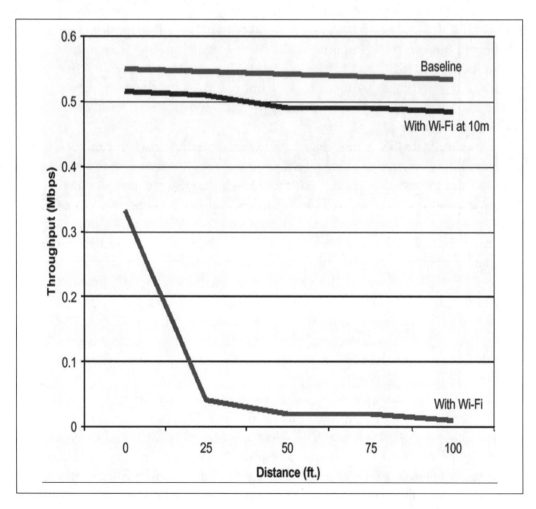

the technologies in our PAN and HAN paradigms intelligently. One way is to not have a Wi-Fi access point, providing us with the high data rate required for video streaming too close to the desk where our PDA and laptop "do their thing"!

Coexisting Piconets

A consideration not yet discussed is Bluetooth devices interfering with Bluetooth devices. How many devices do we need to reduce the data throughput to a trickle?

Figure 1.10 The Effect of Wi-Fi Throughput with Bluetooth Interference (Courtesy of Texas Instruments)

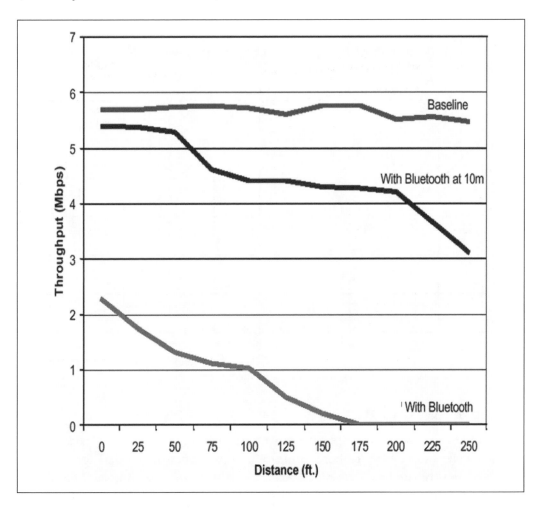

Consider the scenario of having Bluetooth devices in every room, With PANs for each member of the household. The majority of teenagers today have a PC and a mobile phone at the very least. Combine this with the toys of our younger children (and ourselves!) and any "household" Bluetooth devices; access points, control units, security systems, and so on. This adds up to tens of devices operating in the same area. Admittedly, they will not all be operational at the same time, so significant degradation is not likely to occur, But if our product requires dependable data delivery, the retransmission overhead that interference can cause might make Bluetooth technology unviable. Figure 1.11 illustrates how the probability of a packet collision increases with the number of operating piconets.

Figure 1.11 The Effect of Interfering Bluetooth Devices on Each Other

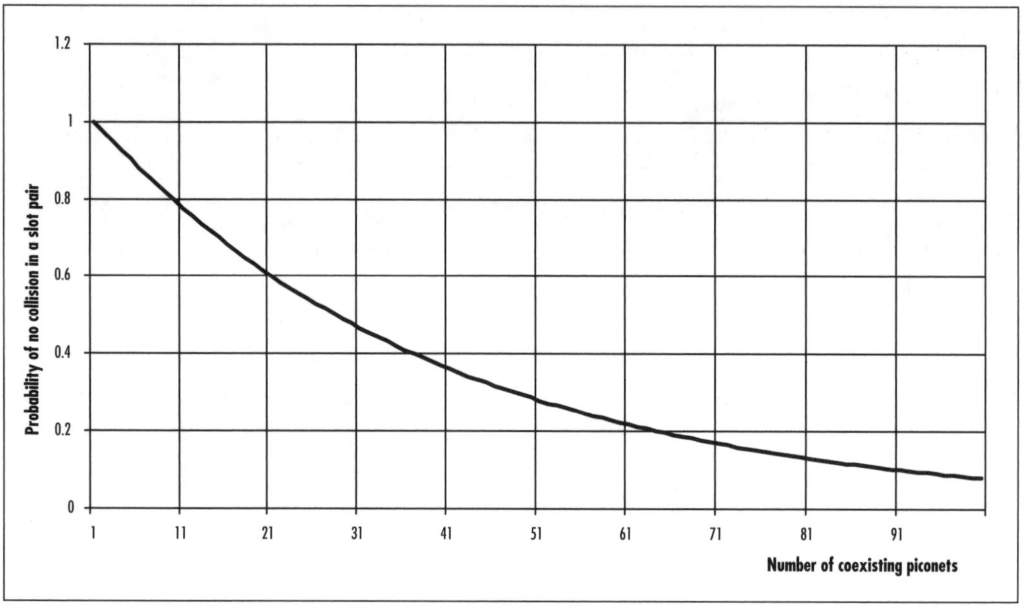

Using Power Control

We must also consider the respective power class of our Bluetooth devices. To enable all classes of device to communicate in a piconet without damage to the RF front ends of the lower power devices, a method of controlling the output power of Class 1 (100mW) devices is required.

Transmit power control is mandatory for Bluetooth devices using power levels at or above 4 dBm. Below this level (i.e., all Class 2 and 3 modules), it is optional. To implement a power control link, the remote device must also implement a Receive Signal Strength Indicator (RSSI). A transceiver that wishes to take part in a power-controlled link must be able to measure its own receiver signal strength and determine if the transmitter on the other side of the link should increase or decrease its output power level.

To set up a power controlled link, the transmit side must support Transmit Power Control and the receive side must support RSSI. Support is indicated in the Locally Supported Features (Bluetooth Spec 1.1 Part C (LMP) Section 3.11). The RSSI need only be able to compare the incoming signal strength to two levels: the Upper and Lower Limits of the Golden Receiver Range. The Lower Limit is between −56 dBm and is 6 dB above the receive sensitivity (0.1 percent BER level) for the particular implementation. The Upper Limit is 20 dB +/− 6 dB above this. The RSSI level is monitored by the receive side's Link Controller. When it strays outside the Golden Receiver range, the Link Manager is notified. A message is sent to the transmit side, requesting an increase or decrease in transmit power to bring the RSSI back in line. If the transmitter is a master, it must maintain separate transmit powers for each slave.

Host Controller Interface (HCI) commands exist to find out the current transmit power and RSSI level, but they are for information only. Layers above the Link Manager are not directly involved in power control. The implication of this is that it is perfectly possible to sit a Class 1 module transmitting at +20 dBm right next to another module which does not support RSSI and not limit the first's transmit power. If the second module's maximum receivable level is the Bluetooth spec of −20 dBm, there is every chance its RF front end will be overloaded. RSSI, although not mandatory, is highly recommended, as is a large power control range implemented on all modules, not just Class 1.

Figure 1.12 illustrates interfering Bluetooth piconets, but the principle holds true for coexisting networks of different technologies. Devices that are close to one another turn their power down and do not interfere with devices at a distance. Devices transmitting a long distance have to turn their power up to reach one another, which generates more interference and affects more devices. The hypothesis for us is ultimately to persuade our consumers to site devices intelligently. The home user is typically unaware of the implications of radio interference and will not position their devices for best performance!

Aircraft Safety

The Federal Aviation Authority (FAA) does not permit "intentional emitters" to be active on planes in flight. Bluetooth technology is an intentional emitter and as such is not legally usable on flights covered by FAA regulations. This means that any systems such as Bluetooth radio tags, which automatically identify baggage for airline baggage handling systems, need to be deactivated in-flight. The inconvenience of deactivating devices may mean that passive radio tags would better suit some applications. Certainly, in-flight deactivation issues *must* be considered by anybody whose products may be used in an aircraft in flight.

Figure 1.12 Interfering Piconets

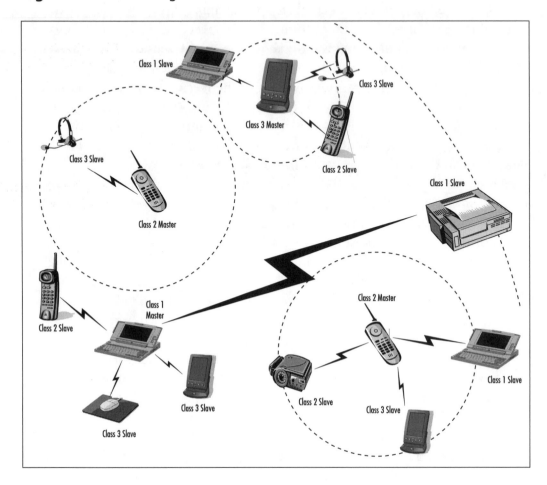

Assessing Required Features

The Bluetooth specification has many optional features, and even if features are mandatory to support, they do not have to be enabled. This section briefly examines a few features of the Bluetooth specification that may affect your product.

Enabling Security

To prevent unwanted devices connecting to our personal devices, or to prevent our personal data from being "snatched" from the air, Bluetooth technology provides security in the form of a process called *pairing*. It utilizes the SAFER+

encryption engine, using up to 128-bit keys. How this provides us with the means to "pair" with another selected device and create a secure link is interesting.

It is possible to "authenticate" a device—this allows a pair of devices to verify that they share a secret key. This secret key is derived from a Bluetooth pass key or Personal Identification Number (PIN). The PIN is either entered by the user or, for devices with no MMI (such as a headset), it will be programmed in at manufacture. After the devices have authenticated, they are able to create shared link keys which are used to encrypt traffic on a link. This combination of authentication and link key creation is called pairing.

Pairing devices allows communication secure from eavesdropping, but enabling security can make it much more difficult to connect with other people's devices, thus security features can seriously compromise usability. For devices where disabling security may be appropriate, the user interface should allow security to be turned on and off simply.

Using Low Power Modes

The Bluetooth specification provides low power modes, *hold*, *sniff*, and *park*. Devices in low power modes can still be connected to another device, remaining synchronized to that specific hopping sequence and timing, even though they do not have to be active. Thus, when they wish to communicate, they do not have to perform the inquiry, page, SDP procedure again—they are effectively just "reactivated."

Hold Mode

The ACL link of a connection between two Bluetooth devices can be placed in hold mode for a specified hold time. During this time no ACL packets will be transmitted from the master.

Hold mode is typically entered when there is no need to send data for a relatively long time—for example, if the master is establishing a link with a new device. During hold mode, the Bluetooth transceiver can be turned off in order to save power.

What a device actually does during the hold time is not controlled by the hold message, but it is up to each device to decide. The master can force hold mode if there has previously been a request for hold mode that has been accepted. The device in hold mode always retains its active member address (AM_ADDR). After the hold period has expired, the slave resynchronizes to the master and the active connection resumes.

This allows for our laptop to place our PDA that it is connected to in hold mode while it establishes a connection to a LAN access point, thus minimizing PDA power consumption when not in use.

Sniff Mode

In sniff mode, the slave remains synchronized to the master, but the duty cycle of the slave's listen activity can be reduced, thus placing the constraint upon the master to only transmit in certain slots. To enter sniff mode, master and slave devices negotiate a sniff interval and a sniff offset, which specifies the timing of the sniff slots and the occurrence of the first sniff slot. After this negotiation, the sniff slots follow periodically according to the prenegotiated sniff interval. In order to avoid problems with a clock wrap-around during the initialization, one out of two options is chosen for the calculation of the first sniff slot. A timing control flag in the message from the master indicates this. In sniff mode, the slave retains its AM_ADDR. This mode is extremely useful if we have our PDA waiting to receive e-mail from our phone. Normally, there will not be any traffic, but the PDA needs to be ready quickly when there is.

Park Mode

If a slave does not need to participate in the channel (that is, it is no longer actively transmitting or receiving data, but needs to remain in the piconet and thus remain synchronized to the master), it must monitor the master's transmissions periodically so that it can keep synchronized. Park mode allows this by having the master guarantee to periodically transmit in a beacon slot. Because the parked slave can predict when a beacon transmission will happen, it can sleep until the master's beacon is due.

In park mode, the device relinquishes its AM_ADDR. Instead, when a slave is placed in park mode it is assigned a unique park-mode-address (PM_ADDR), which can be used by the master to unpark slaves.

Parked slaves must still resynchronize to the channel by waking up at the beacon instants separated by the beacon interval. A beacon offset and a flag are sent in the park message to indicate the instant when the beacon will first happen. A beacon interval is also sent in the park message. Beacons happen periodically separated by the beacon interval.

Park mode conserves the most power and would be appropriate for a device in our PAN that we would only want to randomly access—for example, the printer, which we could un-park when we required its services but not go through the lengthy inquiry procedure each time.

The headset profile allows park mode to be used with headsets, this is so that when an incoming call is received, a cellular phone can rapidly unpark the headset instead of having to wait for a lengthy connection procedure to finish.

Unparking

Via the beacon instant, the master can activate the parked slave, change the park mode parameters, transmit broadcast information, or allow the parked slave's request access to the channel. All messages sent from the master to the parked slaves are broadcasted, and to increase reliability for broadcast, the packets are made as short as possible.

Following the beacon slots, there are a number of access windows defined, through which parked slaves can request to be unparked. The access window that they request to be unparked in is determined by the PM_ADDR assigned to them by the master when they are parked. This allows the parked population to share the access windows, thus reducing the probability of a collision if two slaves require unparking at the same time. Slaves have to be unparked periodically by the master in order to ensure that they are present and that any virtual connections can be maintained.

Which Devices Need Low Power Modes?

In practice, most devices will need to support low power modes. Consider the case of a desktop PC. It is connected to mains power, so it has no need to save power. However, it could communicate with a battery-powered Bluetooth mouse, which will want to use sniff mode to extend its battery life. If the PC does not support sniff mode, the mouse cannot use it, and so its battery life can be seriously compromised by lack of features in the PC.

Similarly the PC may connect with a PDA which wants to synchronize and would like to be put in hold mode if the PC needs to interrupt the synchronizing process to go and service another device.

Park mode might be needed if the PC is connected to a cellular phone so that the PC's microphone and speakers can be used as a hands-free set for the phone.

Do not just consider the requirements of your product—think about the impact your product's capabilities could have on other devices used with it.

Providing Channel Quality Driven Data Rate

The Bluetooth specification provides a variety of packet types—single and multiple slot packets, each coming in medium- and high-rate types.

Multislot packets pack more data into longer packets, and provide higher throughput in noise-free environments, but their throughput is worse than single slot packets in noisy environments because they take longer to retransmit.

Medium rate packets have more error protection. This makes them tolerant to noise, but the space taken up by error protection means they cram less data into each packet. High-rate packets get better throughput in error-free environments, while medium-rate packets get better throughput in noisy environments.

Channel quality driven data rate (CQDDR) allows the lower layers of the Bluetooth protocol stack to measure the quality of the Bluetooth channel, and choose the packets most appropriate to the noise levels. Not all chips/chip sets implement CQDDR, so if you expect maximum throughput in noisy conditions to be an important factor for your product, you should ensure that you choose a chip/chip set which implements this feature.

Deciding How to Implement

Once you have made the decision to implement, what are the available options for Bluetooth technology enabling your products?

There are many options to consider in both hardware and software. Even once you have chosen a chip set and protocol stack, there are different ways that these can be added into your product. In this section, we shall begin by looking at software system architecture, then we'll consider some of the hardware options.

Choosing a System Software Architecture

The choice of system architecture will obviously be determined by footprint, cost, and time-to-market, but the end functionality will have the biggest influence. We will briefly examine the Bluetooth protocol stack as it can have an influence on our product's system architecture.

We will examine the stack in its simplest form—the upper stack and the lower stack. The lower stack controls all of the physical functionality, the radio, the baseband, and the Link Manager (LM) and Link Controller (LC) layers.

The upper stack deals with the channel multiplexing, with the logical link control and adaptation protocol (L2CAP). Serial port emulation and the interface with the application software happens in the RFCOMM layer. A Service

Discovery Protocol (SDP) layer is also essential for all Bluetooth devices, as it allows them to find out about one another's capabilities—an essential facility when you are forming ad-hoc connections with devices you may never have seen before.

There are three implementation models for the stack, dependant upon the functionality or resources the respective product has: *hosted, embedded,* and *fully embedded* (see Figure 1.13).

Figure 1.13 Software Stack Implementations

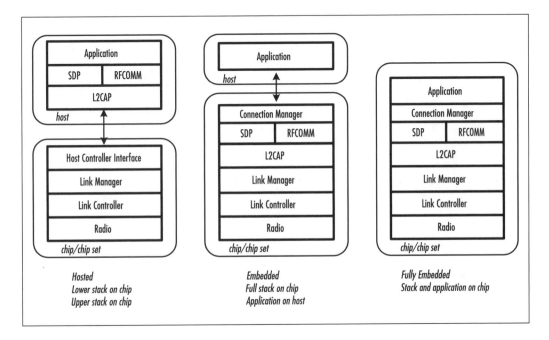

In the hosted model, the lower stack layers reside on the Bluetooth (BT) device, while the upper stack resides on a host (this may be a PC or a micro-controller if the product is mobile or standalone). They communicate via the Host Controller Interface, which sits between the lower layers and upper layers of the protocol stack forming a bridge between them. The two most common physical transports are UART (H4) and USB (H2). The UART protocol was designed for communication between chips on the same board and does not cope well with errors that occur in cables, so there are also proprietary transports which add extra facilities to the simple UART protocol. One example is CSR's BlueCore Serial Protocol (BCSP) which achieves a more reliable form of UART transport with retransmission and error checking. The hosted model is optimum for applications where powerful host processors are already available and there is plenty of memory. Examples of hosted devices include USB dongles, PCMCIA cards, compact flash cards, V90 modems, Internet gateways, and PC motherboards.

In the embedded model, the complete stack resides on a BT device, but a separate user application is running on a host. This model is ideal for 2 and 3G mobile phones, ticket or vending machines, or PC peripherals that have limited processing power and available memory.

In the fully embedded model, the complete stack and the user application are all on the Bluetooth device. There is limited memory resource on the BT device so any application will need to be relatively simple. The best example of a fully embedded device is a headset. It has no need for complex processing, so the whole Bluetooth stack can run on the single microprocessor within the BT chip/chip set.

The lower stack up to the HCI is always provided with the Bluetooth chip/chip set as it is unique to that silicon implementation. With the embedded model, the upper layers are also provided by the chip/chip set vendor—either free of charge if it is their own stack or there may be a license fee per device if they are using another vendor's upper layers. The fully embedded model requires a silicon solution that allows the application code to be written and downloaded to it without compromising the integrity of the Bluetooth stack which should have already undergone the stringent qualification procedure. Any changes to the stack requires it to be requalified!

The upper stack layers, above HCI, can be licensed from numerous vendors. Due to the inherent interoperability requirement of any qualified Bluetooth component, the choice is open. All of the available stack offerings "will" be compatible with the chosen silicon's lower layers. You can, of course, write your own upper layers, but it will be a vast software undertaking—illustrated by the cost of licensing one. Protocol stacks can be expensive, but an expensive stack might just offer you extra features which help to sell your product. Because of this, examine all the available options closely.

NOTE

As the Specification stabilizes, there will be chips entering the market dedicated to a specific purpose only—the headset profile being the primary example. The chips will have all the relevant stack layers and the profile implementation in masked ROM, reducing the cost significantly.

Constraining Implementation Options with Profiles

The Bluetooth profiles deliberately restrict implementation choices. If you are implementing functionality, which is covered by a profile, then you must implement that profile. This is intended to make it easier for devices to interoperate: if everybody implemented their own proprietary methods of communicating, then nobody's devices would ever work together.

You may find that you do not want to follow the profiles, many of them are compromises intended to provide functionality that will address a variety of potential use cases. This means that they may not be optimum for what your application and your product wants to do. This need not be a problem: once you have implemented the relevant Bluetooth profile, you are free to also implement your own proprietary solution.

You may find that having to implement profiles makes Bluetooth technology too burdensome, and this might start to make alternative technologies such as infra-red look attractive. However, you should consider that by implementing a profile, you have vastly increased the number of devices which will interoperate with your product.

Choosing a Hardware Implementation Option

Choosing a software architecture may limit the choice of hardware. Some chips/chip sets can not support the complete protocol stack, so if you do not have a hosted system, you will have ruled out these options. Still, there is likely to be a range of chips/chip sets open to you, each with its own inherent compromises, in time-to-market, cost, and R&D resource.

There are numerous solutions currently available from multiple vendors. Chip sets come as separate radio and baseband devices in a variety of technologies: silicon-germanium, silicon-on-insulator, and CMOS, or as single-chip CMOS device integrating the radio with the baseband. Chip set prices range from $8 to $29, although this will no doubt decrease with large volumes. This option is designed-in directly onto the product's printed circuit board (PCB).

The alternative to buying a chip set is to get a "module." These are PCBs complete with RF deign and antenna, and will be pre-tested and pre-qualified.

Tɪᴘ

All qualified Bluetooth components and products are listed on the "qual-ified products" section at www.Bluetooth.com (the official SIG Web site). Here you will find the manufacturers of chips/chip sets, modules, devel-opment kits, and software components. Data sheets or specifications can then be attained from the respective manufacturer's own Web site, as well as information on how to purchase.

The single chip/chip set approach requires an RF design resource to provide the matching networks, filters, amplifiers, and antennas to the transmitter and receiver paths and will require expensive synthesis and test equipment along with a lengthy qualification process. It will, however, incur a significantly lower finan-cial cost per unit along with a reduced PCB real estate overhead. Many chip/chip set manufacturers will supply you with reference designs. If you exactly follow their instructions, you can get away without designing your own system. You must be very careful if you are following a reference design; apparently insignifi-cant changes can alter the radio performance. For instance, changing the manu-facturer of a capacitor can change its characteristics even though it might be listed as the same value and type.

The module approach is far simpler since the primary RF hardware concern is soldering the module onto your motherboard. Keep in mind that it's larger to integrate onto your motherboard and financially more expensive. Figure 1.14 provides examples of some of the available options and their dimensions. The multiple chip approach separates baseband and radio into two packages, whereas single chip combines both. The single chip approach can also be divided into single chip plus flash (allowing larger flash memory), or single chip with inte-grated flash (for minimum size).

Whichever stack configuration you choose, you will still have to somehow add the hardware. There are two primary options for adding the Bluetooth hard-ware to a product: designing Bluetooth technology directly onto the PCB, or using a pre-qualified complete Bluetooth module. In the following sections, we will briefly question how each method will impact on time-to-market and what the more common risks of implementation are likely to be.

This is by no means a definitive summary. Every individual application will have its own unique implementation issues. You can, of course, employ a third party design house to do it for you and let their designers go through the learning process! The most expensive, yes, but if you have no R&D resource yourself, this may be your only route to joining in the Bluetooth Dream, and it is certainly easier than trying to recruit and manage a complete development team if you don't have one already. There are many design houses that now specialize in Bluetooth design, thus you would get the additional benefit of their experience.

Figure 1.14 Examples of Bluetooth Hardware Solutions

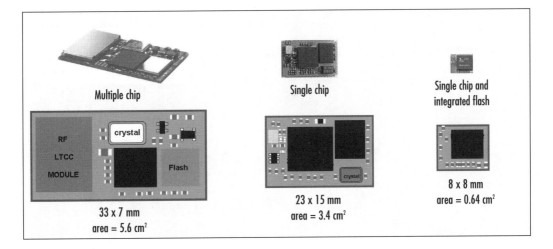

Design Bluetooth Directly Onto the PCB

Designing Bluetooth technology directly onto the PCB is the optimum method if PCB real estate or end unit costs are our primary design constraints. Choose the silicon wisely. Devices are available that have a comprehensive level of integration and do not require difficult-to-source/expensive external components—SAW filters being the obvious example. If we are using a "hosted" stack configuration, we need to ensure that the HCI transport is available and fully functional. As the Bluetooth system has many optional features, we also need to check that our chosen silicon vendors lower stack implementation provides the Bluetooth functionality that we require. PCB real estate needs to be available and thus will affect our choice of solution. A PCMCIA card or PC motherboard, for instance, is a predefined size, irrespective of component population. As a result, the smallest solution is not a primary objective; however for a headset, a compact flash card or a mobile phone size would be a significant determining factor.

PCB structure is an issue if we use this method. Due to the inherent nature of RF striplines and microstrip, a multilayer PCB is needed to give the required power planes, ground planes, and associated dielectrics, and to separate the digital signals to avoid noise pickup in the RF and crystal sections. The PCB is a high proportion of the manufacture cost, if the product typically uses a two layer PCB. This additional cost overhead can impact significantly on the total unit budget. For large PCBs, the cost of a multilayer board may swing the balance in favor of a separate Bluetooth module, allowing the multilayer section to be kept as small as possible. Figure 1.15 is an example of the PCB structure required for a Class1 Bluetooth design.

Figure 1.15 PCB Construction Example for a Class 1 Bluetooth Module

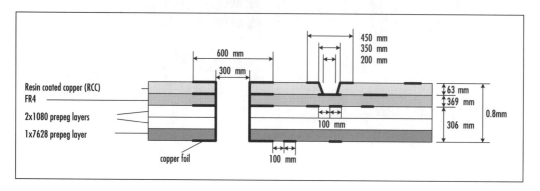

The fastest time-to-market approach if we use this method is for us to use one of the chosen silicon vendor's reference designs. These are normally free-of-charge on purchase of a Development Kit, and will have been proven and qualified. Most vendors provide a schematic and a set of Gerber files that can be imported into our own computer aided design (CAD) packages ensuring exact translation of the crucial PCB tracking layout. Some of us may know better, however, or have our own ideas (for instance, if a lumped balun is recommended in the reference design but you wish to use a printed one as a cost-saving exercise). Experience has illustrated that this can work but may incur repeatability problems with secondary PCB batches. You may wish to use a different power amplifier (PA) for a Class 1 design to the one recommended. Again, cost or a favorite supplier may be an influencing factor. Check with the silicon vendor. They would have evaluated several prior to selecting the one in the reference design. Most chip/chip set vendors work closely with the other Bluetooth component manufactures to provide us with a wide choice of options not all at a cost premium! To get Bluetooth technology into as many consumer products as possible, the ultimate aim is to get the Bill Of Material (BOM) cost on a downward spiral to the now infamous $5 target, which was set during the press hype of the initial technology rollout. This sum represents the cost to replace the average data cable! Figure 1.16 illustrates this method, showing the Bluetooth device and the flash memory (the two chips towards the bottom of the card).

Figure 1.16 Bluetooth Technology Designed Onto the PCB of a Compact Flash Card

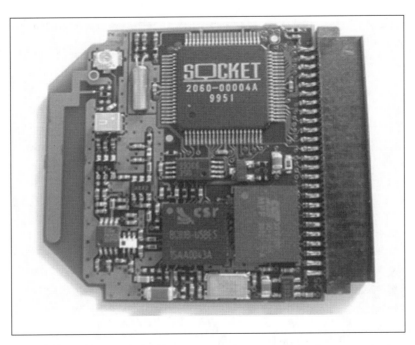

The most common risks associated with this approach can be very simple but add serious time delays to project schedules. A simple component change to improve a matching network between the Bluetooth device's transmitter output and a PA, for example, can incur problems with your manufacturer's component stock and tooling, and cause havoc with any quality assurance (QA) procedures that have been developed concurrently with the design to meet a project production deadline. Examples of the two problems that could have a significant impact on time-to-market are detailed next (design verification and manufacturing). However, test equipment incompatibility, qualification testing, and ultimately, production test development will also have their own impact.

Debugging…

Programming and Upgrading Firmware

How we get the firmware into our chip/chip sets could become a design nightmare considering that the Bluetooth specification is still undergoing revision, and the silicon vendors are still developing their lower stack firmware either for the purpose of adding new functionality or remedying interoperability "bugs." We must have a means to upgrade the firmware in our development labs, our manufacturing sites, or in the "field," if we have put our products onto the market.

All of the silicon available today uses flash technology as the storage media. This enables programmability for upgrades. The ideal scenario would be to program the flash initially via a programming/debug interface. This would require the respective interface pins from the chip to be brought out to pads on the PCB. In a development environment, we could then attach a cable; while in a production environment, we could use a "bed-of-nails" approach.

But what about the "field" products? Do they join the ever increasing pile of technical obsolescence, or do we recall them? Do we really want to put ten thousand or more products straight off the production line back through the same production line for reprogramming? The solution is to follow the example of those clever USB chaps: Device Firmware Upgrade (DFU). A DFU facility allows us to upgrade our products over the standard UART or USB interfaces via software, and requires no soldering of cables or secondary production runs. A "bootloader" is programmed into the chip when it is initially programmed via the methods previously mentioned. The bootloader can be used with upgrade software shipped with our products to provide the "in-the-field" upgrade facility to our customers.

As lower stacks mature and the specification stabilizes, this will not be such a pertinent issue. Nevertheless, before selecting a silicon solution check the programming and upgrade facilities that it offers you, and when designing your systems, consider how you might take care of upgrades both on the production line and in the field.

Design Verification

Design verification can be a problem: despite the most precise synthesis, the prototype may not always exhibit the same RF characteristics in reality. This can involve lengthy diagnosis, component changes, or a board respin if layout issues are suspected to be the cause of the problem.

This can be overcome by the development of several prototypes concurrently, as well as adhering to the selected silicon vendors' design guidelines. If advice states that the device is sensitive to noise, you will know not to run digital lines from the flash next to the Bluetooth device or under the system crystal, and to take de-coupling very seriously! Figure 1.17 illustrates the problems caused if a design routes the address and data bus (or another digital line that changes rapidly) near the crystal or its traces. Any digital signal has fast edges which can easily couple several millivolts into a small signal output from a crystal; this is not helped by the lack of drive you receive from a crystal. As the crystal output passes through the Phase locked loop (PLL) comparator, a slice level is used to determine if the crystal output has changed from a zero to a one, or vice-versa. If there are glitches on the crystal output from digital coupling that are greater than the hysteresis of the comparator, it can result in the square wave output having glitches or excessive jitter. Glitches can confuse the divider and phase comparator and result in excessive frequency deviation at the output, which will cause variations in the RF output.

Figure 1.17 The Effect of Routing Digital Signals Near a System Crystal

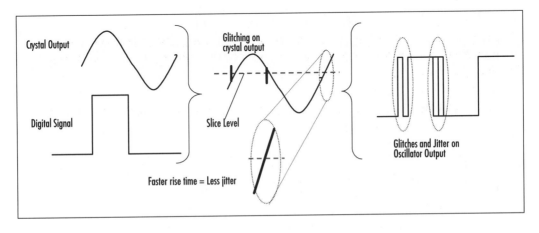

Figure 1.18 illustrates the noise incurred on the output spectrum due to insufficient filtering of the power supply to the BT device, the top trace. This will have a detrimental effect on system performance and will impact negatively on some of the qualification tests for frequency drift and drift rate.

Figure 1.18 The Effect of Poor Filtering on the Bluetooth Output Spectrum

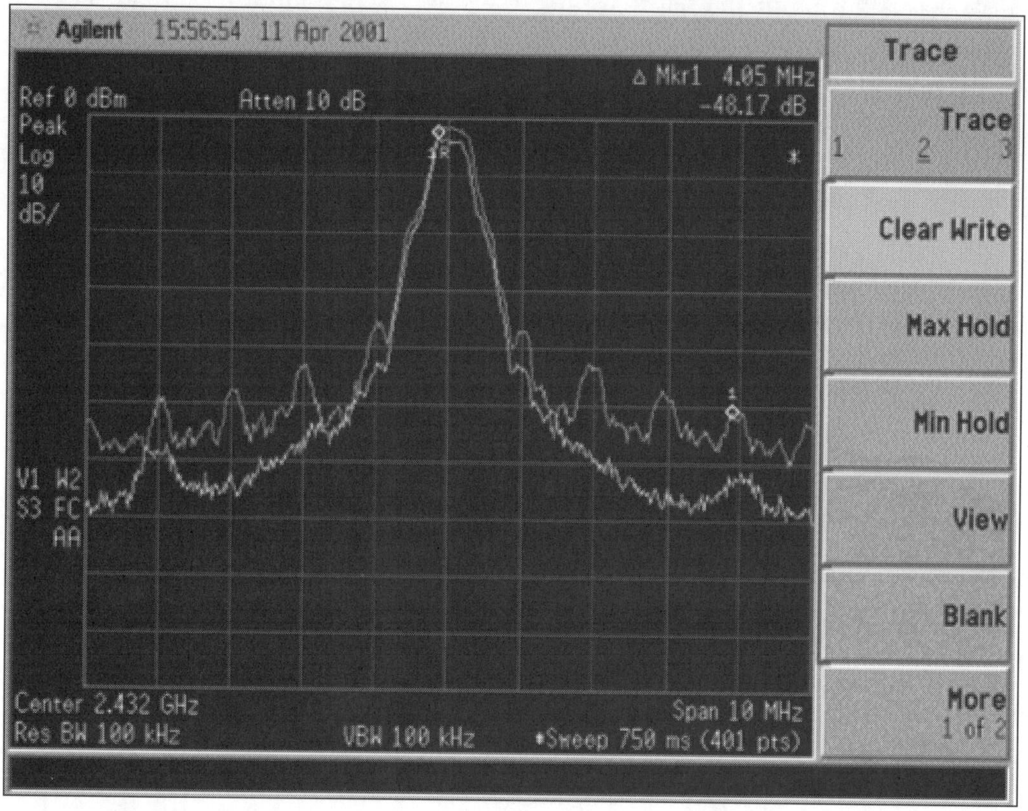

Manufacturing

As previously indicated, the manufacture of Bluetooth PCBs themselves can be problematic. Repeatability of performance with printed RF components and the expense of the multilayer PCB, as well as other problems can be incurred with component placement. As this method of design is optimum for size, the physical dimensions we are working with can be extremely small. This means we have to be precise not only in our layout for noise, feedback, and coupling issues, but also with pad size and component placement.

The Bluetooth chips/chip sets available are mainly packaged as ball grid arrays (BGAs), and the associated passive components have to be the surface mount 0402 type to adhere to the size constraint. There are many factors to take into account when using components on this scale: unless the solder resist finish is of the photo image type with a maximum thickness of 0.025mm, the 0402 resistors and capacitors could be lifted away from the pads on the PCB. A maximum solder resist window around the component pad should be in the region of 0.05mm with an alignment tolerance of 0.05mm to ensure that any tracks or vias between the pads of the BGA are not exposed, reducing the risk of short circuits. Figure 1.19 illustrates some of the problems expected if we get this wrong!

Figure 1.19 PCB Solder Mask Considerations

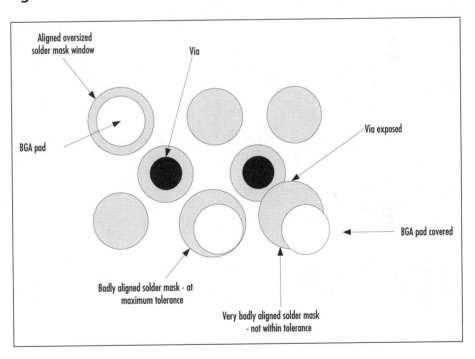

Using a Prequalified Complete Bluetooth Module

Using a prequalified complete Bluetooth module is optimum if time-to-market is our primary design constraint. We have the PCB real estate available and can transfer the additional cost per unit to our end users while remaining competitive.

Modules are available from numerous sources with a choice of Bluetooth silicon. They are available in class 1 or class 2 and can take several forms. Modules are currently being developed that integrate the entire external RF and system components (flash, crystal, filters, and amplifiers) into a single device—predicted sizes being as small as 5mm by 5mm! These modules are all pretested and prequalified, thus simplifying both the production test and qualification required for our end product. Examples of two modules currently available are illustrated in Figures 1.20 and 1.21.

Figure 1.20 An Example of a Class 1 Bluetooth Module (Courtesy of ALPS, Japan)

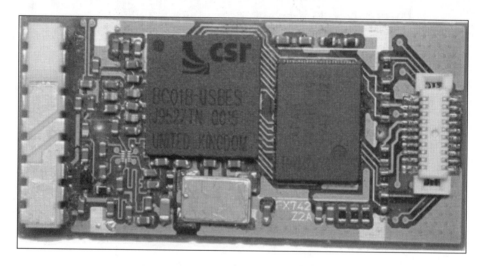

The PCB issues examined in the previous method are irrelevant when using a module since we just solder the module onto our own PCB. There are no new Bluetooth technology-induced RF layout considerations or BGA placement issues. Of course, we must ensure that the antenna is placed in a position where propagation is not adversely affected by surrounding components, and this will require some RF expertise. But antenna siting is really the only RF issue we have to think about.

This method, however, would not suit size-conscious products. The added module will currently increase the overall height of the PCB, which isn't appropriate if your product has to fit in a PC slot where dimensions are predefined and resolute. There are limitations to this method other than cost, size, and supply, although how seriously they affect us will be subjective and dependent upon our own requirements.

Figure 1.21 An Example of a Class 2 Bluetooth Module (Courtesy of Mitsumi, Japan)

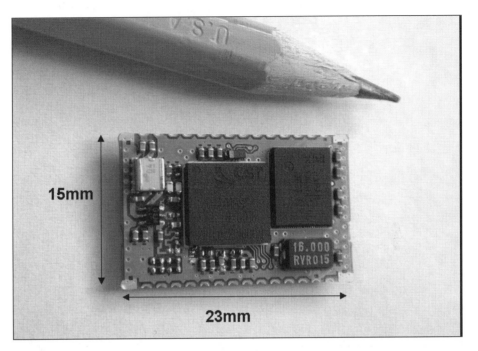

Firmware Versions

The module will be supplied with a version firmware deemed appropriate at manufacture. However, the respective silicon on the module may have undergone many revisions since the module was produced. We are dependant on the module manufacturer to provide us with access to this new firmware and provide us with the means to upgrade it.

Dependant for Functionality

The module we have chosen will be static. It will only provide us with the ability to configure our product designs according to its specification. If we require, for instance, to change the PCM interface to utilize a more price-conscious or better performing codec, we will require access to the Bluetooth chip/chip set that the module is based upon to reconfigure it. If this could affect your product, then ensure that this reconfigure option is available from the module you choose.

Developing & Deploying…

Obtaining Bluetooth Technology Qualification

In order to obtain qualification of a component or product, the manufacturer may use a test house for two services:

- The test house is contracted to make tests to a Bluetooth test specification, and to produce a test report containing the results of the tests.

- An employee of the test house who is appointed by the Bluetooth SIG as a Bluetooth Qualification Body (BQB) reviews evidence submitted by the manufacturer in a Compliance Folder (CF), and if satisfactory, the BQB submits the product or component to the Bluetooth Qualification Administrator (BQA) for listing on the Bluetooth Qualified Products List (BQPL).

A Bluetooth component is an implementation that contains some Bluetooth functionality, and which can be included into another component or product. It can be prequalified so that components or products containing the component do not have to be tested for the prequalified functionality. A Bluetooth product or end product is a device to be sold to the end user, and it can be made up of prequalified components to reduce the testing required by the product manufacturer.

The list that follows gives more details on the tests necessary for qualification:

- RF Tests are required to be made once for each new PCB design. If the same pretested module is reused in other end equipment, no tests need to be repeated.

- USB, UART, or BCSP variants should not need to be retested for RF as the HCI does not affect radio performance. PCB variants where all RF layout and components are identical should not need to be tested, subject to agreement with the BQB.

- The Bluetooth Qualification Body (BQB) may require one or more BB timing tests to be repeated for each new PCB design. This may not be necessary if the crystal is the same as used by the qualified component. If extra testing is required, one timing test needs to be tested at extreme conditions.

Continued

(Currently these tests can be performed by manufacturers using standard test equipment. In the future, there are plans to move this testing into test facilities.)

- Both Module manufacturers and end product users can use a software component that is prequalified at baseband (BB) and LM.

- If the new design includes the upper layer stack components HCI, L2CAP, RFCOMM, Service Discovery Protocol (SDP), or Bluetooth Profiles, these must also be qualified.

- Software components affecting profiles must be qualified. This could be done by developing and qualifying your own profile software components, or by buying in prequalified profile software components and integrating them into the end product.

Considering Battery Limitations

Current handheld PCs offer considerably longer battery life than notebooks because they do not have hard drives, CD-ROMs, or floppy drives. This makes it possible for users to work for hours, and in some case weeks, without having to worry about losing power. Most Palm-size PCs use AAA batteries that last for 20 hours to several weeks, while handheld-size PC batteries last from 8 to 15 hours on a single battery charge. Mobile phone battery technologies offer 130 hours standby and 5 hours talk time as standard. Key consideration when adding Bluetooth technology to any product is the additional power consumption inevitably reducing the overall battery life of the product. This is a serious consideration in products that are normally static, where battery life has not been an issue before and size constraint is predefined.

Due to the expected size constraint within a typical headset mould, a battery with a high charge density/gram would be the most effective solution to employ. A typical application example would be to have a headset capable of 2 hours talk time combined with 100 hours of standby time before recharging. Assuming that the headset has been paired, the RFCOMM connection has been established and the most optimum power configuration is used (see the following section), we can calculate the following:

Codec power consumption = 3 milliAmperes (mA)

SCO connection power = (28mA + 3mA) × 2 [2 hour talk time] = 62mAH

Standby power = (0.6mA) × 100 [100 hours standby] = 60mAH

Therefore we would select a battery capable of delivering 122mA hours of energy.

Table 1.4 illustrates some of the currently available rechargeable battery technologies indicating the respective weight energy density.

Table 1.4 Battery Technologies

Battery Technology	Operating Voltage	Weight Energy Density (WH/KG)	Number of Cells Required
Ni-Cd	1.2V	40 – 60	3
Ni-MH	1.2V	60 – 80	3
Li-ion Circular	3.6V	90 – 100	1
Li-ion Prismatic	3.6V	100 – 110	1
Li-ion Polymer	3.7V	130 – 150	1

Adding Batteries

We are all aware that although the lack of cables makes our lives convenient, the simple act of recharging batteries is tiresome. How many of us have picked up our cell phones to make a call and found it needs recharging? This even with the vast battery life and battery status indicators in current phone battery technology! The ultimate aim with any wireless product is to ensure the time differential between charging sessions will not affect the user's experience in other words, to make sure their products are not connected to the mains longer than they are wireless!

Long battery life means designs with low power as the primary objective. With any low power application, choice of design configuration is crucial in achieving the power consumption targets that you require for optimum use. Initially, there are the hardware configurations relating to choice of processor, design topology, asynchronous (event-driven) over synchronous (polling) designs. Then there is hardware power management and efficient power supply designs. Software considerations include speed gearing, idle, and sleep operations.

Fundamentally determined by the application is the system's design topology. This is the most effective utilization of the hardware and software parameters to achieve a design specifically targeted towards low power operation. Parameters we should consider include:

- Selection of duty cycles for active and passive periods

- Choice of power saving features versus system performance

- Vendor-specific deep sleep modes

Most of the silicon chip/chip sets or modules available offer a wide selection of options to provide for applications where power efficiency may be an absolute necessity, including on chip battery monitors. Check with the manufacturer's data sheets or specifications for information.

Using Power Saving Modes to Extend Battery Life

To appreciate how power saving modes can effect current consumption, we will again take the example of a headset and the audio gateway (AG) of a mobile phone.

The first step in establishing a functional system when both devices are virgin is *pairing*, where both the headset and audio gateway become aware of each other's BT addresses and generate the associated link keys. Generally authentication will be requested through the use of a PIN code which will be built into the headset at time of manufacture. Once paired, there is no need for the audio gateway to do inquiry and SDP searches for subsequent connections to the headset. During the pairing process, the headset must be in page scan mode so as to be able to connect to the audio gateway enquiry. A page scan interval of 800ms with a 12ms window is appropriate here since the connection is not time critical (the typical current figure in this state is 2.5mA). Once the pairing has been completed, the headset must decide to go into page scan again or to go idle.

Once the headset and audio gateway have paired, an RFCOMM link will need to be established before any communication can take place. This is usually initiated by a user action at both the headset and audio gateway. The headset will go to page scan mode where an interval of 800mS is sufficient, and the audio gateway will try to connect to the headset. Once a connection is established, the audio gateway will have control over the headset's power-saving features. Generally, a 40mS sniff mode interval can be set for a period of time in which some action may take place. This will allow acceptable delays for "Ring" commands or "Talk" button pushes while significantly reducing the power consumption figures of the headset (typically 5.5mA). Once it has been deemed that there is no further activity required, then the audio gateway can choose to disconnect altogether, or put the link into park mode.

NOTE

This example is based on CSR's BlueCore2 single chip CMOS device with a recommended operating voltage of 1.8V and optimum device configuration for low power in a Class 2 design. (A class 1 design requires a PA and therefore the power consumption of the PA would need to be considered.)

A beacon interval of about 1 second is appropriate for a parked headset link. This significantly reduces the headset power consumption (typically 1mA) while still allowing a rapid response to an incoming call even when the device is unparked. A rapid response is also possible if the headset initiates a button push: the button push triggers an unpark request on the next beacon, then the headset is unparked by the audio gateway and a SCO connection is established.

The audio gateway will ultimately decide the quality of the audio link and the power consumption of the headset during a SCO link. CVSD is the more appropriate method of encoding for use with speech and is mandatory with the headset profile. There is an option as to what type of packets should be used. HV1 will allow a clearer connection at the expense of increased power consumption. HV3, however, can reduce the power consumption by a third and take advantage of sniff mode. There will be some degradation in the quality of the audio link but the degree of degradation may not be sufficient to warrant the use of HV1. A good design can still give a very clear HV3 packet signal and decent voice intelligibility. For headset applications where voice bandwidth is already limited, HV3 would be the recommended packaging method. Once the call has been terminated, the audio gateway can decide whether to park the link once again or disconnect the RFCOMM connection. Usually, the link is put back into park mode. Table 1.5 summarizes the preceding scenario.

Table 1.5 Typical Power Consumption Figures

Mode	Remarks	Current
ACL Connection	Master [115k2 UART] no power saving	15mA
ACL Connection	Sniff mode, 40ms sniff interval [38k4 UART]	4mA
ACL Connection	Sniff mode 1.28s interval [38k4 UART]	0.5mA
Link Parked	1.28s interval [38k4 UART]	0.6mA
SCO Connection	HV1 packet, CVSD encoded, no sniff interval	53mA
SCO Connection	HV3 packet, CVSD encoded, 40ms sniff interval	28mA
Deep Sleep	CSR proprietary power saving mode	50µA

Figure 1.22 illustrates the scenario just described indicating the complete procedure with current consumption per action.

Assessing Battery Life

As we are now acutely aware, a Bluetooth device consumes current, and thus can have an influence on the battery life of any Bluetooth-enabled product. For products with powerful batteries inherent to their normal use, a laptop being the primary example, it will not be a significant issue. For smaller products like mobile phones and PDAs, it could impact on the overall time available for use. We have examined the power consumption on the headset and AG scenario, which is restricted in its functionality, but a multifunctional product like a PDA will have many varying needs for power, dependant upon what activity it is involved in—exchanging a business card, waiting for an e-mail, or Web browsing will all involve different connection models. We will now consider this in a "real-life" situation. To try to get an objective view of the effect on battery life of the Bluetooth functionality in a PDA, it is necessary for us to make some assumptions. These assumptions are variables but will give us a viable model to consider.

Figure 1.22 The Headset and AG Scenario with Current Consumption

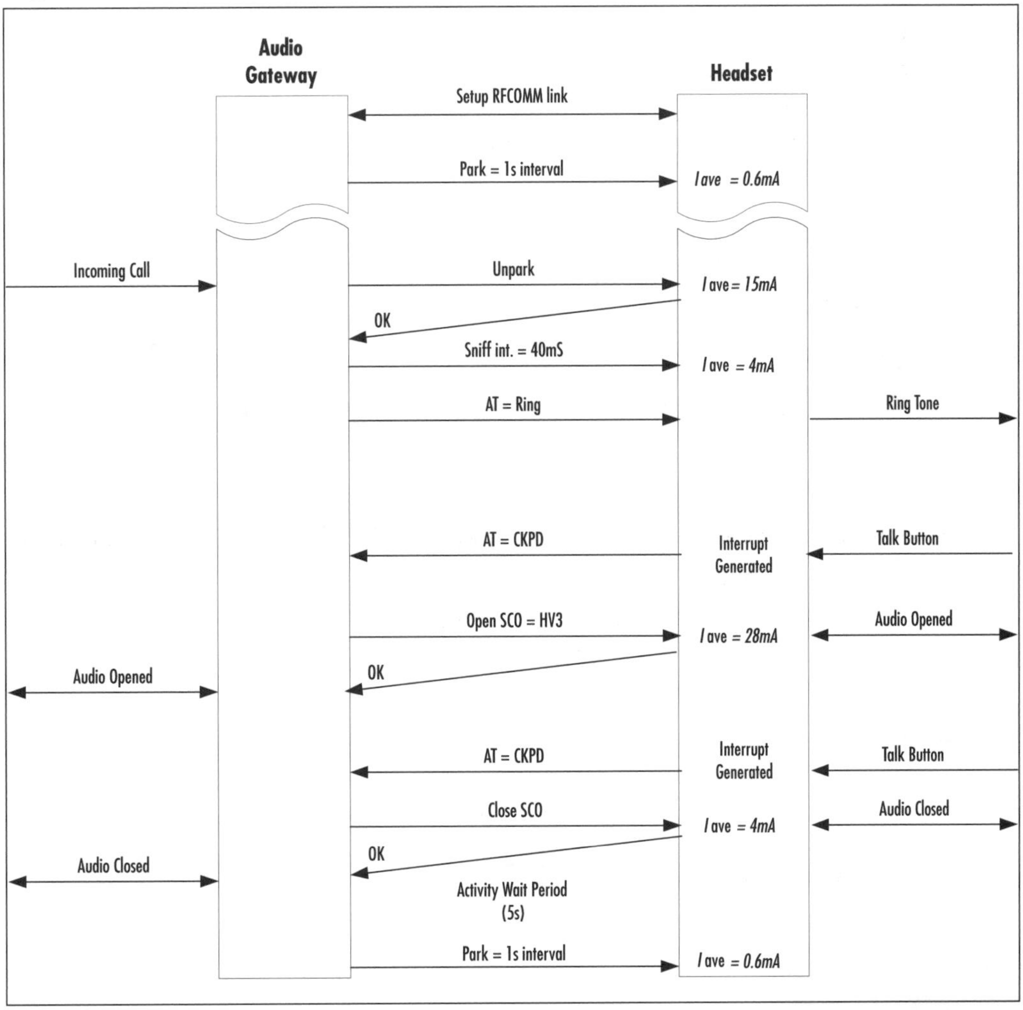

Let us assume that when the PDA is on (with the Bluetooth unit fully powered and operational for eight hours per day), the number of times it is used are limited to:

- Four Web browsing sessions
- The exchange of nine business cards
- Two 30-minute presentations using the PDA as a radio mouse
- Receiving an e-mail every hour
- Using power-saving modes built into the Bluetooth system

For the purpose of modeling power consumption, we need to define a number of states that the Bluetooth device could be in, and for each state, note the power consumption:

- **State 1: Inactive** The device is powered, with clocks running and ready to receive commands over HCI. It has no active connections. Consumption average = 50µA

- **State 2a: Discoverable and Connectable** The device is performing Inquiry Scan and R1 Page Scan every 1.28 seconds. The PDA software will probably require human input to put it into this mode. A timeout will return it to the inactive state after some time if no connection is made, perhaps after 1 minute. Consumption average = 1.3mA

- **State 2b: Connectable but not Discoverable** The parameters are the same as state 2a, except that only Page Scan is enabled. Consumption average = 0.6mA

- **State 3: Paging** The master of the piconet has to page a known slave in order to establish the baseband connection. The time this takes depends on the duty cycle of the slave device. We will assume that the slave is using the parameters described in State 2b, in which case, the mean time to connect is 1.5 seconds. Consumption = 41mA

- **State 4: Connection establishment and parameter negotiation** Once a baseband connection is established, the slave and the master transmit or receive in nearly all slots. There may be power control, authentication, SDP database searches and other management traffic before the link is fully established. This takes on the order of 250 milliseconds (ms), determined by the reaction times of the host. Consumption = 47mA

- **State 5a: Connected, low latency**   The device is a slave in a piconet (in sniff mode). The latency before data flows can be up to 40 ms, but the mean is 20. The latency is programmable. Consumption = 4mA for a 40ms sniff interval.

- **State 5b: Connected, high latency**   The device is a slave in a piconet, in sniff or park mode. The latency before data flows can be up to 1.28 seconds, but the mean is 0.64 seconds. The latency is programmable between zero and 42 seconds. 1 second is a usual compromise. Consumption = 0.5mA for 1.28 second sniff or beacon interval.

- **State 6: Data transfer in progress**   We assume a UART connection. The consumption depends on packet rate and whether the unit is a master or a slave. However, with appropriate choices of sniff parameters, the slave and master will have similar consumption. Consumption = 15mA for an ACl link with a baud rate of 115k2.

With these states defined, we can now examine use of the PDA in specific activities to determine what the power consumption is expected to be:

Web browsing   The user initiates connection to an access point, and the PDA enters State 3 and then State 4. Once connected to the access point, an IP connection is made to the Internet. The slave listens in every slot by default but transmits infrequently. It may request sniff mode; a mean latency of 20ms is appropriate and will dramatically reduce consumption during the time, assume 10 seconds, during which the URL is being searched for. This is State 5a. While data is transferred, assume a mean transfer rate of 24 Kbps, limited by the Internet (the slower this is, the longer it will take, and hence the more pessimistic the result). Assume 90k of bytes transfer, comprising 2 or 3 GIF or JPEG files and one HTML page. Thus, it is in State 6 for 3.84 seconds. Following the data transfer, the device returns to State 5a for a time (e.g., 10 seconds), and then to State 5b if no more traffic is seen. After a further timeout period (for example, 120 seconds), it disconnects and returns to State 1.

Business card or file exchanges with another PDA   The users put one PDA into discoverable mode, State 2a, and the other initiates a connection, entering State 3 and leading to State 4. For this analysis, we will consider the slave only. After connection establishment, the data transfers at the ACL data rate allowed by the UART, the filing systems and the upper layer stacks on the two PDAs. Assume a low speed UART, 38k4 and a small file; 1000 bytes is typical for a business card or diary synchronization. The device is thus in State 6 for approximately 0.3 seconds. Following the data transfer, the connection is broken and the device goes to State 2b for 60 seconds. After 60 seconds, it returns to State 1. Clearly, these timeouts are under the control of the application programmer.

Use as a "cordless mouse" (to control a PowerPoint presentation, for instance)   The user initiates connection to the PC, and the device enters State 3 and then State 4. Typically, the PC will request a role switch, become master and put the device into State 5a. This lasts the length of the session; there is no timeout. Let us assume a 30-minute presentation, after which the user ends the session and the device returns to State 1.

"Unconscious" synchronization   The purpose of this use case is to ensure that the diary or e-mail inbox is always up to date. The PDA runs a daemon. Every so often (5 minutes) it tries to connect to the access point(s) it is paired with, by entering States 3 and then 4. An alternative scenario is for the PDA to do an Inquiry to look for public access points instead, or in addition to the ones it is already paired with. The slight extra traffic required for this is ignored here. Once connected, an IP connection to the appropriate server is made. The slave listens in every slot by default, but transmits infrequently. It may request sniff mode: a mean latency of 20ms is appropriate and will reduce consumption during the time the database is being searched, so the master should put the PDA into State 5a. Let us assume the connection is up for 2 seconds while the server responds, and the data to be transferred, when there is some, is 30K. Further assume that there is new data only once per hour. Thus, when there is new data, the PDA is in State 6 for about 3 seconds, assuming a UART speed of 115k2 baud. After the data, if any, is transferred, the application disconnects and the unit returns to State 1 until the next time it is scheduled by the daemon.

Table 1.6 illustrates the actual power consumption for each of the specific activities previously listed. The model assumes a Class 2 device is used based on CSR's BlueCore2. We must note that there are many variables in each case, and this is only recommended as a model to provide us with some guidance to enable us to determine the effect of Bluetooth technology within a multifunctional device such as a PDA. It is apparent from this table that the proportion of the time that data is being transmitted or received is low, and that the average current consumption is dominated by the time spent in power saving modes.

Table 1.6 PDA Power Consumption for Specific Activities

Use Case	Web Browsing	Object Exchange	Mouse or Keyboard	Unconscious Synchronization
Number of sessions per day	4	9	2	96
Number of pages or files downloaded	3	1	-	8
Time in State 1	27785.92	28210.05	25196	28392
Time in State 2a	0	45	0	0
Time in State 2b	0	540	0	0
Time in State 3	7	0	3.5	168
Time in State 4	1	2.25	0.5	24
Time in State 5a	120	0	3600	192
Time in State 5b	840	0	0	0
Time in State 6	46.08	2.70	0	24
Consumption in mAH per day	65	0.3	4.2	2.7

The application program, which is above the Bluetooth specified profiles, determines the efficiency of the use of power saving modes and will be a very important differentiator between manufacturers or software providers. This clearly illustrates the importance of the application programmer being aware of hardware performance issues.

Summary

We began this chapter by examining the factors that may influence whether a product is a suitable candidate for becoming Bluetooth-enabled. The answer is that a device is suitable if data rates of a few hundred kilobits per second are adequate, if it can tolerate short outages in the communications link, if instant connections are not needed, if it can cope with the power consumption of the Bluetooth system, if a range of 100m or less is adequate, and if Bluetooth technology will add end-user value by increasing usability or functionality.

It is all very well to talk of "adding end-user value" but sometimes it is not obvious how that can be achieved, so it is important to consider how Bluetooth technology can add value to various products. The primary value add is through enabling unconscious connectivity, through the ability to seamlessly connect devices without lengthy software installation and configuration.

A product that misses its market is no good to anyone; time factors must also be examined when implementing a Bluetooth device. There is a significant learning curve, and development takes time. Finally, qualification and type approval are necessary before a product can go to market. These factors may mean that adding Bluetooth wireless technology may not be compatible with your product's development cycle.

Before deciding to add Bluetooth capability to your product, you must be aware of the performance limitations of wireless links. It can take ten seconds to find a Bluetooth device and the same again to connect with it. Once connected, data rates in the hundreds of kilobits are to be expected, but these may be reduced drastically by interference. Latency (delay) on the link is likely to be significantly higher than for wired links.

Before choosing hardware, it is wise to assess the features which Bluetooth technology offers, decide whether you need them in your product, and whether they should be enabled by default. Security features can make it difficult to establish links, but offer privacy when enabled. Low power may not be needed by your product, but you will still need it if you are likely to connect with devices which require low power modes.

Once the decision to implement is taken and you are broadly familiar with the criteria for choosing between Bluetooth solutions, there are many options for hardware and software. The protocol stack on a chip can stop at a host controller interface allowing the higher layers of the Bluetooth protocol stack to run on a separate host processor. Alternatively, the whole stack can be embedded on a

Bluetooth chip/chip set. In the latter case, the application could be run on the Bluetooth chip, or on a host device.

When looking at hardware implementation, there are many more options to consider. Either a single chip or a chipset incorporating multiple chips can be chosen. Factors which can influence chip/chip set choice include available space, power consumption, and, of course, price.

Once the silicon is chosen, you must decide upon a design strategy: whether to design your own PCB, or use a prequalified module. A module is undoubtedly the faster and easier option, but your own PCB can give you more flexibility in component placement, and for very high-volume products will be cheaper in the long term.

Finally, you may have to consider batter technology. Obviously, not an issue for anything connected to the mains, but many Bluetooth devices will be hand-held and will require batteries. Bluetooth subsystems will drain the battery when active, but the good news is that most of the time they are not active, and there are many long life battery technologies available which are adequate for the power requirements of the Bluetooth subsystem.

Many of the issues in this chapter may seem to be the province of the hardware designer, and you might wonder why they are included in a book on applications. We have seen, however, that hardware choices influence the available features used by software, so it makes sense for our introductory chapter to take a holistic view of Bluetooth products.

Solutions Fast Track

Why Throw Away Wires?

☑ You know Bluetooth technology is a good idea if your product satisfies the following six criteria:

1. Adds usability, convenience, or ease-of-use—the Bluetooth Dream!

2. Interference or latency will not affect its primary function.

3. Is tolerant to the connection time overhead.

4. Can afford the limited Bluetooth bandwidth.

5. Battery life or power supply requirements are compatible.

6. The range is adequate.

Considering Product Design

☑ Think about the following items:

- Are you adding end-user value by using Bluetooth technology?

- Does your product's development cycle allow you to add Bluetooth technology to it?

Investigating Product Performance

☑ To know whether Bluetooth technology is right for your product, you must consider:

- Connection times—it can take up to ten seconds to find a device and ten more seconds to connect

- The quality of service—throughput and latency; this will be lower than wired links

- Interference can badly slow down your links, or even cause them to fail

Assessing Required Features

☑ Question whether or not you need to support all the following features:

- Security—you must support it, but will you enable it by default?

- Low power modes—if your product doesn't need them, will it connect with one that does?

- Channel Quality Driven Data Rate—is maximum throughout in noisy conditions important?

Deciding How to Implement

☑ Should your stack be hosted, embedded with application on host, or fully embedded?

☑ Should you design your own PCB (cheap in volume), or buy in a module (faster and easier)?

☑ Battery—if your product is not mains-powered, consider the impact of time spent in different modes on the battery life. Constantly running in scan modes might give you fast connection time, but it will also rapidly drain your batteries. Setting short windows of activity can give almost equivalent performance, and greatly extend your battery life.

Frequently Asked Questions

The following Frequently Asked Questions, answered by the authors of this book, are designed to both measure your understanding of the concepts presented in this chapter and to assist you with real-life implementation of these concepts. To have your questions about this chapter answered by the author, browse to **www.syngress.com/solutions** and click on the **"Ask the Author"** form.

Q: Should I embed the whole stack, or use the host controller interface?

A: This depends on whether you have a host processor with spare resources available. If you have an application which runs on a host device, such as a PC with a powerful processor and lots of memory, then you should run the upper protocol stack on the host and connect to the Bluetooth subsystem using the Host Controller Interface. If you have an application like a headset where your existing device has no processor at all, then you should run the whole Bluetooth solution lower stack, upper stack, and application on one processor to save power, cost, and space. If you have a host with limited resources, such as a mobile phone, you may do best taking an intermediate approach and running the whole stack on the Bluetooth processor instead of running the application on your host processor.

Q: Which hardware solution is for me? A complete prequalified module or a chip?

A: This is dependant upon what your primary design constraint is—cost, time-to-market or PCB real estate—and the recourses you have available. The chip/chip set designed onto your product motherboard will ultimately be the most cost effective option per unit and afford you the smallest footprint but you will require RF design skills and equipment and can encounter significant problems with PCB layout, affecting the performance of your design. This approach also requires that you undergo all of the stringent qualification tests—the chip/chip set you use will ultimately be prequalified, but you will need to perform all the RF tests on your hardware. The module approach offers a faster time-to-market, but the cost overhead per unit will be increased and you will be limited to functionality.

If you need to get to market in a hurry, then a module is probably the way to go. If you have time, development resources with knowledge of radio hardware, and you are anticipating very high volumes for your product, then a chip may be the best option.

Q: Generally, what is the range of battery life?

A: This depends upon the product functionality. Power consumption is much higher when either transmitting or receiving, so the longer you expect your product to be in these states the shorter the battery life. Clever power management design, battery monitoring and use of the Bluetooth power saving modes will all contribute to reducing power consumption.

Chapter 2

Exploring the Foundations of Bluetooth

Solutions in this chapter:

- **Reviewing the Protocol Stack**
- **Why Unconnected Devices Need to Talk**
- **Discovering Neighboring Devices**
- **Connecting to a Device**
- **Finding Information on Services a Device Offers**
- **Connecting to and Using Bluetooth Services**

- ☑ **Summary**
- ☑ **Solutions Fast Track**
- ☑ **Frequently Asked Questions**

Introduction

Bluetooth wireless technology differs from wired connections in many ways. Some differences are obvious immediately: when you are not tied to a device by a cable, you have to find it and check if it is the device you think it is before you connect to it. Other differences are more subtle: you may have to cope with interference, or with the link degrading and dying as devices move out of range.

If you're used to developing applications for static wired environments, all of this may sound daunting, but don't worry—there are simple well-defined procedures for coping with the complexity of Bluetooth connections. This chapter will take you through those procedures step by step, along the way explaining the pitfalls and how to avoid them.

We will start with a review of the protocol stack, and then look at some of the basic requirements of wireless communications the stack cannot hide: finding nearby devices, connecting to them, discovering what services they can provide, and then using those services.

You need to know the basic structure of the Bluetooth protocol stack before reading this chapter.

Reviewing the Protocol Stack

The wide range of possible Bluetooth applications means that there are many Bluetooth software layers. The lower layers (Radio Baseband, Link Controller, and Link Manager) are very similar to the over-air transmissions. They can provide voice connections and a single data pipe between two Bluetooth devices. To ease integration of Bluetooth into existing applications, the specification provides middle layers that attempt to hide some of the complexities of wireless communications. In combination, these layers, when transmitting, can take many familiar data formats and protocols, package them, multiplex them together, and pass them on in a manner that matches the lower layers' capabilities. Matching layers at the receiving end de-multiplex and un-package the data.

At the bottom of the stack are some layers that are fundamental to Bluetooth wireless technology: Radio Baseband, Link Manager, Logical Link Control and Adaptation Protocol (L2CAP), and Service Discovery Protocol (SDP). Above these layers, different applications require different selections from the higher layers. Each profile calls up the higher layers it requires. If you implement more than one profile in your application, you may be able to reuse the common layers. Not all stack vendors support all layers so, if you are buying in a stack, make sure that it supports the layers required for your application's profiles. Figure 2.1 shows the layers defined by the Bluetooth specification (shown unshaded) and some other common layers (shown shaded).

Figure 2.1 Bluetooth Protocol Stack

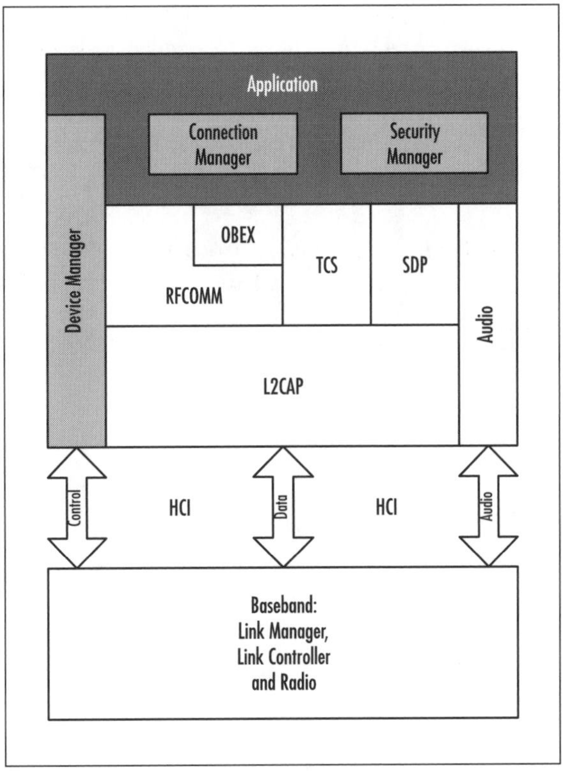

L2CAP

Logical Link Control and Adaptation Protocol multiplexes upper layer data onto the single Asynchronous ConnectionLess (ACL) connection between two devices and, in the case of a master device, directs data to the appropriate slave. It also segments and reassembles the data into chunks that fit into the maximum HCI payload (the HCI is the Host Controller Interface, which connects higher layers on a host to lower layers on a Bluetooth device). Locally, each L2CAP logical channel has a unique Channel Identifier (CID), although this does not necessarily match the CID used by the remote device to identify the other end of the same channel. CIDs 0x0000 to 0x003F are reserved with 0x0000 being unused; 0x0001 carrying signaling information; and 0x0002 identifying received broadcast data.

Debugging...

Reliability of L2CAP

Because of the nature of wireless communications, the links provided by the baseband are not reliable. Errors are caused by radio interference or fading of signals. There is a chance that two or more errors in a packet will combine to give a packet that contains errors but still has a correct checksum. The Bluetooth Special Interest Group (SIG) is considering implementing error correction at L2CAP, which would make such errors less likely to affect applications.

The stack layers that sit above L2CAP can be identified by a Protocol Service Multiplexor (PSM) value. Remote devices request a connection to a particular PSM, and L2CAP allocates a CID. There may be several open channels carrying the same PSM data. Each Bluetooth defined layer above L2CAP has its own PSM:

- SDP – 0x0001
- RFCOMM – 0x0003
- Telephony Control Protocol Specification Binary (TCS-BIN) – 0x0005
- TCS-BIN-CORDLESS – 0x0007

L2CAP only deals with data traffic, not voice, and all channels, apart from broadcasts (transmissions from a master to more than one slave simultaneously), are considered reliable.

RFCOMM

RFCOMM (a name coming from an Radio Frequency [RF]-oriented emulation of the serial COM ports on a PC) emulates full 9-pin RS232 serial communication over an L2CAP channel. It is based on the TS 07.10 standard for a software emulation of the RS232 hardware interface. TS 07.10 includes the ability to multiplex several emulated serial ports onto a single data connection using a different

Data Link Connection Identifier (DLCI) for each port. However, each TS 07.10 session can only connect over a single L2CAP channel and thus only communicate with one device. A master device must have separate RFCOMM sessions running for each slave requiring a serial port connection.

Version 1.1 of the Bluetooth specification has added to the capabilities of the standard TS07.10 specification by providing flow control capabilities. This caters for mobile devices with limited data processing and storage capabilities allowing them to limit the incoming flow of data.

OBEX

The Object Exchange standard (OBEX) was developed by the Infrared Data Association (IrDA) to facilitate operations common to IR-enabled devices like personal digital assistants (PDAs) and laptops. Rather than develop a new standard, the Bluetooth SIG took OBEX largely as is, detailed a few specifics regarding Bluetooth implementation (e.g., making some optional features mandatory), and used it in the File Transfer, Synchronisation, and Object Push profiles. OBEX allows users to put and get data objects, create and delete folders and objects, and specify the working directory at the remote end of the link. IrDA has also provided formats for data objects, while the Bluetooth specification has adopted the vCard format for business card exchange and the vCal format for exchanging calendars.

PPP

The Point-to-Point Protocol (PPP) is the existing method used when transferring Transmission Control Protocol/Internet Protocol (TCP/IP) data over modem connections. The Bluetooth specification reuses this protocol in the local area network (LAN) Access Profile to route network data over an RFCOMM port. Work is already underway on a TCP/IP layer that will sit directly above L2CAP, bypassing and removing the overhead of PPP and RFCOMM. This work is hinted at in some areas of the specification, but in v1.1 PPP, is all that's available.

TCS Binary

Telephony Control Protocol Specification Binary (TCS Binary, also called TCS-BIN), is based on the International Telecommunication Union-Telecommunication Standardization Sector (ITU-T) Q.931 standard for telephony call control. It includes a range of signaling commands from group management to incoming call notification, as well as audio connection establishment and termination. It is used in both the Cordless Telephony and Intercom profiles.

SDP

The Service Discovery Protocol differs from all other layers above L2CAP in that it is Bluetooth-centered. It is not designed to interface to an existing higher layer protocol, but instead addresses a specific requirement of Bluetooth operation: finding out what services are available on a connected device. The SDP layer acts like a service database. The local application is responsible for registering available services on the database and keeping records up to date. Remote devices may then query the database to find out what services are available and how to connect to them. The details of service discovery can be complex and are discussed further in Chapter 5, but each profile describes exactly what information should be registered with SDP based on the application implementation.

Management Entities

Device, Security, and Connection Managers are not protocol layers so much as function blocks. The Device Manager handles the lower level operation of the Bluetooth device. The Connection Manager is responsible for coordinating the requirements of different applications using Bluetooth channels and sometimes automating common procedures. The Security Manager checks that users of the Bluetooth services have sufficient security privileges.

HCI

The Host Controller Interface is not a software layer, but a transport and communications protocol that aids interoperability between different manufacturers' solutions. It is not mandatory to use the HCI interfaces defined in the specification (Universal Serial Bus [USB]; RS232; or a simple Universal Asynchronous Receive Transmit [UART]), or indeed any HCI transport at all, if there are better solutions for your application.

Lower Layers

The lower layers (Radio Baseband, Link Controller, and Link Manager) format the over-air transmissions, handle error detection and re-transmission, and manage the links between devices.

Table 2.1 illustrates which profiles use which layers.

Table 2.1 Stack Layer Requirements by Profile

Profile	Lower Layers	L2CAP	SDP	RFCOMM	PPP	OBEX	TCS-Bin
Service Discovery Application	X	X	X				
Cordless Telephony	X	X	X				X
Intercom	X	X	X				X
Serial Port	X	X	X	X			
Headset	X	X	X	X			
Dial-up Networking	X	X	X	X			
FAX	X	X	X	X			
LAN Access	X	X	X	X	X		
Generic Object Exchange	X	X	X			X	
Object Push	X	X	X			X	
File Transfer	X	X	X			X	
Synchronization	X	X	X			X	

Why Unconnected Devices Need to Talk

As mentioned in the Introduction, not all the details of operating a radio communication link can be hidden from the application by intervening software layers. Some of the basics of wireless communications will be exposed and it is essential to handle these functions correctly if operation is to be as seamless as Bluetooth proponents envisage. With wired connections, the user might check that two devices have the same type of physical interface port, that the ports support the same communications protocol, and that both devices run applications that can use this protocol to talk to each other. If all these checks are passed, the user might then plug a cable into the two ports and expect some useful communication. With Bluetooth devices, the user may not initially know that there are other Bluetooth devices nearby, so a method is required to find them. Then there is the Bluetooth equivalent of plugging in a cable: forming a connection. The checks on communications protocols and applications compatibility are actually done once a basic Bluetooth link is established. They are called *service discovery*.

This is not a book about the details of Bluetooth radio operation, but a little knowledge about a few fundamental principles of the radio and baseband will greatly help you understand what application level decisions are key, why they are key, and how making the wrong decisions could lead to some very undesirable behavior.

First, it is important to understand that Bluetooth radios use a frequency-hopping scheme. When connected, the precise frequency for each hop is selected by a pseudorandom algorithm that depends on the master device's clock and Bluetooth address. Slaves in a piconet synchronize on the master's hopping pattern. However, when unconnected, there is no master to synchronize to. Bluetooth devices need a way to exchange a limited amount of data, allowing them to find and connect to each other before synchronizing on a common clock and Bluetooth address.

The procedure used to find devices is called *inquiry*, and the procedure used to connect to devices is called *paging*. In both cases, one device transmits and receives on special sequences of frequencies that are known to all devices. The other device needs to be listening for the transmissions—if a transmission is received correctly, it sends out a reply. Since it knows the sequences used for inquiry and paging, it can work out the correct frequency on which to send the reply. The key points are:

1. The application must place a device in a listening mode if it is to be found or connected to. The listening mode that allows a device to be found is called *discoverable mode* or *inquiry scanning*. The listening mode that allows a device to be connected is called *connectable mode* or *page scanning*. The terms *discoverable* and *connectable* are used at the user interface, and the terms *inquiry scanning* and *page scanning* are used within the software layers.

2. Whether finding or connecting, for communication to take place, one device must transmit on the frequency that the other is receiving on. This is done by the transmitter changing frequency quickly (1600 times a second) while the receiver changes frequency slowly (every 1.28 seconds). Their frequency hopping is not synchronized, so the procedure must last long enough for the two devices to collide on a frequency that isn't subject to interference. This also introduces a random element to the procedure: how long they take before transmitting/receiving on the same frequency.

3. A Bluetooth device will not reliably find or connect to other devices at the same time as transferring voice. Voice links take priority over everything, while inquiry and page operations take precedence over other data transfers. It is allowed to inquire and page in the gaps between voice transmissions, but because the voice transmission takes priority, often responses will be lost due to a voice transmission, so finding and connecting devices can be slow and unreliable when voice links are in use. You must be aware of these limitations when deciding how your application will behave.

In the following sections, we will discuss the inquiry and page procedures in more detail.

Discovering Neighboring Devices

All Bluetooth devices must be discovered before a connection to them can be initiated. You may not need to carry out a device discovery every time you wish to connect to a device. Instead, you might be able to reuse information gathered from a previous device discovery. There must always be an initial device discovery before the connection, however.

There are two reasons to carry out device discovery. Either you do not know what devices are within range and wish to find out, or you know a device is within range and want to know its details so you can connect to it. In both cases, the procedure is the same and is called an *inquiry*.

Inquiring and Inquiry Scanning

To discover other nearby devices, a Bluetooth device conducts an *inquiry*. The basic command is *HCI_Inquiry* and has three parameters:

- **Lower Address Part (LAP)**
- **Inquiry_Length**   The inquiry will time-out after this period. Note that this parameter is in 1.28s units.
- **Number_Of_Responses**   If the number of responses given here is reached, then the inquiry will end before the *Inquiry_Length* period has elapsed.

The LAP determines the Inquiry Access Code (IAC) used in the transmitted ID message which listening devices respond to.

Debugging...

Messaging across HCI

Some host stacks do not handle multiple simultaneous transactions across HCI. These protocol stacks will wait for one command to complete before sending the next. If you have one of these stacks, then the *inquiry cancel* command will not work: this is because the *inquiry* command will be allowed to run until the *inquiry complete* event returns from the lower layers. Only after the *inquiry complete* has been returned will the next command (*inquiry cancel*) be sent. This means that the *inquiry cancel* is sent after the inquiry has already completed, so the lower layers respond with an error message as they cannot cancel an inquiry which is not in progress.

This is a rare problem as few commercial stacks now available cannot handle multiple simultaneous HCI transactions. But if you find your HCI misbehaving, it is worth investigating whether your stack is one that queues up messages for simultaneous HCI transactions rather than sending them to the lower layers.

There is also the option for the application to use *HCI_Periodic_Inquiry_Mode* and configure the Bluetooth lower layers to conduct periodic inquiry procedures automatically. There are corresponding commands, *HCI_Inquiry_Cancel* and *HCI_Exit_Periodic_Inquiry_Mode*, which cancel the inquiry commands.

The listening mode for inquiry is called Inquiry Scan. Only devices in Inquiry Scan will respond to inquiries and then only to inquiries which contain the correct IAC. This has consequences for your application—you can hide from other devices by not enabling Inquiry Scan; a device which does this is in non-discoverable mode. Conversely, you are not guaranteed to find all Bluetooth devices in an area because devices which are not inquiry scanning are effectively invisible.

Placing a device in Inquiry Scan mode involves setting up the right parameters, then enabling the mode. *HCI_Write_Inquiry_Scan_Activity* is used to set up the scan duration and the interval between scans.

HCI_Write_IAC_LAP is used to define the IAC that the device will be listening for. There are currently only two valid IACs. The General IAC (GIAC), 0x9e8b33, is used by most devices, most of the time. It is the default, the common meeting place for all devices, and must be supported. Some devices may also support the Limited IAC (LIAC), 0x9e8b00, which can be used if you only wish to be discovered for a limited amount of time and in response to a specific event. Instructions and guidelines on their use are provided in the Bluetooth profiles.

The GIAC is most commonly used. All devices that scan will listen for this code. The Limited Inquiry Access Code (LIAC) could be used in crowded environments where many devices are answering inquiries and it can be difficult to select the desired device. The owners of a pair of devices can agree to temporarily put them into Limited Inquiry mode. They will then use the LIAC as well as the GIAC for a short period before automatically reverting back to using only the GIAC. The Generic Access Profile (GAP) mandates that any device listening for the LIAC must also scan for the GIAC. If the Bluetooth hardware supports it, both IACs can be listened for at the same time, in parallel. However, many hardware implementations can only listen for one IAC at a time, so the scanning must be done in series. In this case, it is the application's responsibility to manage the time-slicing between IACs so that GAP requirements are met.

The Limited Inquiry Access facility has not proved popular so far since it requires user intervention at both ends of the link and tends to be seen as an unnecessary complication for the user.

HCI_Write_Scan_Enable is used to both enable and disable the Inquiry Scan mode.

If a device in Inquiry Scan responds to an inquiry this is reported, at the Inquiring device, by an *HCI_Inquiry_Result* event. It is not reported at the Inquiry Scanning device. In fact, the application is unaware that a response has been generated. The *HCI_Inquiry_Result* event is variable in length, depending on the number of responses, and has seven parameters:

- **Num_Responses** The number of responses being reported in this message.

- **BD_ADDR** The Bluetooth Device Address for each device responding.

- **Page_Scan_Repetition_Mode** For each device responding.

- **Page_Scan_Period_Mode** For each device responding.

- **Page_Scan_Mode** For each device responding.

- **Class_Of_Device (CoD)** CoD is a brief description of the type of device responding. Details are in Section 1.2 of the Bluetooth Assigned Numbers document. Again, there is one CoD for each responding device.

- **Clock_Offset**   Since the hop frequency of the responding device is determined by its address and clock, information on the clock offset can be used to predict what frequency it will be listening on and reduce the time to connect to it. Again, one response for each device.

The *Page_Scan* parameters all refer to the frequency, intervals and exact method by which the scanning device allows other devices to connect to it. See the following section for more details.

Since both Inquiring and Inquiry Scanning devices randomly hop frequency, they may end up on the same frequency more than once during an inquiry procedure and several responses may be generated. Whether each response is reported by an *HCI_Inquiry_Result* event is dependent on the lower layer implementation and how many previous responses the lower layers can keep track of. The application must therefore be able to identify duplicate responses and filter them out.

When an inquiry is complete, because either the specified number of responses or duration has been reached, an *HCI_Inquiry_Complete* event is generated. It contains only a *status* parameter.

You can carry out inquiries or inquiry scans as an unconnected device, a master, or a slave. However, a slave's responsibility to regularly listen for master transmissions means it will not be able to devote as much of it's time to the procedure, which may need to continue for longer to compensate. It is also possible to define intervals and windows to allow both operations to run over the same period. See the next section on timing for more detail.

Timing

Since one device needs to be in Inquiry and the other in Inquiry Scan for a successful discovery, it is important for applications to give a high chance of finding devices in a short time. The Generic Access Profile offers guidelines on how to accomplish this. Devices that are generally discoverable (using the GIAC) repeatedly conduct a short inquiry scan over a long period of time while Inquiring devices conduct a long inquiry either once, upon user prompting, or periodically, but with a large interval in-between inquiries.

The actual numbers from the GAP are as follows:

- While discoverable, enter Inquiry Scan for at least 10.625 milliseconds every 2.65 seconds. Remain discoverable for at least 30.72 seconds.

- When inquiring, enter Inquiry mode for at least 10.24 seconds.

- For devices using the LIAC, it is not recommended to stay in Inquiry Scan mode for more than 1 minute.

If there are any voice links present, the data transfer required for them will take priority over both Inquiry and Inquiry Scan operations. You need to consider this when setting up the operations.

- If one HV3 Synchronous Connection Oriented (SCO) link is present, then the inquiry scan period should be extended to 22.5 milliseconds.

- If two HV3 SCO links are present (or one HV2 link), the inquiry scan period should be extended to 33.75 milliseconds.

These rules do not altogether compensate for the effect of SCO links, so you should still consider inquiry and paging procedures to be slower and less reliable if SCO links are in use.

It is often a good idea, if possible, to scale back voice connections to HV3 before entering Inquiry Scan. But note that with three HV3 links present, no inquiry scanning can take place at all: the device is non-discoverable (the same is applied to two HV2 links or one HV1 link; each of these configurations uses up all possible slots and leaves no space for inquiring or scanning).

The inquiry period must be increased to compensate for the presence of SCO connections, or being a slave, in the same way as the inquiry scan period. The Link Controller also makes appropriate changes to the sequence of inquiry transmission frequencies. Again, the presence of three SCO connections would prevent any other operations, including inquiry.

The Bluetooth profiles define which devices within a usage scenario should be discoverable and which should do the discovering.

When to Stop

In an ideal world, once you took the decision to be discoverable, other devices would be able to find you immediately, all the time. In the real world of Bluetooth devices, there are prices to be paid for that level of visibility: power consumption and bandwidth.

Power consumption explains why the default inquiry scan duty cycle is 0.4 percent. For some battery-powered devices, even this may be too high, so dropping into a non-discoverable state may be necessary to save power. Equally, if you are designing a mains-powered device, it may be desirable to increase the duty cycle and thus reduce the time it takes for other devices to find you.

Although transfer of voice (SCO) data takes precedence over Inquiry Scan operations, other (ACL) data transfer does not. In other words, Inquiry Scan uses up bandwidth. If you have chosen a high Inquiry Scan duty cycle, you may need to reduce it, or even disable Inquiry Scan, to achieve a high data rate.

In all applications, there should be an option for the user to manually switch from a discoverable to a non-discoverable mode. The GAP also includes guidelines on how these modes should be described in the User Interface.

Inquiry operations are less problematic. Although the same principles apply as for Inquiry Scan (SCO data has higher priority, ACL data does not), the inquiry operation is normally a one off, and generally triggered by the user. If carrying out an inquiry is going to disrupt a critical data transfer, it might be a good idea to warn the user before proceeding. Automatic periodic scanning should be sensitive to bandwidth use if unexpected drops in transfer rates are to be avoided. Note that if the lower layers are set to periodically inquire, they will schedule inquiries with no allowances for data transfers: intelligent inquiry scheduling is only possible at the application level. The user should also be given the option of disabling periodic inquiry if the feature is offered.

One other consideration for inquiring devices is their effect on other ISM band users. Every inquiry transmission potentially interferes with another piconet, or even with other wireless technologies using the same frequencies as Bluetooth. So, by specifying short inquiry periods the GAP helps Bluetooth devices to be good neighbors, causing the minimum possible interference to nearby devices.

Connecting to a Device

Once a device has been discovered via inquiry, the information gathered can be used to form a Bluetooth connection between devices. At the Bluetooth Radio level, a connection means that the devices in a piconet are all frequency-hopping together, synchronized to the master device's Bluetooth address and clock. Further up the protocol stack, it means that an ACL link has been established that data can pass over. This allows the use of L2CAP and all the other layers that sit above it, including the service discovery layer. The protocol for forming the link is called *paging*.

Paging and Page Scanning

To create a connection between Bluetooth devices one device pages another device, which must be in Page Scan to respond. The terms "create connection" and "page" are often used interchangeably although the latter is more specific since connections can also be created between upper stack layers. A successful page results in an ACL connection between the paging device, which, by default, becomes the master, and the paged device, the slave.

To allow an incoming connection, a device must be placed in Page Scan mode. This is similar to Inquiry Scan in that the mode must be configured, using the *HCI_Write_Page_Scan_Activity*, *HCI_Write_Page_Scan_Mode*, and *HCI_Write_Page_Scan_Period_Mode*, and then activated using the same *HCI_Write_Scan_Enable* command that controls the Inquiry Scan operation. Provided both modes have been configured with timing that allows it (see the following), a device can be in both Inquiry and Page Scan modes at the same time.

HCI_Write_Page_Scan_Activity sets the page scan period and the interval between scans, and hence the duty cycle. *HCI_Write_Page_Scan_Mode* determines if the device scans using the mandatory paging scheme or an optional one. Only one optional scheme is currently defined, although there is a provision for three. It is defined in Appendix VII of the Core Specification and trades an increased level of complexity and a higher duty cycle at the paging device for a lower duty cycle at the Page Scanning device. Few, if any, hardware vendors currently support the optional paging scheme, so a method must exist for hardware that doesn't support it to connect to hardware that does. For this reason, devices in both Page and Inquiry Scan that receive an incoming inquiry must then use the mandatory paging scheme for $T_{mandatory\_pscan}$ seconds following. *HCI_Write_Page_Scan_Period_Mode* sets the number of seconds according to the Page Scan mode (see Table 2.2).

Table 2.2 Relationship between SP Mode and Mandatory Page Scan Period

Scan Period Mode	$T_{mandatory\_pscan}$
P0	>20 seconds
P1	>40 seconds
P2	>60 seconds

To initiate a page, an application issues an *HCI_Create_Connection* command that contains the following parameters:

- **BD_ADDR** The Bluetooth device address of the device you wish to page.
- **Packet_Type** The types of ACL packet the local device will support on this link (i.e. DH/M 1/3/5).

- **Page_Scan_Repetition_Mode**   How often the target device enters Page Scan mode.

- **Page_Scan_Mode**   Whether to use the mandatory Page Scan mode, or an optional mode.

- **Clock_Offset**   The estimated difference between the local device's clock and the target device's clock.

- **Allow_Role_Switch**   Determines whether the local device will accept a request from the target device to swap master/slave roles.

Apart from *Packet_Type*, the first five parameters are provided as part of an inquiry response. The *BD_ADDR* is required to identify the target device. The two *Page_Scan* parameters determine the exact baseband operation during the page. Knowing the *Clock_Offset* of a device is not essential to making a connection—it can still be made if this value is completely wrong—but the better the estimate, the shorter the connection time. The paging device uses the *BD_ADDR* and *Clock_Offset* parameters to calculate the frequency the target device will be page scanning on and starts its paging transmission there. If initially unsuccessful, the paging device then tries other, progressively less-likely frequencies until eventually all possibilities have been covered.

When the target device receives an incoming page, it does not necessarily accept it immediately. The *HCI_Set_Event_Filter* command can be used to switch between three possible behaviors:

- Send an *HCI_Connection_Request* event to the host and wait for an *HCI_Accept_Connection_Request* or an *HCI_Reject_Connection_Request* command.

- Accept the Page automatically.

- Accept the Page automatically *only* if the paging device accepts master/slave role switch.

The last is important for profiles such as LAN access where an access point is discoverable and connectable while being a master of a piconet. A new device, when it connects, becomes, by definition, the master. The new device must allow the role switch so that the access point can become a master again and continue to maintain communications with the existing slaves.

Developing & Deploying…

Masters, Slaves, Role Switches, and Scatternets

To upper stack layers, the only difference between a *master* and a *slave* is that a master can talk to several slaves in a piconet, while a slave can only talk to the master of the piconet.

For some devices, this relationship is important. Take, for example, a PC with a Bluetooth mouse and keyboard already operating. The PC may also wish to allow a PDA to connect and synchronize. Since the PDA initiates the connection, it becomes the master of the new piconet, but the PC will only have allowed this connection if, as part of the connection request, the PDA stated it allows master/slave role switches. As soon as the connection is established at baseband level, the PC requests a switch. If the PDA does not grant it, the PC drops the connection.

Interestingly, for the time between the connection completing and the role switch taking place, the PC is still master of its old piconet even though it's a slave of the PDA's piconet. When a single device is a master of one piconet, and slave of another simultaneously, this is, by definition, a *scatternet*.

Several manufacturers now support the limited form of scatternet required for a master/slave role switch while master of an existing piconet, but maintaining the scatternet for any length of time is still problematic. The Bluetooth specification gives no way for a slave to demand hold, sniff, or park modes from a master; they must always be requested. The master is entitled to refuse such requests, so it is impossible to guarantee that a slave in one piconet will be granted the time required to participate in another piconet as a master or a slave. Even if devices choose to simply switch between piconets as they see fit, ignoring the normal request procedures, there are still problems with how to time these switches in order to maintain multiple connections.

The master of each piconet must periodically poll all its slaves in order to give them an opportunity to transmit (since slaves only transmit data in response to a master transmission). How to cope with the variability of the interval between poll transmissions from a master is particularly awkward. It is possible to devise solutions to these problems, but there are a number of possible solutions and no guarantee that two implementers will choose the same one. A single chip set vendor may be able to demonstrate scatternet operation provided they produce all

Continued

devices in the scatternet, but this provision goes against the fundamental Bluetooth concept of interoperability. Work is progressing in the Bluetooth SIG to devise a standard solution to these problems.

There is an even greater problem with SCO connections in a scatternet, however. The reserved slots for SCO connection in two scatternet-connected piconets are running on different clocks. They will eventually drift, relative to each other, so that the reserved slots coincide, making it impossible for a single device to be part of both piconets. There is no way to renegotiate the SCO timing once the link has been set up.

Fortunately, the problems with ACL scatternets may be resolved soon, but those of SCO scatternets will likely be around for a very long time. For the moment though, no profiles use, let alone require, scatternet operation.

If the page is successful, an *HCI_Connection_Complete* event is generated at both ends of the new link with a "Success" status and other parameters describing the connection. This includes the Connection Handle that, for a master with multiple slaves, is used to route data. A page can fail because it times out or is actively rejected in which case the paging device generates an *HCI_Connection_Complete* event with the appropriate "Failure" status parameter.

Timing

Many of the same principles that apply to inquiry also apply to paging. Where restrictions on inquiry timing are contained in the GAP, the core specification defines restrictions on page scanning. The restrictions on the length of each individual page scan, called the *scan window*, vary according to the number of SCO links present. SCO traffic has a higher priority than page operations, so the scan window must be extended to compensate for the lost bandwidth:

- If no SCO links are present, the scan window must be at least 11.25 milliseconds (ms).

- If an HV3 link is present, the scan window should be at least 22.4 ms.

- If two HV3 links (or an HV2 link) are present, the scan window should be at least 33.75 ms.

Restrictions are also placed on the period between page scans, called the scan interval. The maximum interval between the start of successive scans is 2.56 seconds. If page scanning is continuous (i.e., the scan window is the same length as the scan interval), this is classed as Repetition Mode R0. If page scanning is not continuous, but the interval is less than 1.28 seconds, this is classed as Repetition Mode R1. Intervals between 1.28 seconds and the 2.56 second maximum are classed as Repetition Mode R2. A paging device alters the way it pages depending on the repetition mode of the target device, which is why this information is returned as part of an inquiry response and is a parameter of the *HCI_Create_Connection* command.

There is little point in a device being discoverable via Inquiry Scan but not connectable. Although it is theoretically possible to place a device in Inquiry Scan, but not Page Scan, this mode of operation is not currently used by any profile. Most devices will be in Inquiry Scan and Page Scan at the same time. To do this, the two scan intervals should be set equal, with the scan periods each occupying a maximum of half the scan interval. Like Inquiry Scan, shorter scan intervals can be used to reduce power consumption.

If an inquiry has previously been performed, then there is no need to repeat the process every time a link between two devices is re-established. In fact, placing a device in Inquiry Scan unnecessarily wastes power and allows any other device within range to find it, generating unwanted inquiry responses. The Inquiring device may also attempt to connect—if only to check the device's friendly name—wasting even more power. It is therefore common for devices to be in Page Scan only. This is especially true of devices, like headsets, that are bonded: linked securely to another device. One device of the bonded pair might go into Page Scan when powered on, and the other would page it. The information for the Page operation would come from a single inquiry when the devices first bonded.

As mentioned previously, how long is spent paging before a connection is established largely depends on how accurately the paging device knows the paged device's *Clock_Offset*. If it is exact, then connecting can take as little as 4ms. However, when not in a link, devices' offsets drift. The longer it has been since the last connection between two devices, the less accurate the offset information. It will take longer to connect next time. If one device has been powered off and on between connections, the offset information is useless: no better than a random guess. However, as long as the Bluetooth Address is correct, a connection will still be formed eventually. The theoretical worst-case duration for a page is just over five seconds. Interference or the presence of SCO links may extend this time. The timeout period is set by the *HCI_Write_Page_Timeout* command. The default is 5.12 seconds.

Who Calls Who?

Many Bluetooth profiles don't care which device is the master of a link and which is a slave. For a Point-to-Point Profile, the distinction is meaningless at the higher layers. However, the distinction should be considered, especially for battery-powered devices, as it can have a huge effect on a device's power consumption, for two reasons.

Take, for example, a PDA that wishes to periodically and unconsciously synchronize with a PC. Firstly, if by default, the PC initiates connections, then the PDA must be connectable at all times. Even with an average Page Scan current draw of 0.5 mA, it is still going to use 12 mA-hours of power per day just maintaining the Page Scan mode. It may be more efficient to have the PDA wake periodically and attempt to page the PC.

Secondly, although a slave can request power saving modes such as sniff and park, a master is under no obligation to grant them. If they are not granted, then a slave must listen for a master's transmissions in every possible transmit slot, draining power each time. As a master, a device only needs to transmit enough to maintain a link and there is a better chance that power saving modes can be negotiated and used.

Finding Information on Services a Device Offers

There are many different potential types of Bluetooth device, each with different possible combinations of supported profiles, some of which have not even been thought of yet. All these devices can connect and talk to each other, but they may not support compatible profiles. For example, a headset has little use for Internet access. When initial contact is made, the devices need to ask each other a question. The exact question depends on circumstances. It is a choice between either "Do you provide service X?" or "What services do you provide?"

The first question is appropriate when the device asking the question is only interested in a specific service. Our headset will only be interested in finding devices that can act as an audio gateway. It has no interest in LAN Access Points, so it will ask, "Do you provide an Audio Gateway service?" The second question would be asked, for example, by a PC that wishes to know what devices are in the neighborhood and what services they all provide.

The mechanism to ask and answer these questions is provided by the Service Discovery Protocol, a protocol for accessing a database of the services a device offers. The database also contains the information required to answer the subsequent question, "How do I use service X?" Since the application supplies the services, it is also responsible for maintaining accurate SDP records of them. Remote devices connect to the SDP server as clients and query these records.

A service discovery record contains a number of attributes drawn from 28 possible types. They describe six broad types of information:

- The services on offer (e.g., Generic Audio, Headset Audio Gateway, Handsfree Audio Gateway); their names, availability, and descriptions.

- The protocols used to access the services (e.g., L2CAP and RFCOMM).

- How to connect to these protocols (e.g., the RFCOMM port).

- The supported profiles (e.g., Headset, Handsfree).

- How the service browsing tree is constructed.

- The behavior of the database (e.g., when the service record is likely to change).

Attributes are identified by their own Universally Unique Identifiers (UUIDs). The ideas and mathematics of UUIDs are not unique to Bluetooth. They are designed so that users can generate their own UUIDs with such a low chance of two independently generated IDs being the same that this, in itself, is sufficient to ensure they are not repeated. No central register of new UUIDs needs to be kept. UUIDs in the range 0 to 2^{32} are reserved for SIG-defined attributes, but others can be created by product manufacturers. New manufacturer-created attributes will only be recognized by other products that already know how the related services and protocols work and will not, therefore, experience the high level of interoperability that SIG-defined services enjoy. New services must be different from SIG-defined services, or extensions to them. You are not allowed to create a service that is similar to a headset, but that isn't interoperable with the Headset profile.

The construction of the service discovery record can be complicated, but it is essential if devices are going to interoperate correctly. Fortunately, a majority of attributes that an application should store in the database are exactly specified in each profile.

Every service record browsing tree must have a root named PublicBrowseRoot. PublicBrowseRoot is required as all service browsing trees contain this entry as their root. The presence of PublicBrowseRoot means that all client devices have a known location where they can begin browsing.

Apart from the requirement for a known root, the construction of the service record browsing tree is not defined by the profiles, but by the manufacturer. You should simply try to make the browsing tree logical. For example, a Global System for Mobile Communication (GSM) phone might offer the following services:

- Headset Audio Gateway

- Handsfree Audio Gateway

- Cordless Telephony

- Intercom

With the addition of the Generic Audio service group, Generic Telephony service group, and the PublicBrowseRoot entry, the service record browsing tree shown in Figure 2.2 can be constructed.

Figure 2.2 A Service Record Browsing Tree

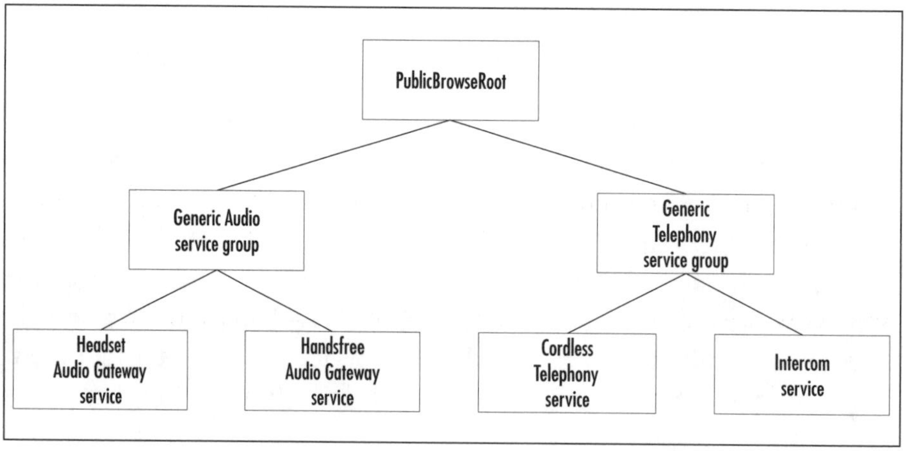

To browse a remote device's service discovery database, a local device must page and set up an ACL connection with it. This means that a device must be in Page Scan mode and accepting connections before information on the services it offers can be gathered. Once an ACL connection is formed, the local device must then open an L2CAP channel and use the reserved PSM (0x0001) to request a connection to the SDP layer. This PSM never changes, and SDP is always present, so you always know where to look for information on a device's services. The L2CAP connection can only be used for service discovery. If you wish to use other services, another L2CAP connection is required. This is important for maintaining security while still allowing service discovery to take place.

The process of service discovery is covered in detail in Chapter 5.

Connecting to and Using Bluetooth Services

Several stages must be completed before you can use a Bluetooth service.

1. Find the device – Inquire.
2. Connect to the device – Page.
3. Discover what services the device supports – SDP.
4. Decide what service to connect to and find out how to connect to it – SDP.
5. Connect to the service.

Stages 3 thru 5 all involve connecting to more than one upper layer. Connections to these upper layers must each be opened separately and in order. The following figures illustrate this process for an Audio Gateway connecting to and setting up an audio link to a Headset. This is a conceptual summary, not a detailed systematic guide. The exact steps an Audio Gateway application will need to go through will depend on how much of the detail is abstracted by a Connection Manager. The following sections give one example sequence.

Stage 1: Finding the device by Inquiring. (See Figure 2.3.) These diagrams are simplified, and omit details of configuration. So, for instance it's assumed that somehow the Audio Gateway has configured inquiry parameters, and that the Headset has been placed in Inquiry Scan mode.

1. The Audio Gateway application sends an inquiry request to the lower layers.
2. The lower layers send inquiry packets to the neighborhood.
3. All Inquiry Scanning devices in the neighborhood, including the headset, reply with inquiry responses.
4. The lower layers send the responses to the Audio Gateway application.

Figure 2.3 Simplified Inquiry Procedure

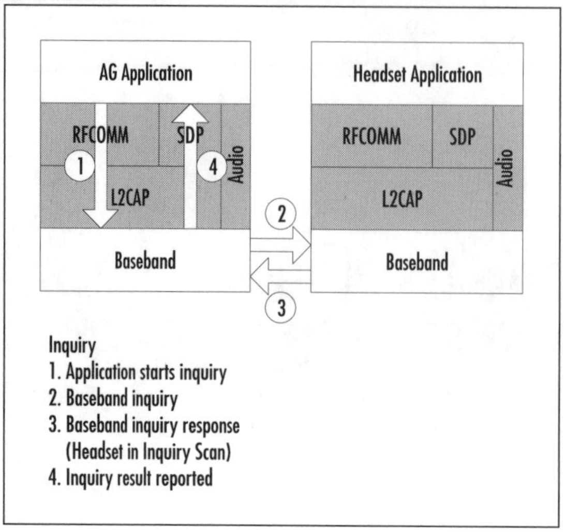

Note that the Headset application is not involved at all: once it has configured the lower layers to Inquiry Scan, it is completely unaware of any inquiry responses they generate.

Stage 2: Connecting to the device by paging. (See Figure 2.4.) Again, these diagrams are simplified, and omit details of the configuration. So, for instance, it is assumed that somehow the Audio Gateway has configured Page parameters, and that the Headset has been set into Page Scan mode.

1. The Audio Gateway application sends a page request to the lower layers

2. The lower layers of the Audio Gateway page the Headset, using its Bluetooth device address to generate ID packets, which only it will be listening for. Other page scanning devices in the neighborhood will not detect the paging or respond to it. At this stage, a series of low-level packets are exchanged. The details are not important except to note that the Headset is passed information on the Audio Gateway device, including its Bluetooth device address and Class of Device.

3. The lower layers on the Headset send a message to the Headset application notifying it of the connection request. This notification will include the Audio Gateway's Bluetooth device address and Class of Device, which were gathered during paging.

Figure 2.4 Simplified Page Procedure

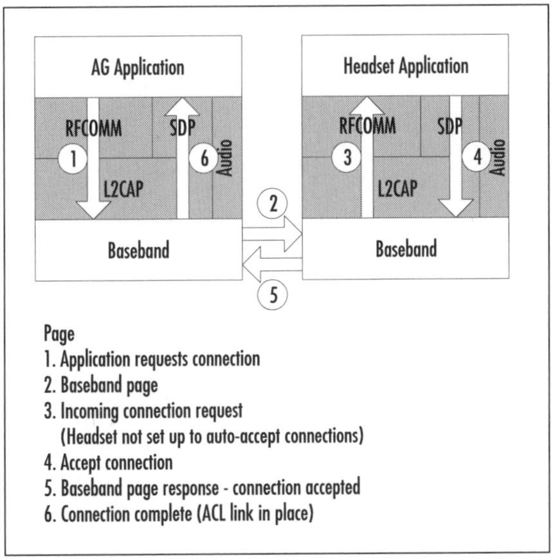

Page
1. Application requests connection
2. Baseband page
3. Incoming connection request
   (Headset not set up to auto-accept connections)
4. Accept connection
5. Baseband page response - connection accepted
6. Connection complete (ACL link in place)

4. The Headset application replies to the lower layers accepting the connection.

5. The lower layers on the headset send the response to the lower layers on the Audio Gateway.

6. The lower layers on the Audio Gateway forward the message, accepting the connection to the Audio Gateway application. The Audio Gateway application now knows it has an ACL (data) connection ready for use.

Stage 3: Discovering what service a device supports through SDP. (See Figure 2.5.) The first thing to do when connecting to SDP is establish an L2CAP connection using the PSM which identifies the SDP layer.

1. The Audio Gateway application sends a request to its local L2CAP layer asking for an L2CAP connection to the PSM for SDP on the Headset.

2. The request is relayed to the L2CAP layer on the Headset, which asks the Headset application if it is willing to accept the request.

3. The Headset application responds that it will accept a connection to the SDP layer.

4. The response is relayed to the L2CAP layer on the Audio Gateway, which informs the Audio Gateway application that an L2CAP connection to the SDP layer on the headset is available for use.

Figure 2.5 Simplified L2CAP Connection to SDP Procedure

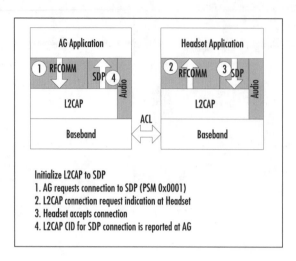

Stage 4: Decide what service to connect to and find out how to connect to it. (See Figure 2.6.) The Audio Gateway application can now send SDP requests and will receive SDP responses from the SDP server on the Headset. Notice that once the Headset application has registered a service record with the SDP layer, it does not need to be involved in SDP transactions—the SDP layer can respond to requests autonomously.

Figure 2.6 Simplified SDP Search Procedure

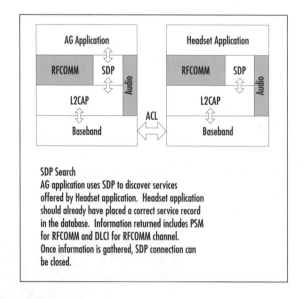

The Audio Gateway will send requests to retrieve the service record for the Headset Service. This checks that the service is really supported, and provides information on how to connect with it.

Stage 5: Connect to the service. (See Figure 2.7.) This stage begins in the same way as connecting to the SDP layer by creating an L2CAP connection. The procedures are exactly the same as those for creating an L2CAP connection to SDP, except that the PSM used this time is the PSM for RFCOMM.

Figure 2.7 Simplified L2CAP Connection to RFCOMM Procedure

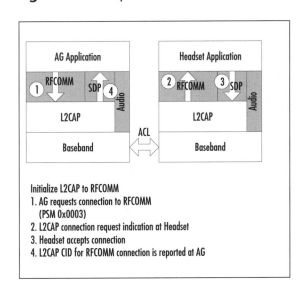

Once the L2CAP connection to RFCOMM is established, it can be used to carry messages between the Audio Gateway application and the Headset application. As we noted in "Reviewing the Protocol Stack," RFCOMM can carry many emulated serial links simultaneously, therefore the Audio Gateway must identify the correct link to use to communicate with the Headset service. This is done by using the DLCI for the Headset service, which was passed to the Audio Gateway in the Headset's service record. See Figure 2.8.

Once the Audio Gateway and Headset are communicating across RFCOMM, the Audio Gateway can send control messages using AT commands (the same command set that is commonly used to control modems). See Figure 2.9. To notify the Headset application that there is a call waiting, and to ask the headset application to alert the user with a ring tone, the Audio Gateway application sends an AT+RING command over the RFCOMM link. If the headset user presses a button to accept the call, the Headset sends this button press in a keypad command: AT+KPD.

Figure 2.8 Simplified L2CAP Connection to RFCOMM Procedure

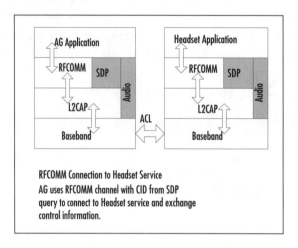

Figure 2.9 Simplified Headset Service Connection Procedure

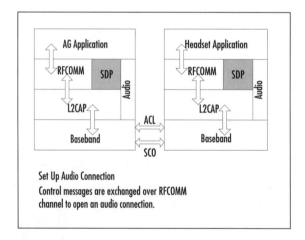

Once the user has accepted the call, a voice (SCO) link must be set up (see Figure 2.10). Although this link is controlled using the RFCOMM link, it is established separately, usually by a separate audio control layer. Once the SCO link is established, it is still controlled by the RFCOMM link. For instance, some headsets support remote volume control using AT commands, and the

SCO link can be destroyed by sending a hang-up command AT+HUP on the RFCOMM link.

The exact procedure for using the service is defined in the appropriate Bluetooth profile. As we have seen, the level of detail in a profile goes to the exact AT command to be sent over an RFCOMM channel when a particular button is pressed. It is this level of detail that allows such a high level of inter-operability. Some procedures, such as those for a Headset, are relatively simple, while others are a lot more complex; the Printer Profile is a good example.

Figure 2.10 Simplified SCO Connection Procedure

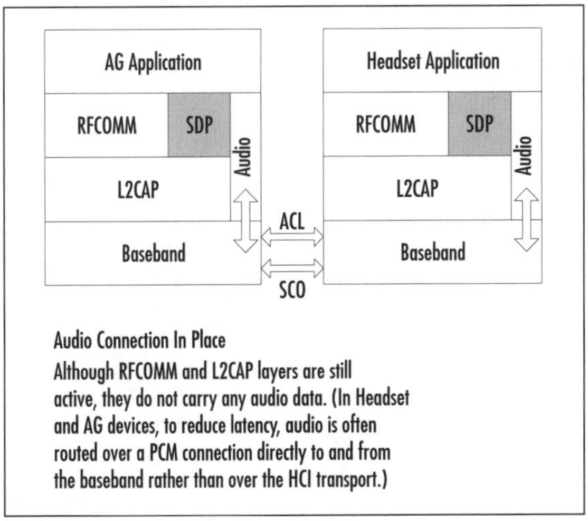

Summary

The Bluetooth stack does a good job of hiding the complexities of a wireless interface, but some peculiarities are still apparent. Before connections can be made between devices, they must find each other. One device discovers another by sending out inquiry transmissions, while the other listens for these inquiries and replies to them. A device must be in Inquiry Scan mode to be discoverable. The specification details timing restrictions on Inquiry and Inquiry Scan designed to ensure that devices have the best chance possible of discovering each other, while still allowing a low duty cycle and hence, minimal power consumption. Increasing the duty cycle reduces latency, but increases power consumption.

Once two devices have found each other, they use a paging procedure to connect. This is similar to inquiry in that one device transmits while the other listens and then responds. Only devices that are in Page Scan mode can be connected to, but devices in Page Scan may reject an incoming connection request if they choose. The Bluetooth specification places limits on Page Scan to allow a good chance of connection while keeping power consumption low.

Devices are usually in Page Scan only (connectable but not discoverable), or Page and Inquiry Scan (connectable and discoverable).

While a Bluetooth service is being used, the complexities of the air interface are hidden by abstracting the interface across a number of software layers. The HCI transport provides a standardized interface to the Bluetooth integrated circuit (IC). Audio is routed directly over the HCI interface. Data traffic from several upper layers is multiplexed through the Logical Link Control and Adaptation Protocol (L2CAP), which identifies upper layer types by their Protocol Service Multiplexor (PSM) values. The actual L2CAP channels each have unique Channel Identifiers (CIDs). The Bluetooth specification describes several different types of layers above L2CAP, including RFCOMM for serial port emulation, and TSC-BIN for telephony profiles.

Different Bluetooth devices support different profiles and offer different services. Each Bluetooth application must maintain an accurate record of the services it offers in a service discovery database. Remote devices can then connect to this database and use the Service Discovery Protocol (SDP) to query it. The SDP layer can always be found in the same place, above L2CAP. Service discovery can be complex, but the Bluetooth profiles detail most of the attributes that should be stored in a service record.

Once a remote device has connected to a local device and found a service in the service database that it wants to connect to, attributes in the service record provide the information on the upper layers required to use the service and how to connect to them. Connections to each protocol layer must be made in turn from lowest to highest.

Solutions Fast Track

Reviewing the Protocol Stack

☑ The protocol stack hides the complexity of the wireless interface and presents, at its highest level, a software interface that resembles that of a wired connection.

☑ Not all the differences between a wired and a wireless interface can be hidden. In particular, the steps required to find and connect to other devices are peculiar to wireless.

☑ Bluetooth devices can contain various combinations of upper stack layers to support various profiles. The Bluetooth specification details a service discovery layer so that devices can find out what services are available and how to connect to them.

Why Unconnected Devices Need to Talk

☑ With Bluetooth devices, the user may not initially know that there are other Bluetooth devices nearby, so a method is required to find them. The Bluetooth equivalent of plugging in a cable is the forming of a connection. The checks on communications protocols and applications compatibility are actually done once a basic Bluetooth link is established, and are called service discovery.

☑ The procedure used to find devices is called *inquiry*, and the procedure used to connect to devices is called *paging*. In both cases, one device transmits and receives on special sequences of frequencies that are known to all devices. The other device needs to be listening for the transmissions—if a transmission is received correctly, it sends out a reply. Since it knows the sequences used for inquiry and paging, it can work out the correct frequency on which to send the reply.

Discovering Neighboring Devices

☑ Only devices in Inquiry Scan can be discovered.

☑ An inquiry is normally a periodic or user-initiated event.

☑ An inquiry response contains all the information required to connect to a device by paging.

Connecting to a Device

☑ Only devices in Page Scan can accept connections, although they may choose to reject incoming connection requests.

☑ If a page and connection request is successful, then the paging device becomes the master of the piconet and the paged device becomes the slave. An Asynchronous ConnectionLess (ACL) connection now exists between the two.

☑ A master can have connections to several slaves, but a slave can only have a connection to a master. For the upper stack layers, this is the only difference between the two.

Finding Information on Services a Device Offers

☑ The application is responsible for maintaining accurate records of the services it offers in a service database.

☑ An ACL and a Logical Link Control and Adaptation Protocol (L2CAP) connection must exist to a remote device before it can browse the service database using the Service Discovery Protocol (SDP).

☑ The service database contains all the information required for a remote device to identify and connect to local Bluetooth services.

Connecting to and Using Bluetooth Services

☑ A remote device must conduct an SDP query before connecting to a local Bluetooth service, and must support a complementary profile.

☑ Connecting to a service involves first opening L2CAP, then higher layer connections in turn, using the information from the SDP query.

☑ The procedure for using a service is detailed in the appropriate Bluetooth profile.

Frequently Asked Questions

The following Frequently Asked Questions, answered by the authors of this book, are designed to both measure your understanding of the concepts presented in this chapter and to assist you with real-life implementation of these concepts. To have your questions about this chapter answered by the author, browse to **www.syngress.com/solutions** and click on the **"Ask the Author"** form.

Q: I don't like the way the Radio Baseband/Link Controller/Link Manager works. Can I change it?

A: No. Interoperability is a fundamental concept of the Bluetooth specification. If you change the way the lower layers function, they will no longer interoperate with other Bluetooth devices. In addition, several core technologies of the Bluetooth specification use Intellectual Property (IP) licensed from Ericsson, or the Bluetooth SIG (depending on which version of the adopter's agreement you signed). The Bluetooth Adopters Agreement gives this license free of charge, provided your products meet the Bluetooth specification. If you change the operation, you would be breaking the specification, the free license would not apply, and you would be using IP without permission. Litigation may follow.

Q: I don't like the way the upper layers work. Can I change them?

A: Yes, up to a point. You can create your own upper layers and profiles, provided the Generic Access Profile (GAP) is still met. The GAP mandates certain minimum functionality, including support for service discovery. This allows other Bluetooth devices to connect and find out what services are offered, even if the devices do not know how to use them: the responses are coherent and sensible. Support for SDP implies the presence of a specification compliant L2CAP layer. New profiles must be different from or extensions to current ones. You are not allowed to create something that is similar to the Headset profile, but will not interoperate with Bluetooth Headset Audio Gateways. However, any stack layer or profile functionality can only be used by an application that knows how it operates. Everyone can read how the Bluetooth specification defined layers and profiles work, so they experience a high degree of interoperability. Manufacturer defined layers and profiles will have a much lower visibility and a correspondingly lower level of interoperability.

Q: What is the difference between an L2CAP PSM value and an L2CAP CID?

A: Protocol Service Multiplexor (PSM) values identify the protocol used to communicate over an L2CAP channel. In effect, this defines the higher layer that uses the channel. Multiple instances of the same higher layer may use different L2CAP channels, but they will all be identified by the same PSM value. Each separate channel is uniquely identified by its Channel ID (CID). A higher layer may request an L2CAP connection to a remote RFCOMM entity by specifying a PSM value of 0x0003. The local and remote L2CAP layers then assign CIDs to this link. The CIDs are used to actually identify traffic sent between RFCOMM layers.

Q: What is the lowest power that a Bluetooth device can draw?

A: This question is only slightly less open-ended than "How long is a piece of string?" The absolute lowest power consumption will be when a device is not doing anything and can drop into a deep sleep mode. Many devices can do this when not part of an active connection; some can also do this in intervals between activity in low-power sniff, park, and hold modes. If low power modes are not used, then slaves can often draw more current than masters, since slaves have to listen in every possible slot for a master's transmission, while masters only have to transmit when they need to. Although page scanning draws a lot less continuous current than paging, if paging is only to be an infrequent activity, the paging device may end up drawing less average current than a device in constant Page Scan mode. In summary, current consumption depends on the mode of device operation, which is determined by the application design. Power consumption implications should therefore be considered carefully when the application is designed. If an application is to be a good neighbor, it should also permit as much flexibility for devices that connect to it as possible (e.g., accept low power mode requests). The actual power consumption during each mode of operation will depend on the Bluetooth hardware implementation.

Chapter 3

Power Management

Solutions in this chapter:

- **Using Power Management: When and Why Is It Necessary?**

- **Investigating Bluetooth Power Modes**

- **Evaluating Consumption Levels**

☑ **Summary**

☑ **Solutions Fast Track**

☑ **Frequently Asked Questions**

Introduction

Bluetooth technology finally makes the mobile application a reality. Not only can users be mobile whilst connected but radio networks can also be used in places where fixed infrastructure is too expensive, dangerous, or difficult to deploy. This, however, leaves you with the difficulty that all these devices must be powered using batteries, which have to be frequently recharged or replaced. If the Bluetooth device uses too much power, this can become a real problem.

As an applications designer, you may think there is nothing you can do about the problem—after all, you have no control over the amount of power your hardware consumes. The good news for Bluetooth applications is that designers *do* have the ability to do something about improving the power efficiency of their application. The Bluetooth specification offers a range of power-saving features, tailored to suit the needs of different applications, which can give your applications a real edge.

The drawback (and there always is one) is that if you use these features badly, you will slow down the response time of your application, making it infuriating to use. This chapter will tell you how to get the best of both worlds: save power while still producing usable applications.

Using Power Management: When and Why Is It Necessary?

Before going further, its worth spending a little time defining what a power managed application actually is and exploring some of the reasons why such applications are necessary. A power-managed application is one that allows the device it is running on to go into sleep mode for significant portions of its duty cycle. Sleep mode need not involve powering down the whole device; in fact, this is highly unlikely, as certain functional blocks will always need to be powered. However, when a device is in sleep mode it should be consuming significantly less power than when it is fully "awake," otherwise power management will be a waste of time.

A further characteristic of application level power management is that it should not adversely affect the performance of the application. In fact, the user should not be aware that your application is using power management and that the Bluetooth device is not constantly powered on. Powering down a device at the wrong time can not only result in almost no energy being saved, but it can also make an application virtually unusable by making it slow to respond. Let's consider the example of a wireless headset and a mobile phone. If the headset is powered down at the wrong time, the phone will not be able to notify it of an incoming call. Even though the headset may be saving significant amounts of power, as far as the user is concerned, it is unusable, because it cannot receive calls in a timely manner.

So, if power management has the potential to make your application unusable or infuriatingly slow, why bother with it? Used in the correct way, the Bluetooth power management modes have the potential to extend the battery life of your device significantly, yet be completely transparent to the user. In general, users do not like having to lug about heavy batteries or recharge their devices frequently. A typical mobile phone has a small battery and yet can last several days without recharging. If adding Bluetooth functionality to such a phone reduces its average battery life significantly, it is unlikely to be popular with the user. Power management at both the hardware and software levels of Bluetooth technology is therefore necessary in order to make these networks viable. A further benefit of application power management is that the energy savings are independent of the underlying technology. This means that if through power management you double the battery life of your device, this will hold true even if the power consumption of the underlying hardware was significantly improved.

A relatively minor, but nevertheless important, point to consider is who owns the devices that are being power managed. Often greater power savings can be achieved by one device at the expense of the energy resources of another. An obvious example would be where a device is powered down for the majority of its duty cycle while another device buffers packets destined for it and therefore must be constantly powered on. Periodically, the first device wakes up to pick up these packets, acts on them if necessary and then powers down again. Thus, the first device can achieve very high power savings at the expense of the buffering device. If the same user owns both devices (and especially if one of those devices can be mains powered, e.g., a PC) then this is a very good approach to achieving high power savings. However, if the devices belong to different users then there is an obvious conflict of interests as both users might be keen to prolong the battery life of their particular device rather than altruistically providing a service for others. In this case, a scheme where both devices achieve some, but not maximal, power savings may be a better compromise rather than having no power saving at all. The anticipated uses of a power-managed application can therefore be important in choosing the power management approach taken.

Having discussed how useful power-managed applications can be, it is worth looking at what types of applications are suitable for these techniques and which ones will have their performance adversely affected by power management. The first thing to remember is that in order to save power, the device must be put into sleep mode. Applications that require large amounts of data to be sent or received, or that need very fast response times, are not suitable for power management. On the other hand, applications requiring small amounts of data to be transmitted or where data transfers are infrequent are very well-suited to being powered down for the majority of the time they are inactive. Similarly, applications where a delay in the response time can be tolerated should also consider power management.

Before choosing a given Bluetooth power management mode to use with your application you should consider the maximum amount of time the device can be powered down without adversely affecting the performance of your application. In general, when using power management, an application designer trades off an increase in latency and a decrease in data throughput for an increase in the battery life of the device running the application. The following sections will discuss the Bluetooth power management modes and the use of each mode in the context of different types of applications.

Investigating Bluetooth Power Modes

For most applications, if a connection exists between two or more Bluetooth-enabled devices, one of the Bluetooth low power modes can be used to extend the battery life of either some or all of these devices. In fact, power-managed devices can be in one of four states, listed in order of decreasing power consumption: active, hold, sniff, and park mode. Each of these low power modes will be described, along with a discussion of what type of applications will and will not be suitable for it.

Active Mode

In active mode, the device actively participates on the radio channel. The master schedules data transmissions as necessary and the slaves must listen to all active master-slave slots for packets that may be destined for them. This mode is a useful benchmark for comparison with the performance of the low power modes since it not only consumes the most power but also has the highest achievable data throughput due to the devices being able to use all available slots. The power consumption of Bluetooth devices is highly dependent on the manufacturer of the device and the application that it is running. Furthermore, as the technology matures, the power consumption of Bluetooth-enabled devices will improve further and hence it is best to compare low power modes relative to the active mode.

We will briefly discuss the type of applications best suited to active mode, which are unlikely to benefit or be able to utilize any of the other low power modes. An application that has very high data rate requirements is unlikely to power save as it will need to have its radio transceiver powered on for the majority of its duty cycle. Similarly, applications that require very low latencies are also unlikely to be able to use the low power modes since they will power down for such short periods that the overhead in powering down the device will be greater than the energy saving made (or powering down for longer periods will mean the application is no longer able to conform to its latency requirements).

Hold Mode

This is the simplest of the Bluetooth low power modes. The master and slave negotiate the duration that the slave device will be in hold mode for. Once a connection is in hold mode, it does not support data packets on that connection and can either power save or participate in another piconet. It is important to note that the hold period is negotiated each time hold mode is entered. Figure 3.1 shows what the interaction between two devices using hold mode might look like. A further important aspect of hold mode is that once it has been entered, it cannot be cancelled and the hold period must expire before communications can be resumed.

Figure 3.1 Hold Mode Interaction

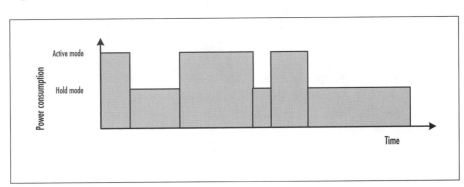

Given these constraints, what type of application would benefit from using hold mode? If your application can determine or control the time of its next data transmission, then it can most probably use hold mode for power management. One example of an application that has some degree of control over when its next data transmission should take place is a wireless e-mail delivery system. E-mail is not a synchronous communications medium and messages can take anything from a few seconds to several hours to be delivered to their destination. More importantly, users do not perceive e-mail delivery to be instantaneous and hence would tolerate a small additional delay in favor of extending the battery life of their device. The following sidebar, "Power Management Using Hold Mode," discusses in more detail how hold mode can be used by such an application, along with power saving techniques available.

Developing & Deploying…

Power Management Using Hold Mode

Given that e-mail is not an instantaneous communications medium and the delivery delays involved can be relatively large, any wireless e-mail delivery system has a lot of flexibility in the way it checks for new messages and sends off ones that have just been written. In fact, if correctly implemented, the delivery delay should not be perceptible to the user.

Let's assume we have a Bluetooth-enabled organizer that periodically communicates with an access point and retrieves newly arrived e-mails as well as sending off ones that have just been written. A simple way of implementing such a service will be to set up an RFCOMM connection between the two devices and have the checking device periodically search for new e-mails. Placing such a link in hold mode is unlikely to have a significant impact on the delivery time of e-mail and can result in power savings at both ends of the link. Furthermore, as each hold interval is negotiated independently of the previous ones, this gives us the opportunity to write an application that dynamically adapts to its usage. For example, successive hold intervals can be increased by a certain factor (up to a particular ceiling, of course) if there are no e-mails retrieved or sent during the previous "active" period. In the same way, successive hold intervals can be decreased if the frequency of e-mail arrivals increases. This approach allows the application to better adapt to the way it is being used and achieve higher power savings when the

Continued

load on the radio is light whilst still being responsive at higher usage rates. However, designers of such applications should be careful not to make such transitions too rapid as this may result in a yo-yo effect with the application swinging from one extreme to the other.

A further power saving technique at the application level, not directly connected with the use of Bluetooth low power modes, may be to compress data before transmitting it. If a high enough compression ratio can be achieved, the time that the transceiver has to be powered on can be reduced enough to justify the extra work. However, this should also be used with caution. A small device with relatively little computation power will use up energy in compressing (or decompressing) a file and this may offset the savings made in transmitting a smaller file. Such power-saving techniques are highly dependent, not only on the type of data being sent, but also on the underlying hardware.

A very different candidate for hold mode is one which relies on the use of a SCO link and does not need to send data packets. Furthermore, if the application can tolerate a poorer audio quality it can use fewer slots and hence power down for longer periods of time. For example, a baby monitor needs to have an active SCO link but does not need the ACL link. Also, given that parents are mainly interested in detecting whether the baby is crying or not, this application could probably get away with a slightly poorer quality of audio. By placing the ACL link in hold mode for relatively long periods of time and reducing the quality of the SCO link, the application can achieve greater power savings.

Having discussed application types able to benefit from using hold mode, we will briefly consider applications that should not use this mode, being it's likely to have a negative impact on performance. Hold mode is not suitable for applications whose traffic pattern is unpredictable and which cannot tolerate unbounded communication latencies. An obvious example is a device that allows a user to browse the Web over a wireless link. Even though access to the World Wide Web is notorious for being slow, if this latency is further increased by using hold mode, the application becomes too frustrating to use. At this point, it's worth remembering that once entered, hold mode cannot be exited until the negotiated hold interval has expired. Furthermore, the traffic pattern of such an application is impossible to predict due to the nature of Web browsing. The user may make a number of page requests in quick succession whilst browsing for a particular page. However, once the page has been found, they may spend considerably longer looking at the page and not need the use of the wireless link for some time.

A very different application type whose performance will be negatively impacted is a network of sensors which need timely delivery of their data—for instance, intruder detection. Once a sensor has been triggered, fast delivery of this information to the control center is imperative. A sensor with a long battery life that spends much of its day powered down may just give an intruder time enough to avoid being caught.

Sniff Mode

This low power mode achieves power savings by reducing the number of slots in which a master can start a data transmission and correspondingly reducing the slots in which the slaves must listen. The time interval, T_{sniff}, between the slots when a master can start transmitting is negotiated between the master and slave when sniff mode is entered. When the slave listens on the channel it does so for $N_{sniff\ attempt}$ slots and can then power down until the end of the current sniff interval. The time of reception of the last data packet destined for the slave is important, as the slave must listen for at least $N_{sniff\ timeout}$ after the last packet is received.

Figure 3.2(A) shows the lower bound of the number of slots that the slave must listen. In this case it just listens for $N_{sniff\ attempt}$. This happens if the last packet for the slave is received when there are more than $N_{sniff\ timeout}$ slots remaining in the sniff attempt. The slave just listens for the remainder of the sniff attempt interval and can then power down.

Conversely, Figure 3.2(B) shows a slave listening for an extended period. In this case the slave listens $N_{sniff\ attempt}$, then receives a packet and listens for a further $N_{sniff\ timeout}$ slots. This shows how the slave must listen for a further $N_{sniff\ timeout}$ slots if the last packet is received when there are less than $N_{sniff\ timeout}$ slots left in its sniff attempt interval. If the slave continued receiving packets it would continue listening for $N_{sniff\ timeout}$ slots after the last packet is received, so if the master kept on transmitting the slave would remain continuously active.

The slave can vary its activity from just $N_{sniff\ attempt}$ slots thru ($N_{sniff\ attempt}$ + $N_{sniff\ timeout}$) slots, and even go all the way to continuously active, all without renegotiating any parameters. You can therefore see that by choosing suitable values for the sniff interval and the number of slots that the slave listens for, power savings can be achieved without adversely affecting the performance of the application.

This section will consider what types of applications are suitable for use with sniff mode and which are not. Sniff mode is more flexible than hold mode since either the master or the slave can request for sniff mode to be exited. However, there is a trade off in the overhead associated with exiting sniff mode and it is more advantageous to choose the sniff mode parameters so as to minimize the likelihood of exit. Since sniff mode requires the slave device to periodically wake-up and listen to the radio channel, it is particularly well-suited to applications where devices regularly transmit (or receive) data. An example of such an application is discussed in the case study that follows. Sniff mode can also be used when there is an active SCO link. Once again, by accepting a slight degradation in the audio quality, power savings can be achieved since SCO links using HV2 or HV3 packets can be placed into sniff mode (note that SCO links using HV1 packets can also be placed into sniff, but in this case it will not have much effect since the device is transmitting in every slot).

Figure 3.2 Sniff Mode Interaction

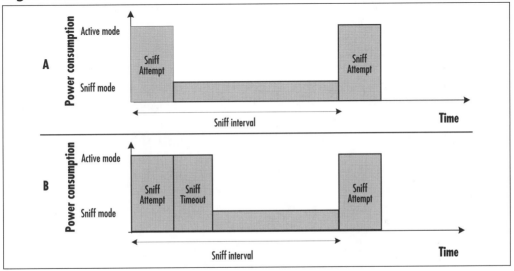

Another set of applications that could use sniff mode are ones where the devices can aggregate data and maybe even do a limited amount of processing before communicating with the master. Thus, not only the frequency of communication can be reduced, but also the actual amount of data transmitted. Once again, sensor networks are an obvious area of application. For example, a traffic monitoring system would be wasting resources transmitting every second the number of cars that have passed through a given point. Since the information is not time-critical, the update frequency can be decreased (i.e., the car count is aggregated at the sensor without affecting the performance of the system). However, this need not be limited to sensor applications—for example, the e-mail delivery system described in the previous example could be implemented using sniff mode instead of hold mode.

Application types not particularly well-suited to using sniff mode are ones frequently requiring relatively large data transfers. In this case, the time necessary to transmit the data is important, because if it takes too much time, your application will not be able to power down for very long, if at all. The application itself will not see a degradation in performance, but it will not achieve any power savings either.

Developing & Deploying…

Power-Managed Sensor Networks

One application that Bluetooth seems particularly well-suited for is sensor networks. As the technology matures, single chip Bluetooth solutions will not only become smaller but also much cheaper, making it feasible to embed them into even the cheapest devices. The number of possible sensor applications is virtually infinite. For this example, we shall consider what a patient monitoring system in a hospital might do and how it can benefit from using sniff mode to prolong the battery life of its sensors. Currently, remote monitoring of patients is limited mostly to intensive care wards and usually only one or two of the patient's vital life signs are monitored. The main reason behind this is that once this information has been collected, it is difficult to disseminate it so that both doctors and nurses have easy access to it. By using wireless sensors, the collected information can be periodically transmitted to a wireless access point and from there stored centrally so it can be accessed from anywhere in the hospital, or even from outside it (e.g., a consultant logging in from home to check up on a patient).

One such system might involve a set of sensors such as heart rate, blood pressure, temperature, and respiration monitors that frequently transmit their readings to a central access point in the ward. This information could then be displayed at the nurses' station so that patients are monitored continuously. In addition, doctors would be able to access the same information from anywhere in the hospital or even from home using their own Bluetooth-enabled organizer and hence be able to react quickly to changes in the patient's condition. To save power, the sensors use sniff mode and during the listen slots are addressed by the access point and transmit their readings. The sensor can then power down for the remainder of the sniff interval. This solu-

Continued

tion has great power-saving potential, but there is one obvious flaw in its design. If a patient suddenly takes a downturn, the sensors might not transmit this information for a relatively long time. This obviously makes the system unusable. However, sniff mode has an important feature in that either the master or the slave can request to exit sniff mode. This would allow a sensor to immediately transmit its readings and the alarm can be raised. Of course, for such safety-critical applications, it is also crucial to include a back-up emergency alert system that does not rely on radio. Adding a small piezo-electric beeper to each sensor will not significantly increase its size, cost, or power consumption. This can then be used in conjunction with the unsniff mode or as an emergency back-up if the sensor is unable to communicate with the master.

Park Mode

Park mode is the Bluetooth low power mode that allows the greatest power savings. However, while parked, a device cannot send or receive user data and cannot have an established SCO link. In this mode, the slave does not participate in the piconet, but nevertheless remains synchronized to the channel. This mode has the further advantage of allowing the master to support more than seven slaves by parking some whilst activating others. A parked slave periodically wakes up in order to resynchronize to the channel and listen for broadcast messages. In order to allow this, the master supports a complicated beacon structure that it communicates to the slave at the time it parks it. However, the beacon structure may change and the master then uses broadcast messages to communicate these changes to the parked slaves.

The structure of the beacon channel is covered in detail in other sources; it is sufficient to say here that every *beacon interval* number of time slots, the master transmits a train of beacons that the slave tries to listen for in order to resynchronize to the channel.

As an application designer, you have to choose the correct beacon interval to save the maximum power whilst maintaining acceptable response times. Response times are governed by how long it takes a slave to request unpark, or how long it takes a Master to unpark a slave, both of which are affected by the park beacon interval.

One factor to consider when choosing the park beacon interval is the clock drift in the devices between successive beacons. If a parked slave loses synchronization, it will stop responding to the master, and may lose the connection altogether. The master will then have to restore the connection by paging it and then parking it again. This is obviously wasteful. Therefore, devices parked for the majority of their duty cycle should have the park beacon intervals set well within the maximum threshold so that if the slave device misses a beacon it can re-synchronize on the next one. So far, park mode sounds very similar to sniff mode. The main difference, however, is that in order to send data packets to a slave, that slave must firstly be unparked (also as mentioned earlier, a slave cannot have an established SCO link when parked). The next section will consider the types of applications suitable for use with park mode.

An application that has been described as being unsuitable for hold mode is one where a Bluetooth-enabled laptop is used for wireless Web browsing. However, the pattern of usage for such an application does make it particularly suitable for park mode. It consists of "bursts" of activity while the user is searching for a particular page, followed by a relatively long period of inactivity while they are reading that page. The slave device can therefore be parked for the majority of the time, while the radio link is not being used. However, when the user needs to send data (assuming the beacon interval is kept relatively short) the slave can be unparked quickly and the request dispatched. Thus, the application can save power whilst keeping response times high. Another advantage of having a short beacon interval is that the slave device has a greater chance of remaining synchronized with the master. As the case study that follows shows, the Headset profile recommends the use of park mode while the headset and Audio Gateway are not actively communicating. This is another good example of an application suited to park mode, since activity is concentrated in bursts, but the response times are bounded by a maximum tolerable latency.

A network of sensors (as discussed previously) is a good example of an application where park mode is not particularly suitable as a low power mode. This is mainly because in order for the sensors to send their data, they would have to be unparked, allowed to transmit, and then parked again. For very short beacon intervals, this is particularly wasteful due to the overhead of the park/unpark procedure. Furthermore, sniff mode perfectly fits the pattern of the application without imposing this extra overhead. This point illustrates quite nicely the conclusion that there is no preferred low power mode. Each of the Bluetooth low power modes is suited to a different class of applications and must be used accordingly in order to achieve optimal performance (in terms of both power consumption and usability).

Developing & Deploying...

Power Management for the Headset Profile

The Headset profile as defined in the Bluetooth specification (part K-6) is designed to provide two-way audio communications between a headset and an "Audio Gateway," allowing the user greater freedom of movement while maintaining call privacy. The profile envisages the user wearing a Bluetooth-enabled wireless headset and communicating with, for example, a mobile phone or laptop computer (the Audio Gateway).

This application is a very good example of what could be termed an *asymmetrically power-managed application*. In this case, the headset has extremely limited energy resources (a coin cell or smaller battery) whose lifetime must be maximized. The Audio Gateway, on the other hand, has considerably greater resources since it is running on a device with a larger battery. The overhead associated with power management should therefore be placed on the Audio Gateway end of the link. By this we mean that not only should the Audio Gateway be responsible for power management on the link but also, if possible, it should use more of its energy resources so that the headset can save more power. Furthermore, as security is an important factor in this application, it is likely that the same user will own both devices and hence it is particularly suitable for asymmetric power management.

A headset must provide *pairing functionality*, allowing it to set up a link key with the Audio Gateway for security purposes. This is not a state that is likely to be entered frequently since once it is paired, the headset will remain so until it is paired again. The headset must also provide *audio transfer functionality* being that is what it is designed to do. Each of these states should be considered with respect to power management.

Whilst pairing, the headset should be in discoverable mode (i.e., it should respond to inquiries and also allow the Audio Gateway to connect to it). In this state, power savings can be achieved by reducing the time the headset spends with its radio transceiver powered on. This can be achieved by setting the page scan and inquiry scan intervals so that the radio is powered on for a relatively small fraction of the time. The downside to this is that the Audio Gateway might take slightly longer to find the headset and pair with it, but this delay is not likely to be significant. Furthermore, given that pairing is performed relatively infrequently, this is not a significant overhead.

Continued

Once the devices have paired and are ready to connect to each other there are two power-saving strategies to be adopted. The first is saving energy while the devices are attempting to establish an RFCOMM connection, and the second is once the RFCOMM connection has been established—an RFCOMM connection must be established in order for "AT" commands to be exchanged so that the audio link (through the use of a SCO connection) can be set up. This is achieved by placing one device into connectable mode (i.e., into page scan mode and letting the other initiate the creation of the connection. According to the Headset profile, either the headset or the Audio Gateway can initiate the connection attempt. If the headset is in slave mode (waiting for the Audio Gateway to connect to it), then it can employ the same technique used in pairing. It can save power by reducing the time it spends scanning (i.e., with its radio transceiver powered on).

Once an RFCOMM connection has been established, it can be placed in park mode until a SCO connection is needed. This avoids the overhead of establishing an RFCOMM connection (and tearing it down) every time a call is placed to or from the headset. Once a connection has been parked, either end is allowed to unpark it. This is to allow an incoming call to be placed through to the headset so the user can utilize voice dialing and dial out. Once the audio call has been completed, the SCO is disconnected and the RFCOMM connection is placed in park mode once more. It is important to note that neither the RFCOMM nor the L2CAP channels are released during park mode, so the connection can be brought up very quickly when required. However, while the connection is parked, data cannot be transmitted or received. Figure 3.3 shows how an example headset application can use both sniff and park to reduce its power consumption. An RFCOMM connection and an ongoing voice call (SCO connection) are assumed to exist between the two devices. The first diagram shows that as soon as the voice call is disconnected the RFCOMM link is placed in park mode. Note that either the headset or the Audio Gateway may initiate park. If at some later time either end wishes to transmit data, the connection must first be unparked. Once again, either device may initiate the unpark. At this point zero or more data packets may be sent and a SCO connection may be initiated. The link cannot be parked until the SCO (if created) has been released and there is no data pending transmission. The second diagram in Figure 3.3 shows how sniff mode can also be used by the headset. If, for example, either device expects to have data to transmit shortly after the voice call is disconnected and does not want to incur the overhead associated with entering park mode, it can place the link

Continued

into sniff mode. In this state, the headset can transmit its button press without exiting sniff. Furthermore, a SCO connection can be set up while still in sniff mode allowing the devices to conserve energy even while there is an ongoing voice call. Figure 3.3 shows that an application is not restricted to using just one of the Bluetooth low power modes, and by using more than one mode it can adapt better to its usage.

Figure 3.3 Headset Use of Park and Sniff Modes

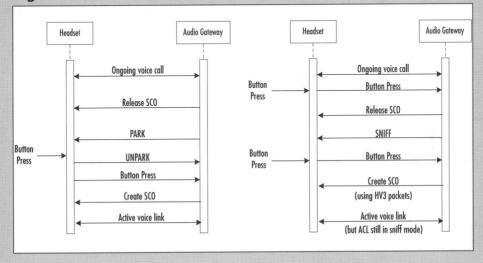

Evaluating Consumption Levels

As discussed earlier, the Bluetooth low power modes have different characteristics and are suited to different classes of applications. Each low power mode also has a different cost in terms of energy consumption. The power consumption of a device is influenced by the hardware used, the low power parameters negotiated, and the type of application it is running. This section will aim to give a very general indication of the relative power consumption characteristics of the Bluetooth low power modes. Absolute values for the average current consumption in each mode are meaningless since it is highly dependent on the underlying hardware. This section will therefore concentrate on the relative power consumption of some of the Bluetooth low power modes.

Figure 3.4 shows a comparison of the average current consumption of a device using different Bluetooth low power modes. Transmission of ACL data has the greatest power cost and will be used as a benchmark against which to compare

Figure 3.4 Relative Current Consumption for Different Bluetooth Low Power Modes

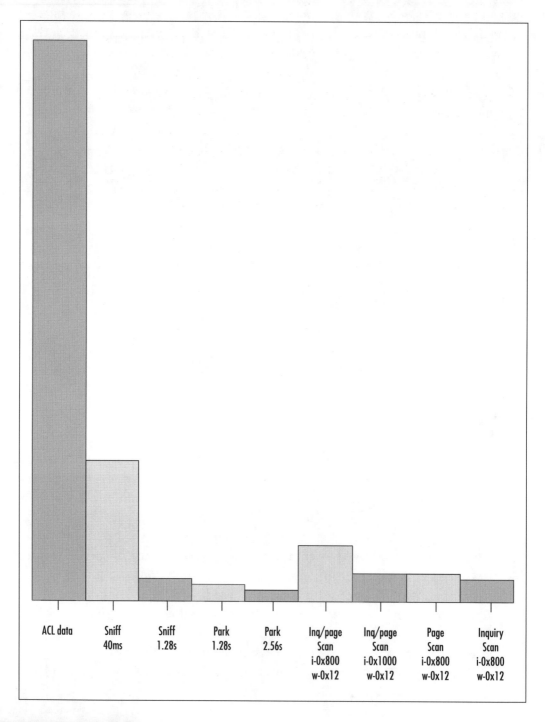

the other modes. As can be seen, a device in sniff mode consumes more current than a parked device. It is also important to note that the interval used while in sniff or park mode also affects its power consumption. The shorter the sniff interval or park beacon used, the more current the device will consume as it has to "wake up" more frequently in order to service that interval. Of course, the trade off is that the shorter the interval, the lower the communication latency. As you can see, there is always a trade off that has to be made between power consumption and latency.

A device must be in inquiry scan mode in order to be discoverable. Similarly, in order to be connectable, the device must be in page scan mode. Of course, both modes can also be enabled simultaneously. As can be seen from Figure 3.4, inquiry and page scan have a current consumption cost associated with them, and as such, should be used only when necessary. For example, if we only need the device to be connectable, then enabling inquiry scan will almost double the current consumption of the device but will not give it the functionality actually needed. Furthermore, as can be seen, the scan interval (denoted by i in the graph) and window (denoted by w in the graph) also have an effect on power consumption, so they should be chosen with care.

Although the graph in Figure 3.4 gives only a very approximate idea of the relative energy consumption costs of the different Bluetooth low power modes, it is easy to see that significant advantages can be gained by having an application use one or more of these modes.

Summary

This chapter has described the properties of power-managed applications and provided a discussion of why applications for Bluetooth-enabled devices can benefit from the use of power management. It has also detailed the different Bluetooth low power modes, illustrating the use of each one with example applications.

Power-managed applications allow the device to power down for a large part of its duty cycle thus saving energy and prolonging its battery life. However, the drawback is that the response time of the application is increased and, if not used correctly, power management can make applications infuriatingly unresponsive. This also means that the application allowing the underlying hardware to power down should be completely transparent to the end user. Bluetooth provides a number of low power modes and each one is suited to a different type of application. Before deciding on the power management mode to use, the maximum allowed latency and expected radio traffic pattern of the application must be considered. Applications with a very low latency or requirements to transmit very frequently might even make it inefficient to use a low power mode due to the overhead incurred in entering and exiting it.

Bluetooth provides three low power modes for application designers to use, hold, sniff, and park. Each mode has different characteristics and is suitable for a different class of application. Hold mode is suitable for applications that can predict or control the time of their next data transmission. As each hold interval is negotiated independently of subsequent ones, this mode is suitable for adaptive power management where the application monitors the usage of the link and increases or decreases its sleep time accordingly. Hold mode cannot be exited and therefore should not be used for applications with hard latency requirements.

Sniff mode allows a Bluetooth-enabled device to save power by reducing the number of slots that the master can transmit in, thereby reducing the slots the slave must listen to. This mode is more flexible than hold mode as it can be exited at any time. The slave listens periodically for a number of slots and this makes sniff mode particularly suitable for use in applications where data regularly requires transmission. Applications that are not suitable for sniff mode are ones that frequently require large data transfers that force the device to remain awake beyond its sniff interval. This does not have a detrimental effect on the application's performance, but it does not allow the device to achieve its full power saving potential either.

Park mode is the mode that allows greatest power savings to be made. This mode is best suited for applications where the radio traffic pattern is unpredictable and the connection establishment latency is bounded by some upper limit. The Headset profile (from the Bluetooth specification) is a good example of such an application. The RFCOMM link must be unparked as soon as possible, once a call needs to be put through from the Audio Gateway to the headset.

The Bluetooth low power modes are different in the power management support they provide and there is therefore no single mode that is best to use. The low power mode used is determined by a wide range of factors dependent on the type of application and its requirements. When considering which Bluetooth low power mode an application should use, the main factors to consider are:

- Whether the application is suitable for power management

- What is the maximum latency the application can tolerate

- What is the expected radio traffic pattern (random, periodic, bursty, and so on)

Solutions Fast Track

Using Power Management: When and Why Is It Necessary?

☑ Consider whether your application is suitable for power-managed operation.

☑ Consider the constraints imposed by the application (e.g., maximum response times, characteristics of the data traffic, and so on).

Investigating Bluetooth Power Modes

☑ **Hold mode** One-off event, allowing a device to be placed into hold mode for a negotiated period of time. Hold interval must be negotiated each time this mode is entered.

☑ **Sniff mode** Slave periodically listens to the master and can power save for the remainder of the time. Important to note that data can be transferred while devices are in this mode and a SCO link may be active. Sniff intervals are negotiated once, before sniff is entered, and remain valid until sniff mode is exited.

☑ **Park mode**   Parked slave periodically synchronizes with the master and for the remainder of the time can power save. Data packets cannot be sent on a parked connection and the devices must be unparked before a SCO connection can be established. Furthermore, there cannot be an active SCO when its associated ACL is parked.

Evaluating Consumption Levels

☑ All other things being equal, the power consumption of a Bluetooth low power mode depends on the parameters negotiated before that mode is entered.

☑ Page and inquiry scan also have a power consumption cost, so these should be entered only when necessary.

Frequently Asked Questions

The following Frequently Asked Questions, answered by the authors of this book, are designed to both measure your understanding of the concepts presented in this chapter and to assist you with real-life implementation of these concepts. To have your questions about this chapter answered by the author, browse to **www.syngress.com/solutions** and click on the **"Ask the Author"** form.

Q: Why don't low power modes work with different version Bluetooth devices?

A: Between version 1.0b and 1.1, improvements were made to the link management protocol messages, which put a device in hold, park, or sniff mode. These improvements made entering the low power modes much more reliable. However, because the protocol messages have changed, devices which have the old version of the protocol cannot work with the new version.

Q: Which versions of the Bluetooth specification are compatible for low power modes?

A: The changes in the link management protocol messages were first introduced as errata to the 1.0b specification. Changes, which were required to interoperate with version 1.1 of the specification, were labeled "critical errata." So:

- "1.0b plus critical errata" should be compatible with 1.1.

- 1.0b is *not* compatible with 1.1 or "1.0b plus critical errata."

- Any version should be compatible with the same version, but there have been interoperability problems with older versions, caused by ambiguity in the specification.

Q: What is the best power saving mode to use?

A: There is no "best" mode, it depends upon the requirements of your application. Look at the case studies in this chapter and consider the requirements of your particular application to decide which power saving mode is best for you.

Chapter 4

Security Management

Solutions in this chapter:

- **Deciding When to Secure**

- **Outfitting Your Security Toolbox**

- **Understanding Security Architecture**

- **Working with Protocols and Security Interfaces**

- **Exploring Other Routes to Extra Security**

☑ **Summary**

☑ **Solutions Fast Track**

☑ **Frequently Asked Questions**

Introduction

As with engineers and administrators whose wired networks provide access to the general public, a very large dose of well-founded paranoia exists in those who want to protect their data as they flow between Bluetooth nodes. There is cause for greater concern when wireless connections are used in establishing peer-to-peer connections, because such communication is easily intercepted. This sentiment has been captured in a statement recently made in the July issue of the technical journal RFDesign, "… any high-school freshman with a scanner and some basic software knowledge can crack a Bluetooth network."

Without considering the implementation of security measures in your product, as outlined in the Bluetooth specification, such beliefs may, in fact, prove very accurate. Presented within this section are very powerful tools that, when properly implemented, can thwart the efforts of those making an attempt to extract information flowing in a completely unprotected public network.

What you need to know before reading this chapter:

- Bluetooth protocol stack component function

- Generic access protocol procedures

- Peer-to-peer protocol connection establishment mechanics

- Host Controller and Host function

- Embedded systems programming

- Familiarity with Bluetooth profiles

Deciding When to Secure

Bluetooth technology is designed to support wireless connectivity inheriting with it a number of unique characteristics associated with this method of invisible communication. For instance, anyone toting a Bluetooth-enabled device could potentially connect to your Bluetooth device, gaining access to data without your knowledge or permission. This should be cause for alarm for two reasons. First, allowing anyone to establish a connection is problematic when your application is to support one specific connection, as is the case in the Headset profile. Secondly, free public access to your data or service can present a problem. Accessing network data and implanting a virus through a local area network (LAN) Access Point (LAP) or having unrestricted access to the telephone network via a wireless telephony gateway are only two examples of applications where the use of security makes sense.

Additionally, once a service is being provided, protecting data being sent wirelessly is necessary for preventing eavesdroppers from intercepting and then interpreting the information.

When to implement security is a related yet different issue. Pragmatically, you will fashion your own security measures around the needs of the application being developed; hints will be provided in this chapter to assist in this endeavour. Reliance upon the Bluetooth specification is obvious for guidance in this matter, but ultimately the decision is yours as a systems designer or application developer.

By offering your end customer the option of enabling or disabling security, you provide them with the option of making your product simpler and easier to use, thereby improving the end users' out-of-the-box experience.

NOTE

Older versions of the protocol stack (pre V1.0B release) have security features incompatible with V1.0B and later releases as a result of changes made to the protocol. To interoperate with earlier versions of the protocol, it is necessary that your device offer the end user the ability to disable all security features.

Outfitting Your Security Toolbox

There are three components that serve as the security "troika" in any network: *authentication*, *authorization*, and *encryption*. Each has a specific function in the scheme of security and can be either enabled or disabled—it all depends on what makes sense for your application.

Authentication is used to verify a device making sure that it is who it says it is. If another Bluetooth device is trying to gain access to your device, either through establishing a radio link or by making a request to use a particular service, you first ask, essentially, "Who goes there?" then "What's the secret password?" In the world of Bluetooth security, you will already have the address of the remote Bluetooth device (from performing the connect procedures), and will use a derivative of a unique secret "link key" stored in your device as the very specific password. If the remote device provides you with the correct password, it is considered authenticated and is free to proceed in accessing all services offered by your device. This process is far more complicated in terms of mechanical operation—something that will be examined in greater detail in the next section.

Authorization has a different function in the security toolbox. It determines if the remote device is to be granted access to specific services offered by your device. Three services, as an example, are supported on your device. They could consist of service discovery, fax, and dial-up modem capability, and have an authorization procedure associated with each. If they do have a requisite procedure, any time a remote device attempts to access a service, authorization is to be triggered. With a remote device requesting access to a service, you would be presented with the name of the remote device, the service it wants to access, and be asked whether you will permit access to this service. Granting permission to a remote device is based upon who it is and the service being requested.

Because authorization depends upon knowing who is asking for access to a service, authentication must be completed successfully prior to entering the authorization procedures.

Encryption protects data by encoding it prior to transmission over the airwaves. The encryption key used is derived from the unique link key associated with the authentication process. To encrypt data, authentication must be triggered and have passed.

A more thorough explanation of each of these security elements is provided in the next sections. Basically, their underlying operation is revealed with an emphasis on the role that the application has in participating in the process.

Authentication

Authentication is the cornerstone of the security paradigm upon which both authorization and encryption depend. Without its successful completion, neither authorization nor encryption will be attempted. The term authentication is somewhat misleading as it refers to only a very specific procedure of verifying a remote device. In the grander scheme of things, other procedures are actually invoked in support of the security measure titled authentication.

Pairing for instance is a procedure invoked when a link key has not been created for the unique connection between devices. (A link key is a secret number associated with a link between two devices.) The pairing procedure requires that an identical personal identification number (PIN) be made available to devices attempting to authenticate for the first time. The PIN is either stored in memory, entered through a man-machine interface (MMI), or changed back to a default value (a byte which is set to the value zero).

Authentication is a very specific procedure used in creating a correct response to a challenge; don't worry, this will be explained shortly. Suffice to say that it follows the pairing procedures in the scheme of things if a link key does not exist.

Bonding refers to the entire process of link-creating, pairing, authentication, link key creation, and semi-permanent storage. Once devices are bonded, pairing does not have to be done again and authentication can proceed without the need for PIN entry. If a device is requested to bond with another device that it already possesses a link key for, this link key is erased. Pairing is then initiated, establishing another link key.

Pairing

Take a look at what happens when successfully traversing the authentication barrier (see Figure 4.1). Let's assume two devices are new to one another, never having gone through authentication before. In this case, the pairing procedure is required for the purposes of creating a temporary link key (Kinit) used by the next process: authentication. In addition to this, Kinit is used in encoding the semi-permanent link key (Ka or Kb) prior to transmitting to the other side for storage and future reference. Here is what happens.

Figure 4.1 The Bonding Process Including Pairing and Authentication

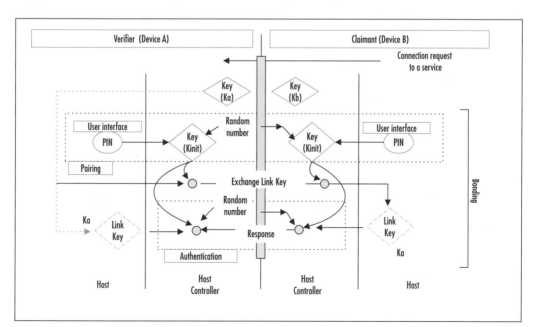

There are two roles: the *Claimant*, which claims to be a particular device, and the *Verifier*, which checks to make sure the Claimant really is who it claims. The Claimant makes a connection request to the Verifier; this can be a request made at the Link Manager level or at upper protocol layers. The trigger point invoking authentication is determined by the application when it configures the service database. Once triggered, however, a PIN is required. The PIN, along with a random number generated within the Link Manager and the claimant's Bluetooth address (not shown) is used in creating the temporary link key, Kinit. This key is created independently by the Claimant and Verifier. Pairing has now completed.

As mentioned, the PIN is furnished either by an MMI, from memory, or provided as a default value by a zero length number. Without an MMI or a stored PIN value, the application should at least try the default PIN value to generate Kinit prior to attempting authentication.

Devices with user interfaces such as phones or laptops will be able to change their PIN numbers. These devices are said to have "variable PINs." Devices such as headsets have no means of entering a PIN, so they have a number programmed in when they are manufactured. This is called a "fixed PIN." Obviously, when connecting a phone to a headset the phone that has the variable PIN must change it's value to match the fixed PIN on the headset.

Link Keys

Authentication is managed by the Link Manager using a *link key*. If a previously stored link key (called a semi-permanent key) exists, it is used to complete authentication. In continuing with the case where a semi-permanent link key does not exist, the next stage is bonding, which creates a semi-permanent key.

Bonding

Kinit is used to encode the unit key (Ka), which is then sent across the airwaves to the other Bluetooth unit for storage. At this point, devices can both exchange unit keys and create a combination key (Kab) which is calculated from both unit keys, or they can agree to just use one device's unit key. A combination key is more secure, but some devices cannot create such a key, so they must use their own unit key as the semi-permanent link key.

This semi-permanent link key is created for future use. With this key now safely stored in memory, the pairing process is eliminated. Now, every time authentication is requested between these two devices, authentication can proceed using the stored link key.

Bonding really refers to the entire process of pairing, authenticating, link key creation, and storage. As shown in this example, Ka became the link key. Kb could have become the link key as well; this is dependant upon the Link Manager and is transparent to the application as far as selection is concerned.

In summary, these are the keys:

- Kinit is calculated from the PIN key and is kept temporarily; it is used to encode unit keys so they can't be read by eavesdroppers.

- The unit key, Ka (or Kb) is derived only once by the Host Controller and stored permanently; this key can be changed but usually isn't. This key can be used as a link key as well (as shown in Figure 4.1). It not only is designated as a link key by the Verifier but is passed to the Claimant and stored as the link key.

- A combination key (Kab) can be created from two unit keys then used as the link key providing even greater security supporting authentication.

The creation of combination keys requires that both Bluetooth devices permanently store this unique key placing a greater burden on Host memory resources required especially when multiple device keys are to be stored. Instead of storing just one key (Ka) as the secret link key to be used for multiple devices, a separate combo key, if used, must be stored for each unique device.

Once the two devices have agreed on a semi-permanent link key, the Verifier begins authentication by issuing a challenge. The challenge is a random number which the verifier sends to the Claimant. A numerical response is calculated by the Claimant (using the link key) and is sent back to the Verifier. The verifier does the same calculations, and compares its results with the claimant's response. If these numbers match, authentication is deemed successful, and the devices are bonded. If the numbers don't match, it will be because one side was using the incorrect PIN key. If this is the case, authentication fails and the devices are not bonded.

At the risk of getting ahead of ourselves, we will briefly mention one last key, Kmaster. This key is temporary, is generated by the master device, and is used to derive an encryption key used in encoding broadcast messages sent to other Bluetooth devices. Each slave also has a copy of the Kmaster, using it to create their own encryption keys, which enables them to decode broadcast messages. Many profiles do not use broadcasts, so some manufacturers have chosen not to implement broadcast encryption.

Debugging...

Security Timeouts: How Long Will the Stack Wait?

During the pairing procedure, there is opportunity for the user to take their time in entering a PIN number. This time period cannot be indefinite as stack timers begin to expire; a connection cannot be established half-way and remain in this state permanently. Interoperability issues have been identified with regard to this situation. Several solutions exist to alleviate the problem. Stack timers can be set not to expire while a PIN is entered. When asking for a PIN at the application level, the amount of time the user has in entering a valid number can also be limited to prevent timer expiry. This situation also presents itself for the authorization procedures since user interaction is required.

Application Involvement

With respect to the procedures necessary in supporting authentication, you can see that there is not that much involvement by the application layer outside of providing a PIN to the Link Manager—this is partially true. Generally speaking, as an applications designer, your responsibility will be to configure your device to instigate security measures as you see fit. Handling PIN entry is an additional interface you will be responsible for (we'll discuss application interfaces later in the chapter).

Also, there are variations on the type of link key that can be created, stored, and used: a unit link key, a combination link key, a master link key, and so on. Each key type has a specific use.

Authorization: How and Why?

Authorization requires that authentication complete successfully. It is then triggered when the remote Bluetooth device makes an attempt to connect to a service. More accurately, this security procedure is invoked when a peer-to-peer protocol connection is requested at the Logical Link and Control Adaptation Protocol (L2CAP) or Radio Frequency Communications port (RFCOMM) layers. We will get to that later, however, when we discuss how to configure security.

Authorization requires that the remote device be identified and that the service being requested be reported to the service provider; this generally happens through an MMI. With this information in hand, the user can choose to permit access to the service requested, granting temporary *Trust*.

Using the Trust Attribute

Trust is an attribute that links authorization permission to a particular service and a device address. When the device is marked as Trusted, the authorization process completes successfully without user interaction. Trust is granted both temporarily, as a result of successful authorization, or permanently. Permanent Trust can be conferred upon any device at any time but is usually done during the initial authorization via the MMI. For Bluetooth devices that do not have a user interface, the Trusted attribute can be granted during an Inquiry session. By simply being within the serving area, remote Bluetooth devices can be labeled as Trusted, tagging their unique Bluetooth address with the Trusted attribute and storing this information in the device database for future reference. Switch into this mode of operation only when you are confident that safe devices are nearby.

A common consideration for devices marked as Trusted is to allow this privilege to expire some time in the future. Expiry of this privilege means that the stored information in the device database remains intact with the exception that the once trusted device is now tagged as Untrusted. Permanently marking a device as Trusted is not a recommended policy as it circumvents the Bluetooth security measures as they relate to authorization. Untrusted devices require that the user intervene on the next attempt to authorize.

Remote Bluetooth devices can also be classified as Unknown. If the device has never been seen before and has no record of existence in the device database, it is referred to as being unknown. If the service being requested by such a device is protected by authorization, then the MMI is used to grant permission. Alternately, a record containing this device's address, the service that it is accessing, along with the Trusted attribute are stored in the device database automatically upon being discovered, bypassing the need for using an MMI.

Enabling Encryption

The last component of security to be described is that of encryption. You really cannot prevent the interception of data that is transmitted wirelessly. What you *can* do, however, is transform the data into something that cannot be (easily) understood. Encryption is the process through which transmitted data is encoded, only to be decoded on the receiving side. When activated, encryption relies upon a special encryption key generated from the stored link key. The encryption key is then used to encode data sent over the airwaves. On the receiving end, the same encryption key (generated from the same link key) is used to decode the data.

Point-to-Point Encryption

Encryption, if used, must be enabled on both sides of the radio link. You cannot use encryption in a unidirectional data transfer. Up until this point, the connection being discussed has been point-to-point (one Bluetooth unit communicating with another unit exclusively). In the case where one unit is broadcasting data to multiple units, there exists a need to distribute an identical encryption key to all other slave units listening in on the broadcast. This scenario is very specific to the master—a slave relationship where the master initiates the point-to-multipoint encryption.

Broadcasting

A new encryption key, briefly mentioned earlier, is based upon *Kmaster*, which is generated using two random numbers. Without going into detail, Kmaster is sent to all slave units that have a need to participate in receiving a broadcast transmission. Once Kmaster is sent to all units, the master device instructs each slave to now use this key in generating a new encryption key, this being now the common denominator allowing all units to decode data originating from the master device. This encryption key is used only while broadcast messages are being sent. Once this activity is no longer required, all units revert to their original link keys under the command of the master. Using point to multipoint encryption is usually temporary and is less secure than point-to-point encryption since it relies upon the lowest common denominator security, that being a common encryption key as shared by a number of different units. For instance, if one unit in a piconet supports 32-bit keys, and all others support 128-bit keys when using broadcast encryption, all units will have to use a 32-bit key.

Under all circumstances, as just described, the application software remains virtually isolated from this process; it does not have to manipulate the link keys used in point-to-point or point-to-multipoint communication. Nor does it concern itself with the operations taking place at the physical layer to manage the use of different link keys. The Link Manager handles the determination of the link key and subsequent use of the encryption keys.

Application Involvement

This brings us to an interesting point in the discussion regarding security. What exactly is the application software responsible for? Thus far, we have examined the basic mechanism used in protecting both a Bluetooth device, or its services from unauthorized access by an unknown and possibly hostile device. Authentication, authorization, and encryption can be considered building blocks on which security rests. Controlling these security instruments, or more accurately, configuring security, is the responsibility of the application developer.

Point-to-multipoint communications can be supported where an encryption key is shared among many different devices—in other words, it is derived from the Kmaster link key. In any event, encryption can be specified for use by the Security Manager and required that authentication be completed successfully.

Understanding Security Architecture

We will now turn our attention toward how security measures are used in the context of a commercial Bluetooth implementation. Figure 4.2 portrays a commercial embedded solution for a Bluetooth device. A Host Controller provides services associated with radio control and is responsible for containing the authentication and encryption engines. When commanded to do so, these engines are fired up and complete the procedures necessary in completing their task: Link key management, random number generation, challenge response routines, and encryption key generation and management. Note that the Unit key (Ka) is permanently stored in the Host Controller, with temporary storage being provided for different types of link keys as required.

The Role of the Security Manager

The Host, on the other hand, is responsible for at least setting up the environment required to start security and in some instances, initiates security itself. A Security Manager module is tasked with many diverse responsibilities, which include providing an application interface to:

- Configure security
- Request PIN entry
- Query the user for an authorization response
- Respond to the Link Manager with PIN information or a link key supporting authentication

Figure 4.2 A Commercial Bluetooth Implementation Showing Interfaces to the Security Manager

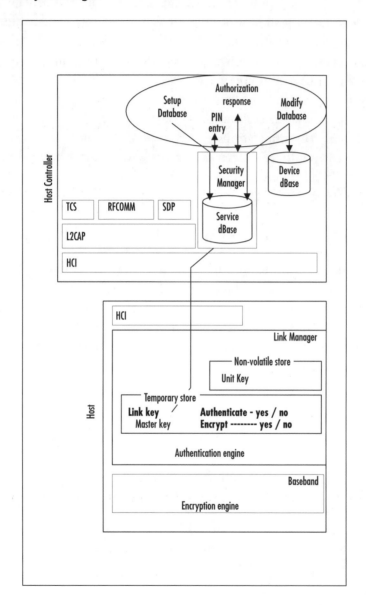

Internal to the Security Manager is a service database, a repository that is configured by the user via application software. As will be explained later, this database is used to implement Mode 2 security and is referenced by the Security Manager to determine which security measures to invoke and when to invoke them. In addition to this, there is a device database which stores link key information, and also keeps tabs on which devices are Trusted and which are not.

Supporting the Security Manager in its responsibilities are three entities:

- The *service database*, which holds the security configuration information as provided by the application software.

- The *device database*, that persistently stores information regarding past sessions with other Bluetooth nodes, allowing quick connections to be established without having to traverse the security barrier again.

- Application software that provides a user interface (UI) for the purposes of entering a PIN or confirming an authorization request and setting up a Trusted relationship. Alternately, in embedded systems where a UI is not to be found, the application will respond to requests in a manner that makes most sense without user intervention.

Two issues loosely related to the Security Manager are:

- Setting up authentication and/or encryption at the Link Manager level; this is done by the application, either indirectly through the Security Manager, or directly by configuring the Host Controller via the Host Controller Interface (HCI) layer.

- The device database, which can be modified by the application code; the time limit associated with a Trusted relationship between two Bluetooth units may expire thereby changing this parameter to Untrusted. The link key can also be erased to force authentication once again.

Before we go any further, we must first understand where triggers can be set to start security procedures. This all begins with defining the three different modes associated with Bluetooth security.

- Mode 1 has no security, obviously making it the least secure mode.

- Mode 2 invokes security when a higher layer protocol or service is accessed.

- Mode 3 invokes security when a connection is requested; this is the most secure mode.

Typically, security is associated not so much with protecting a Bluetooth device as it is with preventing access to services supported by the device itself.

For instance, would it matter that much if another person were to simply establish a radio connection to your device, not invoking peer-to-peer protocol connections at the upper layers of protocol? Or would you be more concerned about the fact that another device could covertly extract files from your device, without your knowledge? More insidious would be the notion that the intruder could plant a virus on your device without your knowledge, then sadistically watch as you frantically tried to prevent your device from self-destructing. The most important line of defense is in protecting services. A close second would be to protect your radio hardware from being tied up by an unwanted intruder, keeping the Host Controller free and available for communication.

Mode 1 Role

Mode 1 security is the simplest of all. It specifies that there are no Bluetooth security procedures at all. Any connection initiated by another device is granted as far as the Bluetooth protocol stack is concerned. Be very careful here as this does not mean that there is no security at all. There is plenty of opportunity at the application layer to implement some level of security, such as the use of a user ID and password in granting access to a network. This can even be done at the object exchange (OBEX) transport layer, which supports the use of authorization independent of the Bluetooth protocol stack. These additional elements of security will be discussed later.

Mode 2 Role

The most common (and useful) form of security is Mode 2 security and is used primarily to protect services being offered by a Bluetooth unit. It is invoked only when a request is made for a specific service, or more accurately, when a connection request is made to establish a connection to a specific layer of protocol.

    With reference to Figure 4.3, you will see that the Security Manager is cognizant of the goings on in both the L2CAP and RFCOMM layers. When an attempt is made to establish a peer-to-peer connection at either of these layers, the Security Manager is made aware of this and acts as an arbiter. It does not matter if the connection is being initiated by your application, or requested by a remote device, the Security Manager has intimate knowledge of what is happening and responds appropriately. It can decide on the course of action, basing its decision on configuration data placed in the service database. The options available to the Security Manager are as follows:

- Do nothing and allow the peer-to-peer connection to establish itself.

- Initiate authentication procedures.

- Initiate authorization procedures.

- Start encryption once a communications link is established.

Figure 4.3 Trigger Points Are Located within RFCOMM and L2CAP to Invoke Mode 2 Security

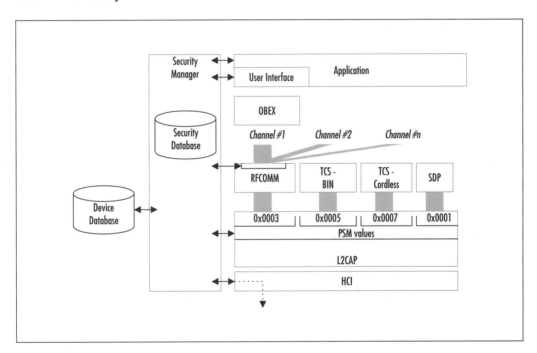

With security being triggered at the L2CAP layer, there is the potential for blocking access to services above this layer. Service Discovery Protocol (SDP), Telephony Control Specification (TCS), RFCOMM, and OBEX functions (and all application profiles relying on these underlying building blocks) can be selectively protected. When an L2CAP connection is established, a value called a protocol service multiplexor (PSM) must be specified, identifying which of the modules above this layer is to be accessed. Table 4.1 lists the PSM values along with their corresponding upper layer connection module to give you a view of services that can be protected if security is linked to L2CAP.

Table 4.1 Associated Protocol Service Multiplexor Specifying the Service It Represents

Service Module	Protocol Service Multiplexor
SDP	0x0001
TCS-BIn	0x0005
TCS-Cordless	0x0007
RFCOMM	0x0003

Usually, when using the L2CAP layer as the security trigger, your intention is to protect either the cordless telephony/intercom profile (TCS) or SDP. Protecting SDP may not be in your best interest as this implies you are not inclined to provide services to other devices that do not know what it is you do. Don't forget that once a remote device passes authentication, and if the link key is stored (bonding completes), authentication will successfully pass in future sessions without user intervention. Perhaps a different strategy would suffice in protecting your device from others that do not know what you do—like configuring your device to be non-discoverable.

In a manner similar to L2CAP, the Security Manager has access to the internal workings of RFCOMM as well and can trigger security based upon connection requests being made at this level. Associating security with the RFCOMM protocol layer protects applications requiring the serial port profile and profiles built upon this foundation such as fax, modem, LAN access, and OBEX.

As was the case with L2CAP, the Security Manager can be selective in determining which applications to protect as well. Peer-to-peer connection establishment at the RFCOMM layer requires that a specific channel (out of a possible 60 channel values) be specified for the connection to complete successfully. This channel number is always associated with a particular service or profile being offered by the Bluetooth server unit. This channel number is made available to client devices through SDP. Therefore, to protect a specific service relying upon serial profile support, you would set up the Security Manager to trigger when a connection attempt is made using RFCOMM and a service-specific channel ID.

There are a few interesting things you should be made aware of. First, server applications (such as a LAN Access Point) relying on RFCOMM must register their use of the RFCOMM interface by entering information into the SDP service database; specifically, this equates to a channel number associated with the RFCOMM module along with the service supported, such as LAN access. Devices interested in using this service must query the service database using the SDP facility, extract this information, then make a request to connect to the specified RFCOMM channel number. The Security Manager detecting this request will make a determination if security is required based upon configuration information contained within its own internal service database. It will then take action and invoke security measures as required.

The Security Manager, in accordance with the Bluetooth specification, can also initiate security measures if a particular type of connection (RFCOMM or L2CAP) is initiated by your own application. For instance, assuming for a moment that as a client application, I want to establish a connection to a server offering "FAX" capability (RFCOMM channel #7 as revealed by an earlier SDP session). After establishing a radio connection at the Link Manager level, a connection request would be made to the server unit at the L2CAP layer. Next, before attempting to connect at the RFCOMM layer, authentication would be invoked by my side. My device would be the Verifier. If successful, a connect request to RFCOMM would then proceed. Note that authentication is supported on outgoing (as well as incoming) connection requests. Authorization and encryption are only triggered on incoming connection requests.

Mode 3 Role

Mode 3 security is the most stringent form supported. When Mode 3 is specified, any radio connection request being made, whether incoming or outgoing, triggers authentication. Optionally, if authentication completes successfully, encryption can be applied to the data link if specified. Authorization is not supported in Mode 3.

Successful completion of authentication results in the establishment of a radio link. For Mode 3 security, the Security Manager remains relatively detached, yet still supports the need for PIN information when required, or link key information, if it exists in the device database. With reference to Figure 4.2, the Host Controller (or more specifically, the Link Manager) has an authentication flag associated with it (Authenticate—yes/no). The application code sets this flag, and if set, authentication is initiated automatically by the Link Manager, allowing the radio frequency (RF) connection to complete once authentication passes. Passing authentication requires the following underlying operations to be managed by the Security Manager running on the Host:

- Getting a PIN if required during the pairing process.
- Providing a link key if one exists as generated from a previous session.
- Storing a link key if one is created by the Link Manager for future reference.

The Link Manager is capable of being configured, initiating authentication procedures independent of the application software. Under this scenario, any attempt to connect at the Link Manager level triggers authentication. As you can see, there is provision to store link key information in the Host Controller as well. The Unit key (Ka or Kb) is usually calculated only one time and stored away in non-volatile store (NVS) for future reference. If you recall, this unit key can be used as a link key only after pairing has been completed. Alternately, the unit key of the other device (Kb) or a combination key (Kab) can also be used as the link key, requiring that it be stored in the Host Controller for use in deriving the encryption key. The link key is also sent via the HCI to the Host for permanent storage as well in the device database. There is also temporary storage available for a master key (Kmaster), which is generated by the Host Controller and used for point-to-multipoint data transfers requiring encryption. The master key is not placed in NVS at the Host Controller level, and as a result is lost once the connection between Bluetooth devices is relinquished.

Mode Unknown

There is one more issue that needs to be addressed and that is the way in which connectionless packets are managed. L2CAP supports connectionless data transfers. Bluetooth supports the notion of datagram transmission—in other words, the ability of one device to send another device a data packet without expecting any type of acknowledgment that the data packet was ever received.

An example illustrating the use of a datagram is in the wireless telephony profile. Multiple terminal units attach themselves to a wireless telephony gateway. Each terminal unit eventually takes on the role of a slave device. With the arrival of an incoming call from the public service telephone network (PSTN), the gateway responds by broadcasting a datagram containing the phone number of the unit being called. All terminal units examine this datagram, and if it contains their phone number, they can then respond by setting up a connection-oriented link. The Security Manager has the ability to block datagrams at the L2CAP layer if it is configured to do so by the application.

So far, the building blocks of security have been presented: authentication, authorization, and encryption. Where and how security is managed has also been covered, yet absent from this picture is how the Security Manager is configured and how it knows what it's supposed to do. This is the next topic of discussion.

The Role of Security Databases

Security management, although automatically administered, depends upon how it is configured, which is the responsibility of the application. There are three ways in which the application participates in setting up the security system. They are:

- Configuring the Host Controller to enforce Mode 3 security.

- Configuring the Security Manager to respond appropriately when L2CAP and RFCOMM layers are attempting to establish a peer-to-peer connection; this is related to Mode 2 security.

- Using the application to command the Host Controller to begin authentication and/or encryption.

In this section, we will examine, from the perspective of the application, how to configure security as it relates to Mode 2.

Service Database Content

Mode 2 security configuration data is stored in a service database under the direction of the application software and through an interface that is supported by the Security Manager. This database is managed exclusively by the Security Manager. The application must access the Security Manager in order to create database records which define the trigger points for security, and identify the components to use in implementing security.

Figure 4.4 illustrates the record content required when characterising Mode 2 security.

First, the trigger point for initiating any security procedure is specified not by specifically referring to a service that requires protection, but rather by the protocol "pipe" leading to this service. Triggering security when a client attempts to attach itself to a Cordless Telephony gateway would have a service definition of:

Protocol level = L2CAP

PSM = 0x0007

Figure 4.4 The Service Database Determines When to Invoke Security

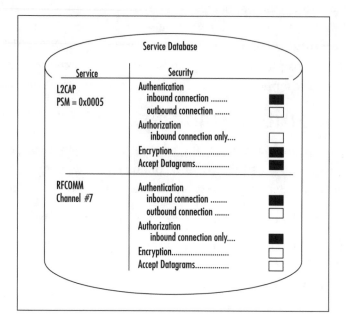

Another example would be a modem server using channel 2 (supported by the RFCOMM module). This would have its service defined as:

Protocol level = RFCOMM

Channel ID = 2

Associated with the service descriptor are security attributes that are exercised prior to allowing the establishment of the peer-to-peer protocol connection. The attributes to be defined are as follows:

- Authentication to be applied (for an outgoing connection) – yes or no
- Authentication to be applied (for an incoming connection) – yes or no
- Authorization to be applied (incoming connection only) – yes or no
- Encryption to be applied (in response to an incoming connection) – yes or no
- Connectionless datagrams to be accepted – yes or no

Service Database Operations

The service database is used only when a protocol event occurs. The Security Manager is activated if a connection is required at the L2CAP or RFCOMM layers; it looks up the corresponding reference in the database. If one exists, it takes action as dictated. The order in which security measures are invoked is:

1. Authentication

2. Authorization

3. Encryption

Attributes in the service database can be modified at any time and must reflect the services offered by the device; in essence, if the SDP database changes in terms of RFCOMM ports being used in supporting services, the same changes have to be taken into account if security is to be applied to the same services. Updates must be reflected in the service database if security is to be effective.

Developing & Deploying…

Mode 1 Security: Configuring for No Security

The absence of a record in the service database for services offered by the device will result in no security measures being executed at least as related to Mode 2. Of course, Mode 3 is different as it is configured by writing to the Host Controller via the HCI; some implementations offer an application programming interface (API) structure associated with the Security Manager that provide commands necessary in configuring the Host Controller.

Authorization is the process whereby permission is granted to the device requesting access to services offered. When the Security Manager determines that authorization is to be invoked, it simply asks the server application the following questions:

- Do you want the device requesting service (as identified by remote username or remote device address) to have access to the particular service being requested (for example, the Fax service)?

- Is this device to be Trusted for future sessions?

In answering yes to both questions, the protocol connections required are completed and the applications' service is offered to the client. The device database is modified to reflect that the remote device or client (as enumerated by its address) is Trusted.

In the future, if authorization is invoked, the device database is consulted. If the Trusted parameter is set for the device requesting access to the service, authorization is deemed to have passed without need for user intervention.

Role of Device Databases

Initiating Mode 2 or Mode 3 security is determined by the application during setup of the service database, or when configuring the Host Controller indirectly through the Security Manager respectively. We now turn our attention to the support activities and structures that need to be managed once the security process is underway. As has been mentioned earlier, there must be a mechanism in place by which historical data is kept for future reference. For example, upon the successful completion of authentication, a link key is created that is unique to the two devices participating in the process. This key must be persistently stored along with the address of the authenticated device for future reference. As equally important as the attribute of Trust, this tag is assigned specifically to devices that have passed the authorization process. It, too, must be stored for future reference. Both entities are placed in the device database, an area that provides persistent storage of information.

Device Database Content

Figure 4.5 illustrates the device database and the content of a record. When authentication is requested, the device database is first accessed to determine whether a link key exists for the device being authenticated. If such a key is available, it is used in calculating the correct response to the challenge issued. If this key is absent, or if it is incorrect, the pairing procedure must begin and a PIN needs to be entered. A new link key will be generated then possibly stored in the device database for future reference. Storage of the key for future use is an option that is managed by the application.

Authorization is very similar in terms of operation. If during the authorization procedure the application determines that the device is to be Trusted (either in response to User input or it is automatically granted without the need for UI), this attribute is stored in the device database as well. Future sessions between the same devices will make reference to this stored parameter, determine that the attribute is Trusted, and bypass the authentication procedure as a result.

Figure 4.5 The Device Database Persistently Stores Data Resulting from Successful Completion of Security Procedures

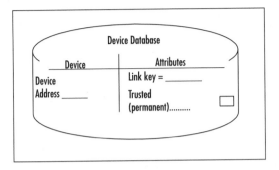

Device Database Operations

This database is accessed by both application and Security Manager. The application can access records for the purposes of changing parameters if required. An example would be in modifying the Trusted attribute to Untrusted upon expiry of a predetermined time period. The Security Manager accesses the device database in response to actions that are dictated by the service database. Extracting a link key in response to authentication activity (as requested by the Host controller), examining the Trust relationship (in response to having to authorize a connection) are two such examples whereby the Security Manager uses information stored in this structure.

Managing the Device Database for Your Applications

Data storage in the device database is persistent to prevent the loss of data as a result of turning the power off. With this in mind, you must be aware of the need to develop your own drivers to manage the device database. Because embedded systems are developed to run on different hardware platforms and to use different operating systems, they require the applications developer to take on the added responsibility of porting the Bluetooth protocol stack to the particular Host target environment. Obviously, you will need to do the work necessary in getting the stack to work with your operating system as well as in developing both transport and hardware drivers required for communicating with the Host Controller. In addition to this porting activity, you must develop drivers that will be used in accessing and managing the device database. Because this database is to be kept in non-volatile store, the hardware implementation could be just about anything from a disk drive to FLASH memory, requiring either a serial interface or parallel interface. Because this is implementation-specific, you will have to assume responsibility for completing this custom work.

Such work is highly dependant upon the protocol stack you are using. Hopefully, your stack vendor has provided an interface that you can write to which supports this activity. The stack can then call the drivers that you have developed in managing the device database. It is desirable to access the device database via an application programming interface (API), provided by the stack itself.

Working with Protocols and Security Interfaces

With all components of security now defined, we are now able to look at the mechanics of how security functions are carried out in an embedded device. Secondly, we will be able to look at how your application is to interface to the Security Manager for the purposes of setting up a proper security regime. Lastly, managing the device database is briefly discussed to complete the discussion of how your application is to treat the issue of security with the intention of jump-starting your design work in meeting time-to-market pressures.

Mode 2 Operation

Figure 4.6 is an illustration of the messaging that takes place when the full complement of Mode 2 security is assigned to a particular service, such as access to the TCS binary group of functions in a wireless telephony profile. In this example, L2CAP is identified as the service-related protocol with the designated PSM of 0x0005; this is the security trigger that invokes the Security Manager. Here is what happens when authentication, authorization, and encryption are required.

Authenticate 1   Commands the Host Controller to authenticate the other device.

Authenticate 2   Host Controller responds, asking for a link key (if one exists).

Authenticate 3   The device database is checked by the Security Manager or a link key associated with the address of the device being authenticated (assume no key exists yet).

Figure 4.6 Operation of Mode 2 Security in Completing the Authentication Procedure as Dictated by the Security Manager

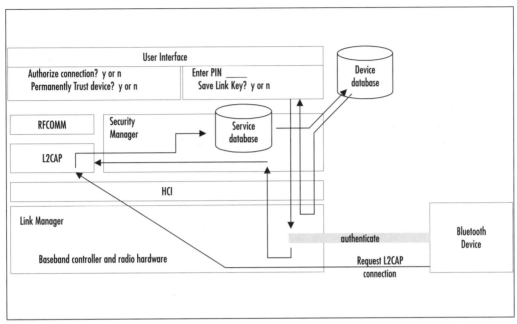

Authenticate 4   The Host responds with "no key."

Authenticate 5   The Host Controller makes a request for a PIN and the Security Manager asks the application for a PIN (either through a UI or from memory).

Authenticate 6   The PIN is returned to the Host Controller and an initial, temporary link key is created (Kinit).

Authenticate 7   A permanent link key (Kab, Ka, or Kb) is created and shared between devices.

Authenticate 8   Authentication proceeds using this permanent link key and passes.

Authenticate 9   The permanent link key is sent to the Host for storage in the device database for future reference.

Authorization 1   The Security Manager examines the device database to see if the device is Trusted (assume it isn't yet).

Authorization 2   The Security Manager presents the name of the device attempting to make a connection, and the service it wants to access to the application software. The application must respond back to the Security Manager if this connection is to 1) be authorized, and 2) if this device is to be Trusted.

Authorization 3   The Trust attribute is entered into the device database by the Security Manager and the peer-to-peer protocol connection is permitted to proceed in establishing itself.

Encryption 1   The Security Manager then commands the Host Controller to invoke encryption, which it does.

During the execution of security measures, there are only two points where the application software is invoked. PIN entry and response to the authorization request are the two elements requiring handlers.

Mode 3 Operation

Mode 3 security is similar in that authentication is initiated by the Host Controller without involvement from the Security Manager; steps Authenticate 2 through 7 are then used in completing the procedure. If encryption is also enabled on the Host Controller, it will automatically be enforced without Security Manager intervention.

Application—API Structure

Application development will now be addressed in terms of implementing security. As was explained throughout the text, there are three application interface points that you will have to concern yourself with after you determine the level of security that you will implement for your device. They are:

- Setting up security (service database for Mode 2 or Host Controller configuration for Mode 3).

- Responding to requests for PIN, specifying permanent storage of the link key, approving authorization requests and allocating semi-permanent Trust (all MMI related).

- Modifying the device database to reflect a change in Trust upon the expiration of a timer or removing link key information if required to do so.

NOTE

The "Bluetooth Security Architecture" white paper currently available through the Bluetooth Web site (www.bluetooth.com) is an excellent reference in how to deal with Bluetooth security.

With an understanding of security as it has been addressed, it is now time to examine the software routines required in supporting security and how the defined interfaces are to be used, as they pertain to developing your application.

We will look first at configuring a system requiring Mode 2 security, the interface routines that are necessary and what you can expect from a commercial Bluetooth protocol stack in terms of implementing your particular solution.

Being able to configure the service database with both service information and the levels of security to be applied when this service is being instantiated is supported by the following routine abstractions supported by the Security Manager API:

- SEC_registerApplication (Name, Security Level, PSM, Protocol ID, Channel ID); this interface configures the service database to trigger security measures when connections are being set up at a particular PSM at the L2CAP layer.

- SEC_registerMultiplexingProtocol (Protocol ID, Lower Protocol, Lower Channel, Security Level); this interface configures the service database to trigger when a link is being requested at a particular channel number on an RFCOMM connection.

In either instance, the parameter governing security being passed into the routine is "security level" and it defines which security elements are to be associated with the specified service.

- Authentication incoming connect request

- Authentication outgoing connect request

- Authorization incoming connect request

- Authorization outgoing connect request

- Encryption incoming connect request

- Encryption outgoing connect request
- Connectionless packets (datagrams) allowed

Commercial implementations may differ somewhat from this description, yet they should provide the same level of functionality in the configuration of security Mode 2. Mode 3 is slightly different, as it is setup by sending commands directly to the Host Controller via the Host Controller Interface. Command abstractions recommended in the security white paper are:

- HCI_Write_Encryption_Mode
- HCI_Write_Authentication_Enable

Again, when you are using a commercially available stack, the command structure made available to the application layer may be slightly different; all you really need is to have the capability to configure the Host Controller to implement authentication and or encryption. Such calls could be made through an API specific to the Security Manager which in turn communicates with the Host Controller. Unlike Mode 2, security measures will be applied to both incoming and outgoing connection requests. You do not have a choice.

Mode 1 is the simplest in terms of setting up security; specify nothing. For those that want to play it safe, simply ensure that the security service database contains no record for the service being protected and Mode 2 will not be used. Also, remember to configure the Host controller to disable authentication and encryption.

In support of the completing authentication or authorization, the application code has to be notified of when a PIN is to be entered or if authorization is to be granted. This is wholly dependant upon the protocol stack, as its architecture will determine how this is to be managed. Two potential ways of handling the required activity are to use a messaging structure and inform another task that information is required, or to make use of callback functions. In the case of either method, the application has to respond and does so by using the following abstractions:

- SEC_PinRequest (Bluetooth address, Name, PIN); this interface returns the PIN, gathered from a User Interface or from memory, to the Security Manager which then passes the PIN to the Host Controller such that it can continue the pairing process.
- SEC_AuthorizationRequest (Service name, Device name, Trusted relationship); this interface presents to the user both the name of the service being requested and the name of the device making this request. In return, the application returns the Trust value that gets written into the

device database. If Trust = TRUE, future sessions will proceed without the need to authorize. If Trust = FALSE, authorization will be mandatory once again. In addition to this parameter, there must also be a way for the application to inform the Security Manager that Trust is granted temporarily, at least for this session. Your protocol stack will have its own way of handling this since it is not addressed in the Bluetooth security white paper.

In the case of responding to a request for authorization, the Security Manager should automatically handle the setup and configuration of the device database to reflect the status of a device. Remember that Trust is a parameter which can be changed from TRUE to FALSE with the passage of time. The application is responsible for keeping track of this and must have a way of modifying the device database to make such changes.

To complete the discussion on the programming interfaces, there is opportunity for the application itself to initiate either authentication or encryption. Supporting this are the following interfaces:

- HCI_Authentication_Request; this interface commands the Host Controller to begin authentication on a specific connection. Remember that if the device is a master, it is capable of supporting up to seven unique data connections to slave units. The Security Manager is used to either respond with a link key to the Host Controller, or to inform the application that a PIN is required and handle the entry (SEC_PINRequest) as described previously.

- HCI_Set_Connection_Encryption; this interface instructs the Host Controller to encrypt a data channel associated with a specific connection that has already been established. Earlier, it was stated that once a device is authorized for one service, it is authorized for all services. If you have a need to re-authorize a device for a service, this is the way you do it. By directly requesting authorization upon the initialization of the service, you are able to protect access to the service by outside users.

Exploring Other Routes to Extra Security

You should now feel very comfortable regarding the Bluetooth security troika and how to apply it in your device. This may not be enough, however. There are a few other tricks you can consider when actually deploying your device to your customer base, as well as a few tricks your customers may have up their sleeves in enhancing system security. Is this being paranoid? You decide.

Invisibility

The ultimate in security is to make your device non-connectable. This is only for the truly paranoid who will go to any measure to protect their services, their data, and their device from hostile as well as legitimate users. Unfortunately, this is not very practical when used as a security measure, even though it is very desirable should the device ever be taken out of service for any reason. (Perhaps the LAN to which a LAN Access Point is connected.)

Less onerous, and quite clever, is to make your device non-discoverable yet connectable. By doing this, your device cannot be "seen" by other devices while they are scanning the vicinity using the Inquiry procedures. By not responding to an Inquiry message, your device will not reveal its presence, nor will it divulge its address, thereby becoming a silent device. Without an address, all other devices will be unable to establish a connection, consequently enhancing security. Users that have been told about the presence of this device can be provided with its address. They can then manually enter this unique address into their Bluetooth device and proceed to connect to the device at will.

An added benefit of configuring your device as either non-connectable or non-discoverable is in saving power consumed by the Host Controller, thereby prolonging battery life (if the device is battery powered).

Application Level Security

Applications themselves often use their own forms of security giving them greater control over the selection of legitimate users. LAN access, for example, relies upon a Point to Point Protocol (PPP) layer which, among other responsibilities, usually asks the client for its user ID and a pre-determined password. When PPP security is in use, network access is granted only after this information is provided and verified by the network, although using the security features at this level is optional. The network manager can dynamically modify network access parameters, providing access to users that are new to the corporation, or restricting access to others that may have left. With reference to the LAN Access profile, there are several different types of PPP that can be supported, each having a similar way of implementing security.

Additionally, network access may have user ID and password requirements that are under complete control of the IT department.

OBEX, although included as part of the Bluetooth protocol stack, can provide a layer of security that acts in a manner similar to that of authorization. When security is used at this level, a connection between OBEX transport layers invokes user interaction generally through a User interface. If the connection is approved, the OBEX transport layer completes the peer to peer connection and application profiles can then be used.

Using application specific security may be preferred since complete control is maintained by the IT department and is not dependant upon Bluetooth security alone.

Implementing Security Profiles

To assist your efforts in developing a strategy for implementing security, a summary of all profiles defined in Bluetooth specification V1.0B and their associated Bluetooth security levels are presented. In addition to this information, which is used to provide guidance as well as to ensure interoperability between different products in the marketplace, different strategies will be presented to provide further assistance toward applying sufficient security to your application.

SDP

We will start by looking at support functions first, that being SDP. Do you really need to protect this feature? The profile specification indicates that authentication and encryption can assume a default value of 'not active', yet authentication and encryption are to be supported. If another device, during the establishment of a connection to SDP, enforces authentication and encryption, then you must reply in kind supporting such requests. It should be obvious that level 2 security is used in this instance as this is the only mode supporting service protection.

Why would you want to protect SDP and would this be a prudent move? Remember, once authenticated, a remote device can then access all services during the same session since the link key is established between devices and is stored temporarily in the Host Controller. (It can also be stored permanently on the Host.) In denying access to information in this fashion may imply that you really don't want people knowing what you do or how to connect to your device. It is better to use a different security measure – perhaps setting your device as non-discoverable to prevent strangers from 'seeing' you. It is probably best to offer unprotected access to SDP providing important connection information, then protecting the actual application that your server provides.

Cordless Telephony and Intercom

Above the L2CAP protocol layer resides the TCS module supporting cordless telephony and intercom profiles. It is mandatory to use security modes 2 or 3; you get to select. Authentication and encryption are to be used and the bonding process is to be initiated by the terminal unit.

In a public environment, a gateway may be provided for users to access the PSTN. Mode 3 may be appropriate, quickly keeping radio connections from being established for unauthorized users. In doing so, you would prevent the loss of an otherwise useful and limited resource: the radio link. Only users that could enter the correct PIN would be able to establish a link with the gateway. Another approach would be to enhance this security by making the gateway non-discoverable; further preventing the occupation of a radio link by casual Bluetooth users. Others that are aware of the gateways presence could connect without having to go through the discovery process.

Mode 2 security is best used in a controlled environment such as an office where users are known. Also, with a fixed number of users known, gateway access may not be a concern. Under this situation, terminal units are able to collect information about the gateway via SDP and choose to continue in establishing a connection. Bandwidth considerations are not that important when compared to the convenience for potential users. Also, being deployed in a friendlier environment, the level of security used can be relaxed to Mode 2.

Placing the device in the non-discoverable mode also limits access to the gateway to those already cognizant of its presence (these are typically regulars that work within the same office space). For larger numbers of users, the address of the gateway could be provided to a fixed number of users. In such a controlled environment, bandwidth considerations (the number of users that can be supported by the gateway) can be managed effectively.

The intercom profile is simpler and does not require security (it is really just an option). Given that a 10-meter distance is not far, one could yell loud enough to overcome the security barrier—unfortunately, your communication would be heard by all!

Serial Port Profile

Security recommendations for this profile are not specific since the applications making use of a simple serial connection are very diverse. As such, I will leave it up to you to decide on what security to use. Suffice to say that you should have a very good idea of what to do after examining security associated with the other profiles that rely upon the serial port profile.

The approach to use is dependant upon the reason for security. If a point-to-point connection (exclusivity) is required, authentication is suggested.

Headset Profile

This is a great example of where a communications link is restricted for use by only a very specific device. A cell phone and headset go through a bonding process—the exchange and storage of a link key. How this is managed is generally up to the vendors of such devices. To date, headset terminals have all been embedded devices incapable of supporting manual PIN entry. Two approaches can be used to accomplish bonding. One approach has the gateway discovering all headset devices in the vicinity and paging at random one of the devices in its headset list of devices. If this is the gateway to which the user wishes to bond with the gateway (cell phone), they acknowledge this connection (by perhaps pushing a button on the headset). The gateway now knows that this device is the correct one. It then begins the pairing process with this unit—using the default PIN. Both devices must use the default PIN (one byte set to the value zero) for this to work. Once authentication passes, a link key is passed between devices (normally from the headset terminal to the cell phone) for storage. With the link key and the address of the headset terminal unit established, authentication can now complete without delay between these bonded devices. Note that authorization is not used in this profile.

A second more convenient approach can be used. A PIN can be programmed into the terminal headset at the factory (and printed on literature accompanying the headset unit). If the cell phone allows it, the user enters this PIN number into the phone. Now bonding proceeds, using this PIN number instead of the default PIN.

Exclusivity in terms of a connection is maintained. Disabling the discoverability of the headset terminal may not be possible given the limited MMI supported, but it is another possibility in supporting an exclusive connection meant to be shared by only two units.

Dial-Up Network and FAX

Access to a service—whether data or the public telephone network (long distance)—must be protected. According to the Bluetooth profile specification, security Modes 2 or 3 are to be implemented for this profile. Also, the client, or terminal unit, is to initiate the bonding process meaning that it initiates authentication, forcing the erasure of its internal link key if one exists. The question now is to identify what security should be used and if it makes sense on the client or the server side of the link.

Clients normally access the dial-up network or FAX server, using SDP to first get a description of the service as well as information required to establish a connection via the RFCOMM interface. Mode 3 security would force any device, either on an inbound connection or outbound connection, to pass through authentication before it was provided with information regarding services offered; this is quite inconvenient. Mode 2 security configured to trigger on an outbound connection attempt at the L2CAP of RFCOMM protocol layers again would protect very little.

Addressing the server (gateway) side, it makes a great deal of sense to trigger security at the RFCOMM protocol layer on incoming connections, allowing client devices access to service discovery information. From this, they can proceed to access FAX or dial-up services. Only then will authentication and possibly authorization be invoked. Typically, either the default PIN (zero length PIN) or one that has been configured into the server will be used.

Bonding is a mandatory procedure initiated by the client (terminal) side of the connection. In essence, the client will initiate this procedure.

LAN Access

Protection of data is the most important consideration when implementing security in a LAN Access Point (LAP). Visibility to potential users can be restricted, as this is an option that is available for use by the security model you use. Restricting access to the LAP is another use of configuring the device as non-discoverable; the notion of exclusivity takes shape when the LAP is perhaps operating to near full capacity. Being non-connectable is a mode that can be configured if the back-end server is down, blocking access to the LAN as a result of equipment malfunction.

Authentication and encryption are to be used in support of connections made to the LAP. Implementing security Mode 3 will force the potential user to authenticate prior to accessing service discovery resulting in tying up an active connection to the LAP. Tying Mode 2 security to RFCOMM allows the potential user to access SDP and determine if an LAP is what they are looking for. Accessing the LAP service will then result in both authentication and encryption to be used in support of the connection. Implied is the need for pairing to take place, as well as bonding; both procedures are to be supported by the LAP.

Client management is not directly addressed by the specification. Security is not critical on the client since information from this device is not made accessible to the LAP unless the user desires to make this data available through their own action.

OBEX

Data transfers and synchronization can be initiated on either the client side or the server side, under the control of the upper layers of the application. Limited discoverable is the preferred mode regarding security on the server side of the connection. Only selected devices are to have direct access to information as provided on the server; non-discoverable is supported to allow the server to completely eliminate others from seeing their device. In configuring the device in this manner, they become completely covert relying on other means to disseminate information. Perhaps this is initially done during conversation, or information is placed into the device manually in order to provide required address information necessary for completing a connection.

Normally, devices providing OBEX services have a user interface of some sort. Computers, cell phones, and PDAs are only a few devices that fall into this category.

Authentication and encryption is supported by both client and server; whether it is used is up to the designer. Where it is used, Mode 2 or 3 is also a design choice. Guidelines that can be applied are dependant on the application supported by the OBEX transport layer.

Object push applications, such as the exchange of business cards over PDAs, could be conducted between users in an area permeated by Bluetooth devices. Use of authentication (and encryption for data that is sensitive) will provide the exclusivity between PDAs required to prevent others from gaining access to the OBEX layer and file information that this layer can provide.

File transfer is similar to object push, and can be treated in much the same way.

Synchronization is slightly different in that this application can be set up to work transparently; the users have no knowledge of the data being synched between a computer and a PDA. In this instance, mutual authentication could be used to protect both devices from establishing connections to a wrong device. Authentication and encryption could be triggered in Mode 2 or 3.

Table 4.2 provides a summary of security attributes for profiles outlined in the Bluetooth specification V1.0B. A mandatory classification indicates that the device must support the corresponding operation, not necessarily use it. For instance, with reference to the LAN Access profile, it is mandatory that the LAN Access Point be pairable. This means that if another device were to begin bonding procedures requiring the invocation of pairing, your device would respond by executing pairing procedures; it does not mean you are required to initiate pairing procedures yourself in support of security. It would be a very good idea, however, to consider using the mandatory features in your security model.

An optional classification indicates that your device can support the security feature, but also has the option of not supporting the feature.

Table 4.2 Summary of Security Attributes Associated with Each Profile

Security Attribute	SDP	Cordless Telephony	Intercom	Headset	Dial-Up Networking and FAX	LAN Access	OBEX
Non-discoverable		Gateway: mandatory	Mandatory	HS: mandatory	Gateway: mandatory	LAP: optional	Server: mandatory
Limited Discoverable		Gateway: optional	Optional	HS: optional	Gateway: optional		Server: 1st choice
General Discoverable		Gateway: mandatory	Mandatory	HS: mandatory	Gateway: mandatory	LAP: mandatory	Server: 2nd choice
Non-connectable						LAP: optional	Server: optional
Pairable		Terminal: optional Gateway: mandatory	Mandatory if bonding used, otherwise optional	HS: optional AG: optional	Terminal: optional Gateway: mandatory	LAP: mandatory	Server: mandatory
Non-pairable		Terminal: mandatory Gateway: mandatory	Optional	HS: optional AG: optional	Terminal: mandatory Gateway: mandatory	LAP: optional	Server: mandatory
Bonding		Terminal: initiates Gateway: accepts	Optional	HS: accepts AG: initiates	Terminal: initiates Gateway: accepts		Optional
Authentication	Mandatory	Mandatory	Mandatory		Mandatory	Mandatory	Mandatory
Encryption	Optional	Optional	Optional			Mandatory	
Security Mode 1							
Security Mode 2		Mandatory			Mandatory		Mandatory
Security Mode 3		(2 or 3)			(2 or 3)		(2 or 3)

Case Study

One of the most popular profiles being pursued by many companies is the Headset profile. The audio gateway resides on a cellular phone and the actual headset rests in the human ear. Incoming calls can be answered by the headset, either automatically or by using manual intervention. How does the cell phone know that it is actually communicating with the correct headset? Security procedures are used in ensuring this connection using the following strategy.

The process of bonding the cell phone and the headset is required in establishing and storing a common link key for the purposes of future authentication. If the headset is within range of the cell phone and an incoming call, the cell phone immediately establishes a radio connection with the headset. Relying on Mode 2 security, the cell phone initiates authentication procedures which, in using the stored link key, pass. The headset application then responds and is ready to accept an audio connection to support the call.

Setting this situation up is of great interest. For instance, bonding requires that a PIN be entered during the pairing procedures. This PIN can be managed in two ways. For headset devices that are manufactured to use the default PIN, the bonding procedures would proceed as follows. The cell phone would issue an inquiry, collect addresses of all Bluetooth devices within range, perform service discovery to isolate all headset applications and then attempt to access each headset. This requires that pairing takes place; the default PIN is then used. Authentication is then completed successfully since the cell phone also uses the default PIN. The headset is then paged and if it responds (because the user pushes a button to indicate it is willing to accept the connection), the cell phone knows that this is the headset to be bonded with the cell phone. If for instance there were several headsets in range and the incorrect headset was accessed, the user should not respond. The cell phone will then know that this is not the device to bond to and will connect to the next headset device in the list of headsets discovered.

Alternatively, the user can be presented with a list of possible headsets and choose which one to connect with, thereby avoiding a query for every headset in range.

Headsets that have a PIN programmed in them (identifying this PIN on the packaging) are bonded differently. If the cell phone permits it, this PIN number is entered into the phone. Pairing continues using this PIN, authentication completes, and bonding is established.

In either case, now that bonding has completed, the headset is now accessible for use by the cell phone.

Summary

Bluetooth security is used to protect services offered by devices as well as enforce exclusivity, permitting only very specific devices to connect. In accomplishing this end, the security troika was introduced consisting of authentication, authorization, and encryption. Specific use of these fundamental building blocks was then discussed in context of three different security modes; Mode 1 was the easiest to understand as it refers to no security, Mode 2 enforces the security troika at the L2CAP and RFCOMM protocol layers, while Mode 3 enforces authentication and encryption at the Link Manager level.

With this basic architecture defined, a commercial implementation of how security was to be configured by using components such as the Security Manager, service database, and device database was shown. Dataflows, although transparent to the application, were discussed to complete the picture. Application interfaces were then introduced to assist the developer in understanding how to implement the security levels required for their particular application. For those developers requiring assistance on this front, a table summarizing Bluetooth profiles and the security measures to be used was provided.

Finally, additional security measures that form part of a larger security strategy were addressed, including the configuration of the Host Controller to remain non-discoverable or non-connectable. Additionally, authorization at the PPP level, as well as that supported by OBEX, were also briefly mentioned.

Practical examples of implementing security features capped off the discussion, introducing real-world solutions to the reader, hopefully providing them with a greater sense that developing applications relying on Bluetooth security is not as complicated as it appeared prior to reading this chapter.

Solutions Fast Track

Deciding When to Secure

☑ Secure for protection of data from eavesdroppers.

☑ Create exclusive links between devices.

Outfitting Your Security Toolbox

☑ Authentication verifies that the other Bluetooth device is the device you believe it is, using a link key as the secret password.

☑ Authorization grants permission to a device making a request to use a particular service.

☑ Encryption encodes data being passed between two devices; it requires successful authentication.

Understanding Security Architecture

☑ The Security Manager, which resides in the protocol stack, manages Mode 2 security transparently to the application.

☑ The Host Controller manages Mode 3 security if configured to do so by the application software.

☑ The Security database is configured by the application and specifies when to trigger Mode 2 security procedures as well as which security measures are to be taken.

☑ The device database offers persistent storage for parameters created during the successful completion of security and makes these available for future sessions to reduce security procedures required.

Working with Protocols and Security Interfaces

☑ Mode 2 security is invoked when a client application attempts to establish a connection with the server application and can use authentication, authorization, and/or encryption.

☑ Mode 3 security is triggered by the Host Controller when either an incoming or outgoing request for a radio connection is made. Authentication and/or encryption can be specified.

☑ Application Programming Interfaces support the configuration of the type of security to use and offer a way to insert user input (PIN entry) when required.

Exploring Other Routes to Extra Security

☑ Security measures are to be supported in many profiles, such that if another device wants to invoke a component of the security troika, it will be met with an appropriate response.

☑ In many instances, implementing security is not made mandatory since this is left up to the discretion of the system designer. What is made mandatory in many instances is supporting security as mentioned previously.

☑ Non-discoverable mode as configured into the Host Controller can prevent device detection during the Inquiry process.

☑ Non-accessibility can prevent any device from establishing a radio connection, thereby preventing access.

☑ Applications often have associated with them User IDs and passwords as further measures toward protecting information resident on a server. Authorization, the act of granting permission to a service, is another application-based security measure used by the OBEX transport layer.

Frequently Asked Questions

The following Frequently Asked Questions, answered by the authors of this book, are designed to both measure your understanding of the concepts presented in this chapter and to assist you with real-life implementation of these concepts. To have your questions about this chapter answered by the author, browse to **www.syngress.com/solutions** and click on the **"Ask the Author"** form.

Q: What happens if authentication fails? What could be the cause of such a failure?

A: When authentication fails, the connection is rejected. If the connection is repeatedly attempted, perhaps because a hacker is trying to penetrate the security shield, the authentication procedures will respond by delaying a response at ever-greater time intervals, allowing authentication to be attempted repeatedly whilst still hopefully discouraging hackers.

Q: Can I prevent the storage or even removal of a link key as stored in the device database, ensuring that each encounter with another Bluetooth device will result in the need to re-enter a PIN?

A: The link key is stored in the device database which should be made accessible to the application; this is dependant upon the implementation of the particular stack you are using. You have direct access to records in the device database, allowing your application to find a record, modify it, then return it to the database for reference by the Security Manager.

Modification of the *Trust* parameter as well as complete eradication of the stored link key is supported.

Q: If I am developing an embedded device without a User interface, how can I use authentication or authorization when I cannot enter a PIN or respond to granting either temporary or permanent Trust?

A: PIN information can be stored in memory and accessed by the application when a request for this data is made. If you use this strategy, you must reveal the stored PIN to the user allowing them to enter this same PIN information into another device to successfully complete the pairing procedure. Authorization can be managed transparent to the user as well. By earmarking every device as Trusted that comes into range of a Bluetooth unit (as determined by the Inquiry procedures), authorization will be successful. Another method that can be used is in parsing out the name of the remote device, and if this is recognized by comparing strings, authorization will successfully complete; note that this requires the entry of valid device names implying that there is some user interface available. Keep in mind, this method is open to spoofing, as eavesdroppers can read the name, too.

Q: Do I have to use Bluetooth security even when I can rely upon legacy security already built into the profile?

A: The simple answer is yes. Support for security, as determined by the specification, is mandatory in many instances, yet its use is optional. Your device may not instigate security procedures, yet another device may (and could) request you participate in traversing the security boundary. The ability to participate in this exercise means you will ultimately have to implement security just in case another device wants to use it.

Q: Do I have to implement the device database in non-volatile store? What about the service database configuration? Do I have to be concerned about its contents being erased after powering down the device?

A: Using NVS is convenient as it allows the retention of device information (link key and Trust) even when the device is powered down. Volatile storage can also be used, but requires that the user enter data back into this database for future reference. The service database is generally managed in RAM; its contents are determined by application code as it initializes data structures (like the service database associated with SDP) prior to offering services.

Q: Who determines which key (Kinit, Kmaster, Kab, Ka) to use and when to use it?

A: The Link Manager makes this decision, generating keys and storing them when required. The Link Manager only communicates with the Security Manager to get PINs and store link keys as necessary. The application has minimal involvement with link key management.

Chapter 5

Service Discovery

Solutions in this chapter:

- **Introduction to Service Discovery**

- **Architecture of Bluetooth Service Discovery**

- **Discovering Services**

- **Service Discovery Application Profile**

- **Java, C, and SDP**

- **Other Service Discovery Protocols**

- **The Future of SDP**

☑ Summary

☑ Solutions Fast Track

☑ Frequently Asked Questions

Introduction

Computing is part of almost everyone's daily routine. From communicating via e-mail and mobile phone to shopping online, computing has found its way into mainstream living. As more people use mobile phones, personal digital assistants (PDAs) and laptop computers to perform daily tasks, it becomes critical that people be able to find services in their local area in a standard way that makes them easy to connect to and use.

The evolution of networking parallels the evolution of computing. As computers evolved from special-purpose, high-cost devices to general-purpose, low-cost devices, so too have networks evolved from single-function and limited-access (university and military networks), to open, multifunction platforms built around core standards (Transmission Control Protocol/Internet Protocol [TCP/IP], Hypertext Transfer Protocol [HTTP], HyperText Markup Language [HTML]). But the very success of such open and truly global networks can create its own problems. A key problem is one that every Internet user has experienced: the "finding stuff" problem. We know the information or service we need is out there, but we don't know how to find it. Most of our first online experiences were slightly overwhelming as we grappled with quantities of information presented to us. Hence, the rise of search engine technology (such as Google) and specialized portals that categorize information for us (such as Yahoo!). The more information there is out there, the more help we need finding it.

As computers became smaller and more powerful, a new category called *information appliance* emerged; it includes PDAs, ultra-light laptops, high-end phones, and Web tablets. These devices are typically used in many different scenarios—at home, at the office, and on-the-move. New types of connectivity available on these appliances is creating a new kind of networking: spontaneous and instant (ad-hoc) networks of consumer devices that join and leave a network at will. Much of the power in this new wave of appliances lies in their potential to connect to other devices, similar to or different from themselves. The purpose of connecting is not just to form a network, but to *do* something, like send a file, print a file, access a Web page or perform a transaction.

As these networked appliances become more popular, a problem emerges: to benefit from this kind of connectivity, the appliances need to work together. The appliances and services must be able to discover each other, negotiate what they need to do and proceed with business—with no intervention from the user. In corporate networks, the problem of finding services is often handled by a directory service. A directory-centric approach relies on the availability of a centralized or federated directory of available services. A given member of the network (a client) finds a service by asking the directory to look it up. The client sends an input query (name, address, or other wide-ranging criteria) to the directory, which then responds by sending a list of matching services back to the client.

For this system to work, the directory must be configured with information about available services that are updated either by an administrator, or by new services registering directly with the directory as they become available. This approach is common in traditional wired (or enterprise) networks. For example, the Domain Name Service (DNS), Lightweight Directory Access Protocol (LDAP) and the Common Object Request Broker Architecture (CORBA) Naming Service all provide directory services where a client queries the directory using some criteria. These systems work well for relatively stable environments—where the available services change relatively infrequently compared to the overall set of services. However, these systems are not ideal for ad-hoc networks, where no centralized services (such as directory services) may be present, where the resources of the appliances are themselves limited, and where the network itself is unreliable. This problem led to the development of less directory-centric approaches to the "finding stuff" problem, and, in particular, to the use of service discovery protocols and frameworks, which allow participants in a network to co-operate in advertising and using services with minimal external infrastructure.

Before reading this chapter, you should have a basic understanding of the layers of a Bluetooth stack, in particular Logical Link Control and Adaptation Protocol (L2CAP) and the Radio Frequency Communication (RFCOMM) protocol. You will also need a good understanding of the C programming language, along with some knowledge of Java.

Introduction to Service Discovery

The term *service discovery* is used to describe the way a networked device (or client) discovers available services on the network. The emphasis is on being able to discover at runtime what services exist, and how to talk to those services. Service discovery makes it possible to have zero configuration networks—where the user doesn't have to manually configure the network. Instead, the network configures itself as it discovers new available services. The ability to self-configure is critical to ad-hoc networks because:

- There is no other infrastructure available, such as a directory service.

- The network is unreliable, so connections will appear/disappear.

- Nodes themselves—such as the supplier of a service—will move in and out of the network.

Discovery protocols specify the "rules of engagement" between those seeking a service (clients) and the service provider (servers). Discovery protocols aim to minimize the configuration required in the system and to maximize the system's flexibility. Key features of a discovery protocol are:

- "Spontaneous" discovery and configuration of network services

- Low (preferably zero) administrative requirements

- Automatic adaptation to the changing nature of the network: addition or removal of nodes, or services

- Interoperability across platforms

Service Discovery Protocols

There are several discovery protocols available, each with different characteristics and a different focus (see Table 5.1 for a summary of service discovery protocols). We will examine these protocols in more detail at the end of this chapter.

Table 5.1 Summary of Service Discovery Protocols

Protocol	Originator	Comment
Salutation	Salutation Consortium	Originally designed for printers, faxes, copiers
Service Location Protocol (SLP)	Sun, IETF RFC 2608	Generic service discovery protocol intended for corporate networks
Jini	Sun/JavaSoft	Extends the Java platform and language to allow dynamic, self-configuring networking
UPnP and Simple Service Discovery Protocol (SSDP)	Microsoft, IETF Draft	Extends Microsoft Plug and Play to a wider, networked world
Service Discovery Protocol (SDP)	Bluetooth SIG	Designed for Bluetooth ad-hoc networks

Bluetooth SDP

It should be no surprise to discover that service discovery is fundamental to the architecture of the Bluetooth standard. Given that Bluetooth is explicitly designed to facilitate ad-hoc networking between a wide variety of devices, it places a strong emphasis on how those devices discover and use services in the network. The standard does not assume that any form of centralized or federated directory service exists, and so is one of the few discovery protocols that is truly peer-to-peer in nature (see Figure 5.1 for a comparison of service discovery protocols).

Figure 5.1 Comparison of Service Discovery Protocols

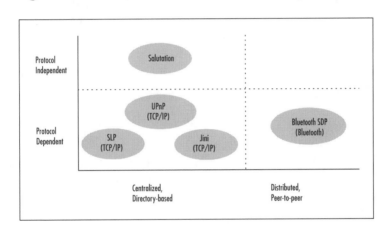

The standard defines a Service Discovery Protocol (SDP) that enables a client to directly query a device it detects on the network about the services offered by that device. We have characterized Bluetooth Service Discovery as being protocol-dependent, in that it mandates the use of the underlying Bluetooth communication protocol as the basis for service discovery. However, it's important to note the following:

- Bluetooth SDP could indeed be implemented using other underlying transport mechanisms.
- Higher-level protocols (such as TCP/IP) may be run over Bluetooth.

The latter attribute allows Bluetooth clients to use other forms of service discovery (for example, Jini) once they have bootstrapped themselves with initial services in the Bluetooth network. It also means that Bluetooth SDP may be integrated with a number of the other service discovery protocols. We will discuss some examples of this at the end of the chapter.

Architecture of Bluetooth Service Discovery

To understand the architecture of service discovery in Bluetooth, three key elements need to be considered: Service Discovery data structures, the Service Discovery Protocol, and the Service Discovery Application Profile (SDAP). The SDP, a part of the Bluetooth specification, describes both the data structures that represent information about services and the protocol used to communicate between SDP components. SDAP stipulates how SDP can, and should, be used by Bluetooth applications. Next, we'll discuss the high-level architecture of each of these elements.

The Structure of Service Records

A Bluetooth application user will need to access an entity on a remote device that will do something for the user. The remote entity is called a *service*. A service might provide information, carry out an action, or access a resource. In order for a user to find information about what services are provided by a device, the device must have an SDP server. The SDP server contains enough information about each supported service to allow it to be accessed by the user (or client). For a particular service (and there may be many services on one device) a *service record* contains a description of that service. The description takes the form of a sequence of *service attributes*, each one describing a piece of information about the service. Within the SDP server, each service record is uniquely identified by a *service record handle* (a 32-bit number). This handle is unique only within the scope of the SDP server.

A *service class* defines the set of service attributes that a particular service record may have. In other words, a service record is a particular instance of a class of services. For example, a service record whose service class is *PrinterClass* is a collection of attributes that describe a specific printer service. In fact, a service record may be an instance of multiple different service classes, each with their own set of service attributes. This is useful for building hierarchies of service types. A service class B can be said to be a subclass of service class A if it contains all of the service attributes of A and also adds its own attributes. You can tell what service classes a particular service record instance belongs to by looking at a particular attribute of the record, namely the *ServiceClassIDList* attribute. The Bluetooth specification defines 15 service attributes that are common to all service records. They're not mandatory, but when used they have to conform to the definition in the Bluetooth specification. These are the Universal Attribute Definitions, and they include attributes like *ServiceClassIDList*, *ServiceRecordHandle*, and *ProtocolDescriptorList* (a list of protocol stacks that may be used to access the service).

A service attribute is a name-value pair that includes an *attribute ID* and an *attribute value*. The attribute ID uniquely identifies the attribute within the scope of the service record. The attribute ID also identifies the type of the associated attribute value (for example, whether the attribute value is a text string, an unsigned integer, a Boolean, and so on). Since an attribute ID is unique only within the scope of a service record, the same ID can be used in different service records to represent different attributes of different types.

An attribute value can contain data of arbitrary complexity, rather than just simple types. This is accomplished using *data elements*. A data element is made up of a header and a data field. The header field includes a size descriptor and a type descriptor. The size descriptor identifies the size (in bytes) of the data in the data element. The type descriptor identifies the type of data stored in the data element, such as:

- Nil, the null type
- Unsigned integer
- Signed twos-complement integer
- Universally Unique Identifier (UUID)
- Text string
- Boolean
- Data element sequence
- Data element alternative (a sequence of data elements from which one element is selected)
- Uniform Resource Locator (URL)

One of the valid types for a service attribute ID is a UUID, as defined by the International Organization for Standardization (ISO) [in ISO/IEC 11578:1996 "Information technology - Open Systems Interconnection - Remote Procedure Call (RPC)"]. These 128-bit numbers are guaranteed to be unique across all space and time (actually, unique until A.D. 3400, based on the UUID algorithm).

One of the key uses of UUIDs is as a type for the members of the *ServiceClassIDList*. That is, each service class is uniquely identified by a UUID. A set of pre-defined service classes is provided in the Bluetooth Assigned Numbers specification. Another use of UUID is as a unique identifier for a particular service instance. This identifier is the *ServiceID* service attribute. Later, you'll see that UUIDs play a key role in searching a service discovery server.

The basic structure of the datatypes used by Bluetooth Service Discovery is summarized in the sample SDP server shown in Figure 5.2. For simplicity's sake, the service class identifiers are shown as text strings rather than UUIDs.

Figure 5.2 The Data Structures of a Sample SDP Server

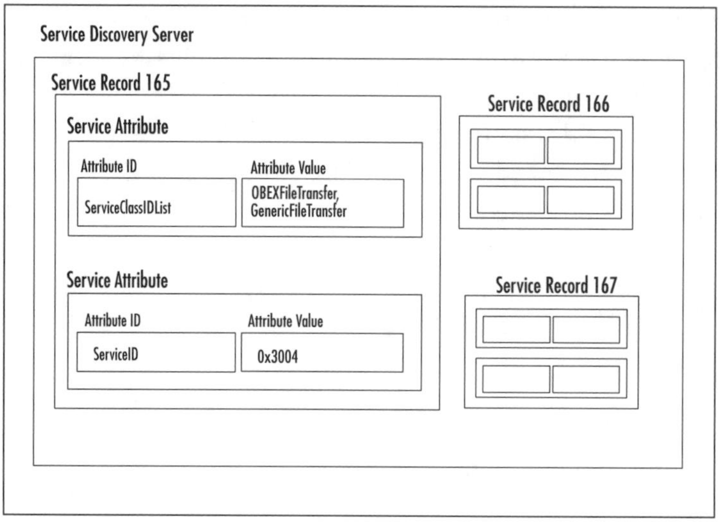

A client wanting to access the service records of a service discovery server can do so in one of two ways: they can search for a particular service record or they can browse the available service records. The search capability of the Service Discovery Protocol is simple but effective. It allows a client to specify a list of UUIDs and then retrieve a list of service record handles for service records, whose attributes contain all of the UUIDs specified by the client. Later in this chapter, you'll see how this mechanism is used in practice.

To support the browsing of service records, Bluetooth Service Discovery uses special service attributes and service classes that allow for the construction of a browseable hierarchy. A service class called *BrowseGroupDescriptor* is defined. A service record that is an instance of this class is analogous to a directory in a file hierarchy—it's a place in a hierarchy where related services can be stored, or where child *BrowseGroupDescriptor* records can be stored. The *BrowseGroupList* attribute of a service record specifies the list of *BrowseGroupDescriptors* that a service record instance is a member of (it may be in more than one). The members of this list attribute are the UUIDs of the *BrowseGroupDescriptor* records. So, a client can browse the Service Discovery Server by specifying the UUID of the Browse Group of interest as a search pattern to the server. This search will match all service records that have specified this *BrowseGroupDescriptor* UUID in their *BrowseGroupList* attribute.

Before looking at the Service Discovery Protocol, it's worth considering the semantics of a service attribute value. Although the Bluetooth specification says that an attribute ID describes both the type and the semantics of an attribute value, this is somewhat sketchy. The semantics of an attribute value are not, in fact, codified within a service attribute. Instead, the meaning of a particular attribute value is understood by the client application once it knows what service class the attribute's service record belongs to. For example, a client accessing a service record of service class 0x1113 (the last 16 bits of the UUID for the Wireless Application Protocol [WAP] service class) must know, at application development time, that the service attribute with attribute ID 0x0306 is the attribute that identifies the Internet Protocol (IP) network address of a WAP Server. This information is not made available to it at runtime for presentation to the end user, for example. If you're familiar with richer software abstractions for discovering network services, this example illustrates the opportunities for an abstract layer of primitives to hide some of the programming detail from an application developer.

The Service Discovery Protocol

So, how exactly do clients discover services in their local areas? Services are discovered using the Service Discovery Protocol, a simple protocol that communicates between SDP clients and servers. It can be implemented over any reliable packet-based transport layer, though it's typically implemented over the Logical Link Control and Adaptation Protocol. The Service Discovery Protocol includes a set of Protocol Data Units (PDUs) that contain the basic requests and responses needed to implement the functionality of Bluetooth Service Discovery. The actual PDU format and protocol are not directly relevant to an application programmer who will work exclusively through the API of a Bluetooth stack. But it's worth summarizing the protocol here since the stack API is usually derived from the structure of underlying PDUs.

An SDP PDU contains a PDU ID, a transaction ID, and a parameter length in its header. Its body contains some number of additional parameters—what these parameters are depends on which type of transaction the PDU contains. The PDU ID identifies the type of transaction. The following are transaction types supported by the protocol:

- SDP_ErrorResponse
- SDP_ServiceSearch
- SDP_ServiceAttribute
- SDP_ServiceSearchAttribute

With the exception of SDP_ErrorResponse, the transaction types are Request/Response pairs. For an SDP implementation to match an incoming Response with a previously issued Request, a number is assigned to the Request that is unique among currently outstanding Requests. This is the Transaction ID. The SDP_ErrorResponse PDU is generated if a Request PDU is improperly formatted, or if some other error has prevented the generation of an appropriate Response PDU. The parameters of this PDU will give you some information about the nature of the error. The ServiceSearch transaction, embodied in a Request/Response pair, searches for services containing service records that match a submitted search pattern. The search pattern (of UUIDs) is passed as a Request PDU parameter; the service record handles of the matching service records are then passed in a Response PDU parameter. The ServiceAttribute transaction retrieves particular service attributes from a specified service record. The parameters of the Request PDU specify the service record handle of the target record, as well as the list of attribute IDs to be retrieved. A list of attribute values is passed in a parameter in the Response PDU. The capabilities of the two preceding transactions are combined in the ServiceSearchAttribute transaction. This transaction retrieves attributes matching the specified Attribute list from the service records matching a specified search pattern.

Developing An Abstract C API for SDP

The Service Discovery Protocol of the Bluetooth specification identifies the protocol data units exchanged between protocol peer entities. Ultimately, it's not the role of the specification to provide an API. So, we start here by providing an API in C that covers the low-level functionality of the protocol. Coding examples in the rest of the text reference this API.

The API uses an "object-oriented" flavor with liberal use of opaque types. All memory management is performed by the API implementation.

First, we look at the API needed from the server point of view—in other words, an API allowing for the creation and advertising of service records.

```
//The basic types are opaque
typedef implementationHandle SDP_SERVICE_RECORD_t;
                                        // service record
typedef implementationHandle SDP_DATA_ELEMENT_t;
                                        // Data element
typedef short SDP_ATTRIBUTE_ID_t;
                                        // attribute
typedef unsigned short SDP_DE_TYPE_t;
                                        // Data element type bitmask
typedef unsigned short SDP_DE_SIZE_t;
                                        // Data element size bitmask
//Used to create a service record
status_t sdp_create_service_record(SDP_SERVICE_RECORD_t *srh);
//Used to free a previously created service record
status_t sdp_free_service_record(SDP_SERVICE_RECORD_t srh) ;
//Create a basic data element from its given type and value
//Type is constructed by ORing a type and size bitmask
//size is ignored for String, URL and sequence types.
//For String, URL types, the given value must be a char*,
//from which the size is calculated.
//For a sequence type the size is calculated directly from the
//list of elements added into the sequence.
//For integer types greater than 32 bit, and for 128 bit UUID
//types, the value is given as a byte array.
status_t sdp_create_data_element(SDP_DE_TYPE_t type,
                                void *value,
                                SDP_DATA_ELEMENT_t *elem);
//These are the bitmask values for the type and size,
//derived directly from the specification
//[SPEC] part E, section 3
```

```
#define SDP_DE_TYPE_NIL     0x00  /* Nil, the null type        */
#define SDP_DE_TYPE_UINT    0x08  /* Unsigned Integer          */
#define SDP_DE_TYPE_STCI    0x10  /* Signed, twos-complement
                                        integer                */
#define SDP_DE_TYPE_UUID    0x18  /* UUID, a universally
                                        unique identifier      */
#define SDP_DE_TYPE_STR     0x20  /* Text string               */
#define SDP_DE_TYPE_BOOL    0x28  /* Boolean                   */
#define SDP_DE_TYPE_DES     0x30  /* Data Element Sequence     */
#define SDP_DE_TYPE_DEA     0x38  /* Data Element Alternative  */
#define SDP_DE_TYPE_URL     0x40  /* URL, a uniform resource
                                        locator                */


#define SDP_DE_SIZE_8       0x0   /* 8 bit integer value       */
#define SDP_DE_SIZE_16      0x1   /* 16 bit integer value      */
#define SDP_DE_SIZE_32      0x2   /* 32 bit integer value      */
#define SDP_DE_SIZE_64      0x3   /* 64 bit integer value      */
#define SDP_DE_SIZE_128     0x4   /* 128 bit integer value     */


//Used to create a data element sequence or data element
//alternative
status_t sdp_create_data_element_sequence(
        SDP_DATA_ELEMENT_t *head);
//Used to add a data element to a previously constructed data
//element sequence or alternative
status_t sdp_add_element(SDP_DATA_ELEMENT_t head,
                        SDP_DATA_ELEMENT_t elem);
//Used to free a previously created data element
status_t sdp_free_data_element(SDP_DATA_ELEMENT_t elem) ;
//Used to add an attribute to a previously constructed service
//record
status_t sdp_add_attribute(SDP_SERVICE_RECORD_t srh,
                          SDP_ATTRIBUTE_ID_t attrId,
                          SDP_DATA_ELEMENT_t attribute);
```

```
//Used to advertise a previously constructed service record
status_t sdp_register_service(SDP_SERVICE_RECORD_t srh);
//Used to stop advertising a previously advertised service
//record
status_t sdp_unregister_service(SDP_SERVICE_RECORD_t srh);
```

Next, we present the API from the client's point of view—in other words, an API for the retrieval of service records and their attributes in order to use the information.

```
//The basic types are opaque
typedef implementationHandle SDP_DEVICE_t;
typedef implementationHandle SDP_CONNID_t;
typedef short SDP_COUNT_t;
//Used to create an SDP connection to a remote device's SDP
//server.
status_t sdp_open_connection(SDP_DEVICE_t device
                               SDP_CONNID_t *sdpConnID);
//Used to close an SDP connection to a remote device's SDP
//server.
status_t sdp_close_connection(SDP_CONNID_t sdpConnID);


//Used to retrieve a list of service records that match
//the given list of UUIDs. Adhering strictly to the protocol
//only the service record handles are retrieved.
status_t sdp_service_search(SDP_CONNID_t sdpConnId,
                            SDP_DATA_ELEMENT_t[] searchPattern,
                            SDP_COUNT_t searchPatternCount,
                            SDP_COUNT_t maxRecordCount,
                            SDP_COUNT_t *numFound,
                            SDP_SERVICE_RECORD_t **res);
//Used to retrieve a list of attributes from the remote SDS
//for the given service record handle. Note that the remote
//nature of the api is explicit, reflecting the SDP protocol
status_t sdp_get_attributes(SDP_CONNID_t sdpConnId,
                            SDP_SERVICE_RECORD_t srh,
```

```
                                SDP_ATTRIBUTE_ID_t[] attrIds,

                                SDP_COUNT_t attributeIdCount);
//Used to retrieve the attribute value (as a data element)
//corresponding to the given attribute ID from the
//given service record. If the attribute value was not
//previously retrieved by the sdp_get_attributes function
//this function will return null.
status_t sdp_get_attribute(SDP_SERVICE_RECORD_t srh,

                           SDP_ATTRIBUTE_ID_t attrId,

                           SDP_DATA_ELEMENT_t *attrValue);
//Used to parse the attribute values (as data elements)
//retrieved from the service record by the preceding api.
//The type, size, and value are returned. For most types (except
//the sequence types), the value can be cast to the appropriate
//C type as given by the type parameter (see the notes for
//sdp_crete_data_element)
status_t sdp_parse_data_element(SDP_DATA_ELEMENT_t dataElement,

                                SDP_DE_TYPE_t *type,

                                SDP_DE_SIZE_t *size,

                                void **value);
//Used to retrieve successive data elements from a data element
//sequence. This function will only work on data elements of
//type sequence.
status_t sdp_get_next_element(SDP_DATA_ELEMENT_t sequence,

                             SDP_DATA_ELEMENT_t *nextElement);
```

Discovering Services

We've put together a practical guide to help you make sense of using SDP to advertise and discover services within a network. Following on the previous section, we'll create and advertise a service record on a server device using the API in the earlier section titled "Developing An Abstract C API for SDP." We'll then connect to the SDP server and find a specific service record or browse service records from a client device. But first, let's discuss how to use the Class of Device (CoD) to assist in short-circuiting the service discovery process.

Short-Circuiting the Service Discovery Process

Every Bluetooth device can contain a Service Discovery Server (SDS) that advertises the services available on that particular device, be it a mobile phone, PDA, or something else. It can do this by making available the service records that describe those services. A client starts by finding a Bluetooth device. Then they use the SDS to pinpoint a service or to browse available services. *Bluetooth device discovery can help short circuit this service discovery process.* During the device inquiry process (before any ACL connection is made between devices), the low-level Frequency Hopping Synchronization (FHS) packet is exchanged between discovering and discovered devices. One of the pieces of information in the FHS packet is the Class of Device. The CoD is a 24-bit value composed of three parts: Major Device Class, Minor Device Class and Major Service Class. Checking these values can be beneficial when determining if a connection should be opened to the device. For example, if a PDA is looking for a printer, it can tell immediately from the CoD if a discovered device can print. It doesn't have to open a connection to the SDS and check the Service Discovery Database (SDDB) of the discovered device. So, a client will know if a device hosts the required service before a connection is made. This "short-circuiting" of service discovery is powerful and increases the speed and efficiency of service discovery. The Bluetooth SIG controls the values of the three CoD attributes. For further information on the CoD, see [SPEC], part B, section 4.4.1.4, and [ASSN] section 1.2.

Creating and Advertising a Service

If the CoD indicates that a service or category of service is available, then a connection can be opened to the SDS on the discovered device. This connection can be used to find an exact match service or to determine the precise mechanism to interact with a service. In general, the service record should only be advertised when the service is available, and the service itself should be responsible for this. (The service is advertised as part of a service bootstrapping process, and conversely, advertising the service is stopped as part of service termination.)

To create a service record, individual data elements that correspond to the attribute values of the service attributes need to be constructed. They are then added into the service record. The following piece of code in this section creates a service record for an Example service. The Example service belongs to the Example service class. This service class has a class description that defines the contents of the service record that defines the Example service. The service description in Table 5.2 lists each of the attributes contained in an Example service record, including the name, ID, value type, and meaning.

Table 5.2 Service Attributes Example

Attribute Name	Attribute ID	Attribute Value	Attribute Semantic
ServiceClassIDList	0x0001	Sequence	a & b in list
ProtocolDescriptorList	0x0004	Sequence	a & c in list
LanguageBaseAttributeIdList	0x0006	Sequence	a & d in list
ServiceName	offset (0x0000)	String	a & e in list

a) This service attribute has the definition as given by the corresponding universal attribute definition, available in the SDP protocol specification [SPEC] part E, section 5.1.

b) This service attribute provides a list of UUIDs that identify the classes (or class definitions) of which this service is an instance. In this case, the class list contains the single ID for the Example class.

c) This service attribute provides a list of the protocols and protocol attributes needed for a client to access this service. In this case, the protocol list contains the single Bluetooth protocol L2CAP, and its attribute is the Protocol Service Multiplexor (PSM) value for the service (this PSM value is assigned dynamically at runtime by the L2CAP implementation).

d) This service attribute contains a list of natural languages supported, and for each language a triple: the ISO language identifier, the encoding used for attributes in this language, and the base ID to be used for all attributes that encode natural language strings in this language (see ServiceName).

e) This service attribute contains the name of the service in a natural language. The offset is added to the base language ID as given in the LanguageBaseAttributeIdList to give the ID for the *ServiceName* attribute in the given language.

The code samples that follow are pseudo-code samples that use our abstract C API. Variables are typically declared close to their first use rather than in an initial declaration block. This is illegal in C (though not in C++), but it improves readability and is an aid to understanding.

```
//Create an element for the service class identifier, which is a
//UUID that uniquely identifies the service class description that
//describes the service record contents for this service
char exampleServiceClassUUID[32] = 0x12672536752ABBC12612AB12BC125A7F;
SDP_DATA_ELEMENT_t exampleServiceClassID;
sdp_create_data_element(SDP_DE_TYPE_UUID _ SDP_DE_SIZE_128,
                        exampleServiceClassUUID,
                        &exampleServiceClassID);
//Create the element sequence for the mandatory attribute
//ServiceClassIDList, which lists the service class IDs of
//all the service classes to which this service belongs
SDP_DATA_ELEMENT_t serviceClassIDList;
sdp_create_data_element_sequence(&serviceClassIdList);
//Add the one service class ID to this list
sdp_add_element(serviceClassIdList, exampleServiceClassID);
//Create the element sequence to describe the access paths through
//the protocol stack, and the element sequence to describe the access
//path through L2CAP
SDP_DATA_ELEMENT_t protocolList, l2capList;
sdp_create_data_element_sequence(&protocolList);
sdp_create_data_element_sequence(&l2capList);
//This Example service is accessed through the L2CAP transport on a
//dynamically assigned PSM (imagine this code is being executed as the
//service is bootstrapping)
//Create the individual elements that constitute the access through
//L2CAP, i.e. the UUID for L2CAP, and the PSM value
SDP_DATA_ELEMENT_t l2capId, psmValue;
sdp_create_data_element(SDP_DE_TYPE_UUID _ SDP_DE_SIZE_16,
                        0x0100, &l2capId);
sdp_create_data_element(SDP_DE_TYPE_UINT _ SDP_DE_SIZE_16,
                        0x1001, &psmValue);
//Add the elements to the sequence
sdp_add_element(l2capList, l2capId);
sdp_add_element(l2capList, psmValue);
```

```
//Add the L2CAP access to the general service access path list
sdp_add_element(protocolList, l2capList);

//Create the attribute ID for LanguageBaseAttributeIdList
SDP_ATTRIBUTE_ID_t langBaseAttributeId = x0006;
//Create the element sequence to describe the main human readable
//language base, i.e. English
SDP_DATA_ELEMENT_t englishLanguageBase;
sdp_create_data_element_sequence(&englishLanguageBase);
//Create the individual elements that constitute the members of the
//language base element sequence, i.e. the ISO language identifier, the
//ISO character encoding of strings in this language, and the base
//attribute ID that all human readable attribute IDs will be added to,
//to determine the actual attribute ID.
SDP_DATA_ELEMENT_t enLangId, enLangCharSet, enLangBaseID;
//For simplicity 'en' and 'fr' are used to represent 'English' and
//'French', as specified by ISO 639:1988(E/F), rather than converting to
//a 16 bit integer, as specified in the Bluetooth specification
sdp_create_data_element(SDP_DE_TYPE_UINT _ SDP_DE_SIZE_16,
                        'en', &enLangId);
sdp_create_data_element(SDP_DE_TYPE_UINT _ SDP_DE_SIZE_16,
                        UTF-8, &enLangCharSet);
sdp_create_data_element(SDP_DE_TYPE_UINT _ SDP_DE_SIZE_16,
                        0x0100, &enLangBaseID);
//Add the elements to the sequence
sdp_add_element(englishLanguageBase, enLangId);
sdp_add_element(englishLanguageBase, enLangCharSet);
sdp_add_element(englishLanguageBase, enLangBaseID);
//Create an element sequence for each human readable language that will
//be supported, e.g. French
SDP_DATA_ELEMENT_t frenchLanguageBase;
sdp_create_data_element_sequence(&frenchLanguageBase);
SDP_DATA_ELEMENT_t frLangId, frLangCharSet, frLangBaseID;
sdp_create_data_element(SDP_DE_TYPE_UINT _ SDP_DE_SIZE_16,
```

```
                              'fr', &frLangId);
sdp_create_data_element(SDP_DE_TYPE_UINT _ SDP_DE_SIZE_16,
                        UTF-8, &frLangCharSet);
sdp_create_data_element(SDP_DE_TYPE_UINT _ SDP_DE_SIZE_16,
                        0x0200, &frLangBaseID);
sdp_add_element(frenchLanguageBase, frLangId);
sdp_add_element(frenchLanguageBase, frLangCharSet);
sdp_add_element(frenchLanguageBase, frLangBaseID);
//Finally, create the element sequence to hold all the language
//lists and add them in
SDP_DATA_ELEMENT_t languageList;
sdp_create_data_element_sequence(&languageList);
sdp_add_element(languageList, englishLanguageBase);
sdp_add_element(languageList, frenchLanguageBase);
//Now create the element to define the service name in both English and
//French
SDP_DATA_ELEMENT_t enServiceName;
sdp_create_data_element(SDP_DE_TYPE_STR, 'Service Name',
                        &enServiceName);
SDP_DATA_ELEMENT_t frServiceName;
sdp_create_data_element(SDP_DE_TYPE_STR, 'Nom de Service',
                        &frServiceName);


//We can now create the service record and add all the attributes
SDP_SERVICE_REC_t exampleServiceRecord;
sdp_create_service_record(&exampleServiceRecord);
sdp_add_attribute(exampleServiceRecord,
                  ServiceClassIDList,
                  serviceClassIdList);
sdp_add_attribute(exampleServiceRecord,
                  langBaseAttributeId,
                  languageList);
sdp_add_attribute(exampleServiceRecord,
                  0x0100,
```

```
                        enServiceName);
sdp_add_attribute(exampleServiceRecord,
                        0x0200,
                        frServiceName);
//Finally we can advertise the service

sdp_advertise_service(exampleServiceRecord);
```

As you can see, creating and advertising individual service records can be an involved process. In an upcoming section, we will explore how the API can be improved with "helper" functions based on the use of the Bluetooth profiles. Now, we'll look at the client side of service discovery and the two ways a service can be discovered: by looking for a specific service or by browsing.

Discovering Specific Services

The Bluetooth Service Discovery Protocol allows for services to be discovered on the basis of a series of attributes with values of type UUID. In reality, when talking about discovering specific services, one of the most important attributes of a service, if not *the* most important, is the *ServiceClassIDList*. It provides a list of the classes to which the service belongs. For example a Headset service as defined by the Headset profile belongs to ServiceClass Headset and ServiceClass Generic Audio. The following code is used to search for an instance of the Example service, as defined in the previous section.

```
//We assume here that the device is obtained through the device
//discovery procedure, and is not discussed here
SDP_DEVICE_t device;
//The SDP connection to the peer device
SDP_CONNID_t connection;
//The search pattern, containing the list of UUIDs to be used. Each
//service record must contain every UUID given in order to qualify.
//In this case we will only have one UUID – the UUID of the Example
//service class.
SDP_DATA_ELEMENT_t searchPattern[1] = {exampleServiceClassID};
//The number of service records found as a result of the search
SDP_COUNT_t numberFound;
//The service records found
```

```
SDP_SERVICE_RECORD_t[] found;
//Open an SDP connection to the device.
sdp_open_connection(device, &connection);
//Do the search for the specific service, specifying a maximum of one
//result to be returned. In this instance numberFound will be one or 0.
sdp_service_search(connection,
                   searchPattern, 1, 1,
                   &numberFound,
                   &found);
```

If the service class ID used to perform the search represents the most specific class needed, then any service represented by the returned service records can be used. Individual attributes which further refine the search may be given, but with our C API, they must be attributes whose values are of type UUID. To provide a search facility using non-UUID type attributes would mean writing this code yourself. This could be done by performing a base search with the UUID types, and then accessing the appropriate non-UUID attributes and comparing them with the values given. The next section shows how this could be done, by discussing how individual service attributes are examined.

Using Service Attributes

Once a client has retrieved service records, the service record's attributes can be examined. The client can retrieve the service name attribute for displaying to the user in the language of the Locale of the user machine. For example, this is how a user in a French Locale would do it:

```
//We assume here that the service record has been returned by the
//previous code. We describe a C function to return the Service name
//as a char*.
char* getServiceName(SDP_CONNID_t connection,
                     SDP_SERVICE_RECORD_t serviceRecord) {
  //The name as a char*
  char* serviceNameString;
  //Utility variables for type and size
  SDP_DE_TYPE type;
  SDP_DE_SIZE size;
  //Get the value of the LanguageBaseAttributeIdList attribute from
```

```
//the remote device
sdp_get_attributes(connection,
                    serviceRecord,
                    &langBaseAttributeId, 1);


//Retrieve the value of the attribute - the sequence of supported
//languages
SDP_DATA_ELEMENT_t langaugeList;
sdp_get_atribute(serviceRecord,
                  langBaseAttributeId,
                  &languageList);
//Iterate through the sequence of languages looking for French
//as given in the language ID - the first element in the language
//sequence
SDP_DATA_ELEMENT_t langauge;
unsigned short langBaseId = 0;
while (sdp_get_next_element(languageList, &language) == SUCCESS) {
  SDP_DATA_ELEMENT_t langaugeId;
  sdp_get_next_element(language, &languageId);
  //Parse out the type, size, and value from the element
  //we know the value should be an unsigned short
  unsigned short id;
  sdp_parse_data_element(languageId, &type, &size, &id);
  //If this is the French language sequence, then parse out the base
  //attribute ID.
  if (id == 'fr') {
    SDP_DATA_ELEMENT_t languageEncoding, baseAttributeId;
    sdp_get_next_element(language, &languageEncoding);
    sdp_get_next_element(language, &baseAttributeId);
    sdp_parse_data_element(baseAttributeId, &type,
                                 &size, &langBaseId);
    break;
  }
}
```

```
    if (langBaseId != 0) {

       //The attribute ID for the service name in French is given by the
       // langBaseId, since the ServiceName attribute has a 0x0000 offset.
       sdp_get_attributes(connection,
                          serviceRecord,
                          &langBaseId, 1);
       SDP_DATA_ELEMENT_t serviceName;
       sdp_get_attribute(serviceRecord,
                         langBaseId,
                         &serviceName);
       sdp_parse_data_element(serviceName,
                              &type, &size,
                              &serviceNameString);
    }
    return serviceNameString;
}
```

Browsing for Services

If the service Class ID for a particular service is unknown, or if a client wants to browse the services on a device, the service discovery protocol provides a way to do this. To be "browseable," a service must be explicitly marked as browseable with a *BrowseGroupList* attribute in its service record. If the service record doesn't have this attribute, it can't be browsed. The *BrowseGroupList* attribute contains the list of UUIDs that identifies the groups that a service belongs to. A well-known root browse group UUID (called *PublicBrowseRoot*) is defined by the SIG (see the [ASSN] section 4.4). Because the root is a well-known UUID, a client knowing nothing about services always has a place to start browsing. A group is defined by a *BrowseGroupDescriptor* service record. This service record has two attributes of interest: the *GroupID* (whose UUID value is contained in a service's *BrowseGroupList*), and the *BrowseGroupList* attribute, which specifies the list of browse groups to which this group itself belongs. The *BrowseGroupDescriptor* service class definitions are given in [SPEC], part E, section 5.3, and its service class ID is defined in the [ASSN], section 4.4.

If you want the Example service to be in a Sample Services group—a group available from the root browse group—you would define a Browse group with this name and some GroupID UUID to tag the group. You'd then insert this tag into the *BrowseGroupList* of the Example Service. Of course, the *BrowseGroupList* of the Sample Services group must contain the root browse group. The following code shows how the Sample Service browse group is created and how the Example service is put into that group.

```
//Create an element for the service class identifier, which in this
//case is a well known UUID for the BrowseGroupDescriptor service class
//ID (defined by the SIG as a 16 bit UUID of value 0x1001)
SDP_DATA_ELEMENT_t browseGroupDescriptorServiceClassID;
sdp_create_data_element(SDP_DE_TYPE_UUID _ SDP_DE_SIZE_16,
                        0x1001,
                        &browseGroupDescriptorServiceClassID);
//Create the element sequence for the mandatory attribute
//ServiceClassIDList, which lists the service class IDS of
//all the service classes to which this service belongs
SDP_DATA_ELEMENT_t serviceClassIDList;
sdp_create_data_element_sequence(&serviceClassIdList);
//Add the one service class ID to this list
sdp_add_element(serviceClassIdList,
                browseGroupDescriptorServiceClassID);
//Create an element for the GroupID attribute, which is a
//UUID that uniquely identifies the group defined by this browse
//group.
SDP_DATA_ELEMENT_t sampleBrowseGroupID;
sdp_create_data_element(SDP_DE_TYPE_UUID _ SDP_DE_SIZE_128,
                        0x87634324b34232cb434d43a43d3444dd,
                        &sampleBrowseGroupID);
//Create an element for the root browse group ID, which is a
//well known UUID defined by the SIG
SDP_DATA_ELEMENT_t rootBrowseGroupID;
sdp_create_data_element(SDP_DE_TYPE_UUID _ SDP_DE_SIZE_16,
                        0x1002,
                        &rootBrowseGroupID);
//Create the element sequence for the BrowseGroupList attribute
```

```
//which lists GroupID of all the groups that this record is
//browsable from.
SDP_DATA_ELEMENT_t sampleGroupBrowseGroupList;
sdp_create_data_element_sequence(&sampleGroupBrowseGroupList);
//Add the one UUID to this list - the well-known root browse group
sdp_add_element(sampleGroupBrowseGroupList,
                rootBrowseGroupID);


//Now create the service record and add all the attributes
SDP_SERVICE_REC_t sampleGroupServiceRecord;
sdp_create_service_record(&sampleGroupServiceRecord);
sdp_add_attribute(sampleGroupServiceRecord,
                ServiceClassIdList (0x0001),
                serviceClassIdList);
sdp_add_attribute(sampleGroupServiceRecord,
                GroupID (0x0200),
                sampleBrowseGroupID);


sdp_add_attribute(sampleGroupServiceRecord,
                BrowseGroupList (0x0500),
                sampleGroupBrowseGroupList);
//Finally we can advertise the service
sdp_advertise_service(sampleGroupServiceRecord);
```

The Example Service (as defined in the previous section) needs to have the following code added in order to be included in the Sample Group. The code should be added just before the service record is advertised.

```
//Create the element sequence for the BrowseGroupList attribute
//which lists GroupID of all the groups that this record (the
//Example Service) is browsable from.
SDP_DATA_ELEMENT_t exampleServiceBrowseGroupList;
sdp_create_data_element_sequence(&exampleServiceBrowseGroupList);
//Add the one UUID to this list - the UUID of the sample group
//GroupID attribute
sdp_add_element(exampleServiceBrowseGroupList,
```

```
                        sampleBrowseGroupID);
sdp_add_attribute(exampleServiceRecord,
                        BrowseGroupList (0x0005),
                        exampleServiceBrowseGroupList);
```

This code makes the Example Service browseable from the Sample Browse Group.

Clients can now discover the service by browsing on their mobile devices. The specific client code for doing this is not given as it will follow the template given already in the earlier section "Discovering Specific Services.", but it employs the following algorithm:

A service search is performed using the UUIDS for both the Public Browse Group (defined by the SIG as a 16-bit UUID of value 0x1002), and the *BrowseGroupDescriptorServiceClassId* (defined by the SIG as a 16-bit UUID of value 0x1001). This specific search should yield only those *BrowseGroupDescriptors* service records that are browseable from the public root. In this instance, given the preceding Example code, this search would yield one record, the *SampleGroup record*. From this, we extract the Group ID, and perform another search using this UUID as the sole UUID in the search pattern. This will yield any service records that are members of the group—in other words, which have the Group ID in their *BrowseGroupList* (in addition to the *BrowseGroupDescriptor* service record itself). In this instance, the Example service record will be returned.

Service Discovery Application Profile

Bluetooth profiles define usage scenarios for Bluetooth devices as well as the functionality that should be available from the underlying protocol stack. The profiles don't present individual programming interfaces (which would be platform-dependent), but instead present a platform-neutral description of functionality to be provided by an application that realizes the profile.

In the previous section, we presented a C-based API for service discovery. If you are familiar with the SDP protocol, you'll notice that the API is based on the description of the protocol PDUs exchanged between the protocol's client and server entities. It's not based on the Service Discovery Application Profile, for reasons that will become clear shortly. The SDAP is a usage scenario describing the functionality a Service Discovery Application (*SrvDscApp*) should provide to an end user on a Local Device (*LocDev*) so that user can discover services on a

Remote Device (*RemDev*). The SDAP doesn't specify an API that will provide this functionality, but suggests primitives that can be mapped to an API. This differs from most other profiles that describe functionality without using primitives. The primitives are:

a) **Enumerate Remote Devices** This primitive is used for device discovery and would likely be implemented by the baseband inquiry mechanism.

b) **Search Services** This primitive is used to search for specific services based on the class of the service or the class of service and some specific attributes of the service. It would likely be implemented by the searchServices functionality (shown in the previous section).

c) **Browse Services** This primitive is used to browse services according to the browse groups. It would likely be implemented by functionality (as shown in the Browsing Services section).

d) **Terminate Primitive** This primitive is used to terminate a previously started primitive.

The *SrvDscApp* is only necessary on the *LocDev* device—the client device. Though the profile says devices without user interfaces are not candidates for *LocDev*, devices can still use the procedures defined by the profile to exercise the SDP protocol. For instance, where another application profile (such as Serial Port Profile) is using SDP to recover applicable service records. We look at this scenario in the next section, "Service Discovery Non-Application Profiles." Primitives *c* and *d* give the necessary procedures for this usage (which are covered by the API in the previous section). Adding APIs to cover the first two primitives creates an interface that achieves the functionality of the SDAP.

Service Discovery Non-Application Profiles

No, it's not a misprint. The title is deliberately jarring to draw your attention to the fact that most profiles detailed in the Bluetooth specification have a service discovery component. This component specifies the structure and content of the service record that accompanies the service (or application) that realizes the profile. The SDAP (in addition to dealing with application functionality for service discovery) specifies the procedures that an application realizing a profile must use to perform service discovery.

If these procedures are upheld, interoperability is ensured. For example, an application that realizes a profile should be able to advertise its service via the

Service Discovery Server and be found by any client on any device that accesses the profile's SDP record—according to the service discovery procedures described by the SDAP. This example of an individual profile's service discovery component (see Table 5.3) describes the Serial Port profile's service record.

Table 5.3 Serial Port Profile Service Record Example

Attribute Name	Definition	ID	Type	Value
ServiceClassIdList	List of services supported	0x0001	Sequence	N/A
ServiceClass0	Serial Port	N/A	UUID	Assigned Number
ProtocolDescriptorList	List of protocols supported	0x0004	Sequence	N/A
Protocol0	L2CAP	N/A	UUID	Assigned Number
Protocol1	RFCOMM	N/A	UUID	Assigned Number
ProtocolSpecificParm0	Server Channel	N/A	UINT8	2
ServiceName	Text name	0x0000	String	"Com1 as example"

The serial port profiles describe a usage scenario where two applications, A and B, are communicating via a serial cable emulation. Device B, which acts the role of the server, must register the previous record with the SDDB. As the profile states, this is the most generic type of service, which indicates nothing of the application functionality. So, additional service class IDs can be inserted into the *ServiceClassIDList*. As you saw in the previous section, the amount of code needed to create and advertise a service record can be extensive. The API offered to the developer can be improved by providing an API for the serial port profile itself:

```
status_t sdp_create_serial_port_record(SDP_TYPE_t UUIDType,

                                       void *UUID,

                                       SDP_SERVICE_RECORD_t *srh);
```

This function performs most of the drudgery of the previous section, and provides a service record ready to be registered with the SDDB. Of course, any updates or extra information needed can be added with the usual API.

Java, C, and SDP

The Bluetooth Service Discovery Protocol doesn't prescribe an API for programmers to use. Although both the SDP transactions and data representation imply the structure of an API, Bluetooth stack implementations vary widely in the APIs and programming abstractions they provide. Some stacks represent SDP transactions asynchronously, through a function call for making a request and a separate callback for replies. Others provide one synchronous function that blocks the caller while waiting for a reply. Stacks also differ in the level of abstraction of their function calls. Some stacks provide functions that return, in essence, raw SDP PDUs that the programmer must then disassemble and interpret—for example, the abstract C API examined earlier. Others return structured data from which the relevant data elements are more easily extracted. Some stacks provide richer abstractions that allow a programmer to carry out simple, routine tasks in fewer steps (for particular profiles, for example). When choosing a stack, it's wise to consider the design and richness of an SDP API to ensure that you can write readable, maintainable code as efficiently as possible, without giving up access to all the features and flexibility you need. Is it more important for you to be able to create, populate, and advertise a service record in one or two function calls, or to have full control over each PDU element in minute detail?

When considering abstraction levels, programming language is a key choice. Most stacks expose C APIs, while others provide Java or C++ interfaces. Service Discovery is arguably the Bluetooth component best placed to take advantage of the richness and usability of the Java programming platform. Java, in particular the Java 2 Platform Micro Edition (J2ME), is rapidly becoming the platform of choice for developing embedded wireless applications. This is evidenced by its adoption by industry heavyweights Nokia, Motorola, Siemens, Matsushita, Sharp, and others. It provides a level of portability, maintainability and ease of programming that languages such as C do not. Of particular relevance here is the potential for rich SDP abstractions that can largely remove the programmer from the detail of PDUs and completely remove them from error-prone pointer and memory manipulation.

As part of Java Community Process (JCP)—the vehicle for standardizing the Java platform—a set of standard Java APIs for Bluetooth is being developed. The Java Specification Request (JSR) 82 Expert Group is carrying out this work. Motorola chairs the group, with contributing experts from a number of companies, including Rococo Software. At the time of writing, the first full version of this specification is due for publication at the end of 2001. Implementations of this standard will allow programmers to implement Bluetooth applications within the J2ME environment in a standard and portable way.

Historically, Java as a programming language for embedded applications has suffered most from one criticism—it was too slow and bulky. This was true in its early versions, primarily since it is an interpreted language and the Virtual Machines in which applications ran weren't optimized, but this has changed. Many developments contributed to Java becoming a key open platform for embedded application development in general, and wireless development in particular. Virtual Machines have been optimized for such environments—for example, the "KVM" in Sun's J2ME Connected Limited Device Configuration (CLDC). Virtual Machines have found their way into silicon, with Java bytecodes being interpreted directly on the chip. The Jazelle product suite from ARM and the MachStream platform from Parthus are good examples of this. Java has also been tailored for particular platforms, with precompilers providing the performance power required by embedded applications without sacrificing the advantages of the Java platform.

In addition to the abstractions possible for SDP implementations in Java, the J2ME platform provides a useful Input/Output (I/O) framework that can be applied to Bluetooth application development. A key element of the J2ME specification is the Generic Connection Framework (GCF). It's a mechanism that allows a programmer to create different types of networking connections through a standard Connector interface. In a Bluetooth extension to the GCF, a Connector could create instances of Bluetooth-specific connection classes, say *RFCOMMConnection* or *L2CAPConnection*. Since this is a standard networking framework used by all J2ME applications, programmers can quickly produce Java Bluetooth applications by applying existing techniques and design patterns.

Rococo Software (www.rococosoft.com) provides an implementation of the standard Java Bluetooth APIs, along with a simulator that allows programmers to run their applications and test their use cases without the need for underlying Bluetooth hardware or stacks.

Other Service Discovery Protocols

Let's elaborate on some other discovery protocols: the Salutation Consortium's Salutation service discovery protocol, the Internet Engineering Task Force (IETF)'s Service Location Protocol (SLP), Microsoft's Universal Plug and Play (UPnP), and Sun Microsystems' Jini.

Salutation

Formed in 1995 by a group of U.S. and Japanese companies, the Salutation Consortium defines an architecture for networking devices, applications, and services. The core focus of the group (and most implementations of the standard to date) has been to enable seamless access to office equipment such as fax machines, printers, copiers, and so on. However, the standard has evolved to include phones, PDAs, and general electronic equipment. The Salutation architecture defines a uniform way of labeling devices with descriptions of their capabilities and with a single, common method of sharing that information.

The architecture is composed of Salutation Managers (SLMs), which coordinate all aspects of registering new services and searching for services on behalf of clients. It also contains Transport Managers (TMs), which sit between the SLMs and the rest of the system (see Figure 5.3 for an illustration of the Salutation architecture). This architecture allows Salutation to be "transport independent." That is, a separate TM may be written for each underlying transport required, and the SLM, which provides the core functionality of the system, remains transport neutral. SLMs act as repositories for local service information as well as brokers who seek services on behalf of clients. SLMs periodically check available services to update their repositories. Table 5.4 outlines the functions of the Salutation protocol.

Figure 5.3 The Salutation Architecture

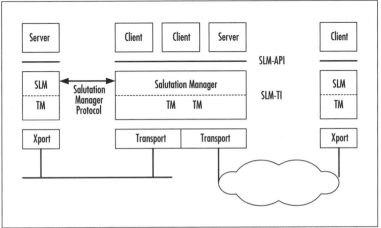

Table 5.4 Salutation Highlights

Function	Description
Announcing Presence	Through cooperation between Salutation Managers (SMs). Register with a known, probably local SM.
Discovering Other Services	Send queries to the local SM. SMs coordinate and return results.
Describing Services	Structured description of services as functional units, which in turn contain attribute records. Functional units identify the "type" or "features" of a service. Attributes provide much more detail. Standard functional unit definitions exist for well-defined services (print, fax).
Self Configuration	Salutation does not address this issue.
Invoking Services	Flexible. Provides for vendor-specific protocols, SLM-managed sessions providing transport independence, as well as defined (standard) data and protocols for selected functional units. The defined APIs can be implemented on most platforms.
Transports	Transport independent architecture
More Information	www.salutation.org

Service Location Protocol

Service Location Protocol (SLP) originated from a working group of the Internet Engineering Task Force (IETF). It's a language-independent protocol for automatic resource discovery on IP-based networks. SLP is designed to be lightweight and decentralized with minimal administration requirements. SLP (like some of the other service discovery protocols) makes use of UDP/IP multicast functionality in TCP/IP. This makes it particularly useful for networks where there is some form of centralized administrative control, such as corporate and campus networks. The discovery mechanism is based on service attributes, which are used to characterize a service. The SLP architecture has three main components:

- **User Agent (UA)** Performs service discovery on a client's behalf (which might be a user or an application).

- **Service Agent (SA)** Advertises the service's location and characteristics on behalf of services, and registers this information with the Directory Agent.

- **Directory Agent (DA)** Accumulates service information received from SAs in its repository and responds to service requests from UAs.

User Agents send a Service Request describing the service they seek to one or more Directory Agents. The Directory Agents respond with Service Replies describing services that match the query (see Figure 5.4).

Figure 5.4 SLP Service Discovery

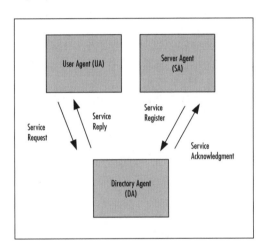

Services are located by their address, the so-called *service:URL*. The address format is composed of the prefix *service:*, the service type, the network address and, optionally, a path. Service types can be of concrete or abstract type. For example, they may either name a particular service type (which is usually a particular protocol), or name a family of service types. For example, in the *service:URL:*

```
service:printer:lpr://www.rococosoft.com/laserprinter
```

the service type *is service:printer:lpr*, a service type name with abstract type printer and concrete type *printer:lpr*.

SLP doesn't mandate the presence of a DA. Users Agents will try to locate a DA when they first start up, but if they don't find any, they will try to operate directly with service agents. When a DA starts to operate on the network, it advertises its presence and all agents that receive the advertisement can start using the DA. Small networks with few services and users may not require a DA on the network. The DA is designed to allow the system to scale in larger networks without imposing undue network traffic. Both Sun Microsystems and Hewlett Packard, among others, have implemented SLP in their products.

Table 5.5 outlines the functions of SLP.

Table 5.5 SLP Highlights

Function	Description
Announcing Presence	Register with DA.
Discovering Other Services	Query DA. Can also multicast a service request in the absence of a DA.
Describing Services	Attribute value pairs.
Self Configuration	Does not address this area. An IP device when plugged onto a network will have to be configured with an IP address, subnet mask and optionally a gateway and DNS server.
Invoking Services	Does not address this area.
Transports	TCP/IP
More Information	www.srvloc.org

Jini

Jini is a distributed service-oriented architecture developed by Sun Microsystems. Jini is considered an extension of the Java language and platform. The key concept in Jini is the *service*, which can be almost anything: a process, a piece of hardware, a communications stream, or a user. Services can be collected together to achieve a task. A collection of Jini services forms a Jini federation: services coordinate with each other within the federation and can join and leave a federation dynamically. Services communicate with each other using a service protocol, which is defined as a set of interfaces in Java. The standard itself provides a base set of interfaces to facilitate core interaction between services—a given implementation of the system may extend these as needed.

A key component of Jini is the *lookup* service. Services are found and resolved by a lookup service. The lookup service is the central bootstrapping mechanism for the system and provides the major point of contact between the system and the system's users. The lookup service maps interfaces indicating the functionality provided by a service to sets of objects that implement the service. Additionally, descriptive entries associated with a service allow more fine-grained selection of services based on properties people understand. A service is added to a lookup service by a pair of protocols called *discovery* and *join*—first the service locates an appropriate lookup service (by using the discovery protocol), then it joins it (by using the join protocol). Having joined, a service is now a member of a federation.

Communication between services occurs using Java Remote Method Invocation (RMI). RMI is a Java-based extension to traditional remote procedure call (RPC) mechanisms. One important extension is that it enables actual code, not just data, to be exchanged between services.

This allows services to provide not only a description of the service they offer to the lookup service, but also the actual client-code (called a service object) that is configured to access the service (see Figure 5.5). Clients can then receive this service object as part of the lookup, and access the service directly.

Figure 5.5 Using a Service in Jini

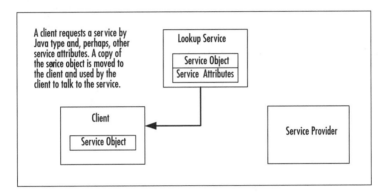

Table 5.6 outlines the functions of Jini.

Table 5.6 Jini Highlights

Function	Description
Announcing Presence	Unicast/Multicast to Jini lookup services and subsequent registration.
Discovering Other Services	Query lookup service(s) with properties of services of interest.
Describing Services	Registration information composed of attribute/value pairs.
Self Configuration	Does not directly address this area. An IP device when plugged onto a network will have to be configured with an IP address, subnet mask, and optionally a gateway and DNS server. From then on, the lookup services can be used.
Invoking Services	Download service proxy and use proxy to access service.
Transports	TCP/IP and proxies to other transports.
More Information	www.jini.org

Universal Plug and Play (UPnP)

In January 1999, Microsoft announced its Universal Plug and Play (UPnP) initiative. The UPnP initiative seeks to extend the original Microsoft Plug and Play peripheral model to a highly-dynamic world of many network devices supplied by many vendors. UPnP defines a set of lightweight, open, IP-based discovery protocols that allow appliances (telephones, televisions, printers, game consoles, and so on) to exchange and replicate relevant data between themselves and the PCs on the network. UPnP is a "wire-only" protocol—it defines the format and meaning of what is transmitted between members of the network and says nothing about how the standard is actually implemented. It requires TCP/IP and HTTP to be present to operate.

UPnP uses the Simple Service Discovery Protocol (SSDP) to discover services on IP-based networks. SSDP can be operated with or without a lookup or directory service in the network. SSDP operates on the top of the existing open standard protocols, using the HTTP over both Unicast UDP and Multicast UDP.

Table 5.7 UPnP Highlights

Function	Description
Announcing Presence	Use SSDP and Directory service proxies (optional).
Discovering Other Services	Listen to SSDP multicast channel directly or contact a directory service proxy.
Describing Services	XML description of the service is made available at a specified URL.
Self Configuration	DHCP (if available) or AutoIP, and multicast DNS.
Invoking Services	UPnP does not address this area.
Transports	TCP/IP and proxies to other transports
More Information	www.upnp.org

When a service wants to join the network, it first sends out an advertise (or announcement) message notifying the world about its presence. In the case of multicast advertising, the service sends out the advertisement on a reserved multicast address. If a lookup or directory service is present, it can record the advertisement. Meanwhile, other services in the network can directly see these advertisements as well. The "advertise" message contains a URL that identifies the advertising service and a URL to a file that provides a description of the advertising service. Devices can also cancel advertisements in order to leave a network.

When a service client wants to discover a service, it can either contact the service directly through the URL provided in the service advertisement, or it can send out a multicast query request.

Table 5.7 outlines the functions of UPnP.

The Future of SDP

The SDP protocol is a low-level, lightweight, compact, and efficient service discovery protocol. Its inclusion in the Bluetooth protocol stack was considered critical to Bluetooth technology's success as its use spread across many types of devices exporting varied services. But, as you've seen, SDP is one of many protocols that deal with the concept of service discovery. One of the key issues is interoperability of the various protocols. One of the Bluetooth white papers [Mill99] deals with the mapping of the SDP protocol to the Salutation service discovery architecture. In the immediate future of SDP, the Bluetooth SIG is defining the Extended Service Discovery Protocol. This "new" protocol is expressed as a profile (dependent on the Generic Access Profile) and allows the Universal Plug and Play (UPnP) protocol suite to run over a Bluetooth stack. The suite runs directly over L2CAP using a connection management layer (to provide flow control, and so on), or over IP, either as currently defined by the LAN Access profile or using the new Personal Area Profile (PAN). As such, the core SDP protocol remains unchanged, but it is used to discover the UPnP service that can then be used. Though not proposed at present, a similar profile could be developed for the Jini service discovery protocol.

Summary

The problem of how a device locates useful services and applications in a distributed network is common in many domains. In Bluetooth, it is the Service Discovery Protocol (SDP) that addresses this problem. Unlike many other lookup or discovery protocols, SDP is a true peer-to-peer protocol that does not rely on centralized, third-party infrastructure. The service record is the unit used to describe a Bluetooth Service. Service records are made up of attributes that capture information about a service. These attributes may contain data that is reasonably complex in structure, through the use of data elements, in addition to simple types.

There are a number of ways to query the services that a particular Bluetooth device supports. The first approach is to use the Class of Device (CoD) which may be extracted from the Frequency Hopping Synchronization (FHS) packet. The CoD contains, among other information, the Major Service Class of the device. This may be used to decide if a remote device is of interest to the inquiring device, and helps to short-circuit the service discovery process. Secondly, a client may search the service discovery server. They may search for specific attributes—most importantly the *ServiceClassIDList* attribute. A client may also search for service records containing attributes with values that match a specified list of UUIDs. Finally, a client may browse a hierarchy of service records by searching for a particular BrowseGroupDescriptor (or "directory" in the hierarchy).

Bluetooth SDP does not mandate a particular programming interface or set of programming abstractions. We presented an abstract C API that exposes the functionality of SDP to the programmer. We examined how, using this API, we would create and advertise a service, discover specific services, use service attributes and browse for services. There are opportunities for richer APIs that provide "helper" functions based on the use of Bluetooth profiles. Such functions could take the drudgery out of some of the coding effort.

The Service Discovery Application Profile (SDAP) is a usage scenario describing the functionality of a Service Discovery Application. It consists of suggested primitives that may be implemented in terms of the underlying SDP API. These primitives are used both by local devices discovering services on remote devices, and also by other Bluetooth profiles that need to advertise their services via SDP.

Though many Bluetooth stack implementations expose a C language API, Java is gaining ground as a platform for developing embedded wireless applications. As part of the Java Community Process, standard Java Bluetooth APIs are being defined. They will be components of the Java 2 Platform, Micro Edition (J2ME).

Future developments in Bluetooth SDP include the definition by the Bluetooth SIG of the Extended Service Discovery Protocol. This Profile will provide a mechanism for integrating the Universal Plug and Play (UPnP) protocols with Bluetooth SDP.

Solutions Fast Track

Introduction to Service Discovery

☑ The term *service discovery* is used to describe the way a networked device (or client) discovers available services on the network. Service discovery makes zero configuration networks possible—the user doesn't have to manually configure the network.

☑ Key features of a discovery protocol are: spontaneous discovery and configuration of network services, low (preferably zero) administrative requirements, automatic adaptation to the changing nature of the network (addition or removal of nodes or services), and interoperability across platforms.

☑ Bluetooth Service Discovery is protocol-dependent; it mandates the use of the underlying Bluetooth communication protocol as the basis for service discovery. However, Bluetooth SDP could indeed be implemented using other underlying transport mechanisms, and higher-level protocols (such as TCP/IP) may be run over Bluetooth.

Architecture of Bluetooth Service Discovery

☑ For a particular service (and there may be many services on one device) a *service record* contains a description of that service. The description takes the form of a sequence of *service attributes*, each one describing a piece of information about the service.

☑ Within the SDP server, each service record is uniquely identified by a *service record handle*. A *service class* defines the set of service attributes that a particular service record may have. In other words, a service record is a particular instance of a class of services.

☑ A service attribute is a name-value pair that includes an *attribute ID* and an *attribute value*. The attribute ID uniquely identifies the attribute within the scope of the service record.

☑ An attribute value can contain data of arbitrary complexity, rather than just simple types. This is accomplished using *data elements*. A data element is made up of a header and a data field.

☑ The Service Discovery Protocol includes a set of Protocol Data Units (PDUs) that contain the basic requests and responses needed to implement the functionality of Bluetooth Service Discovery. An SDP PDU contains a PDU ID, a transaction ID, and a parameter length in its header. Its body contains some number of additional parameters, depending on which type of transaction the PDU contains.

Discovering Services

☑ Every Bluetooth device can contain a Service Discovery Server (SDS) that advertises the services available on that particular device, be it a mobile phone, PDA, or something else. It can do this by making available the service records that describe those services.

☑ The Bluetooth-defined Class of Device (CoD) value can tell a discovering device if a connection should be opened to the discovered device—it doesn't have to open a connection to the SDS and check the Service Discovery Database (SDDB) of the discovered device, "short-circuiting" service discovery.

☑ The Bluetooth Service Discovery Protocol allows for services to be discovered on the basis of a series of attributes with values of type UUID. In reality, when talking about discovering specific services, one of the most important attributes of a service, if not *the* most important, is the *ServiceClassIDList*.

Service Discovery Application Profile

☑ The SDAP is a usage scenario describing the functionality a Service Discovery Application (SrvDscApp) should provide to an end user on a local device (LocDev) so that user can discover services on a Remote Device (RemDev). The SDAP doesn't specify an API that will provide this functionality, but suggests primitives that can be mapped to an API.

☑ Most profiles detailed in the Bluetooth specification have a service discovery component that specifies the structure and content of the service record that accompanies the service (or application) and which realizes the profile. The SDAP (in addition to dealing with application functionality for service discovery) specifies the procedures that an application realizing a profile must use to perform service discovery. If these procedures are upheld, interoperability is ensured.

Java, C, and SDP

☑ As part of Java Community Process (JCP), a set of standard Java APIs for Bluetooth is being developed and is due for publication at the end of 2001. Implementations of this standard will allow programmers to implement Bluetooth applications within the J2ME environment in a standard and portable way.

☑ A key element of the J2ME specification is the Generic Connection Framework (GCF), a mechanism that allows a programmer to create different types of networking connections through a standard Connector interface. This would allow programmers to quickly produce Java Bluetooth applications by applying existing techniques and design patterns.

Other Service Discovery Protocols

☑ The Bluetooth SDP may be integrated with a number of the other service discovery protocols, including Salutation, UPnP, Service Location Protocol (SLP), and Jini.

☑ The Salutation architecture defines a uniform way of labeling devices (fax machines, printers, copiers, and also phones, PDAs, and general electronic equipment) with descriptions of their capabilities and with a single, common method of sharing that information.

☑ Salutation is "transport independent," that is, a separate Transport Manager may be written for each underlying transport required, and the Salutation Manager, which provides the core functionality of the system, remains transport neutral.

☑ SLP is a language-independent protocol for automatic resource discovery on IP-based networks. Like some of the other service discovery protocols, it makes use of UDP/IP multicast functionality in TCP/IP. This makes it particularly useful for networks where there is some form of centralized administrative control, such as corporate and campus networks.

☑ Jini is a distributed service-oriented architecture, considered an extension of the Java language and platform. Services communicate with each other using a service protocol, which is defined as a set of interfaces in Java. The standard itself provides a base set of interfaces to facilitate core interaction between services. A key component of Jini is the *lookup* service.

☑ Communication between services in Jini occurs using Java Remote Method Invocation (RMI). RMI is a Java-based extension to traditional remote procedure call (RPC) mechanisms. One important extension is that it enables actual code, not just data, to be exchanged between services.

☑ Universal Plug and Play (UPnP) defines a set of lightweight, open, IP-based discovery protocols that allow appliances to exchange and replicate relevant data between themselves and the PCs on the network. UPnP is a "wire-only" protocol—it defines the format and meaning of what is transmitted between members of the network and says nothing about how the standard is actually implemented. It requires TCP/IP and HTTP to be present to operate.

☑ UPnP uses the Simple Service Discovery Protocol (SSDP) to discover services on IP-based networks. SSDP can be operated with or without a lookup or directory service in the network. SSDP operates on the top of the existing open standard protocols, using the HTTP over both Unicast UDP and Multicast UDP.

The Future of SDP

☑ SDP is one of many protocols that deal with the concept of service discovery. One of the key issues is interoperability of the various protocols.

☑ In the immediate future of SDP, the Bluetooth SIG is defining the Extended Service Discovery Protocol. This "new" protocol is expressed as a profile (dependent on the Generic Access Profile) and allows the Universal Plug and Play (UPnP) protocol suite to run over a Bluetooth stack. Though not proposed at present, a similar profile could be developed for the Jini service discovery protocol.

Frequently Asked Questions

The following Frequently Asked Questions, answered by the authors of this book, are designed to both measure your understanding of the concepts presented in this chapter and to assist you with real-life implementation of these concepts. To have your questions about this chapter answered by the author, browse to **www.syngress.com/solutions** and click on the **"Ask the Author"** form.

Q: What is Bluetooth SDP?

A: The Bluetooth Service Discovery Protocol (SDP) is a distributed, peer-to-peer lookup mechanism for discovering which services are supported by in-range Bluetooth devices. It is defined in the Bluetooth Specification.

Q: How are services represented in SDP?

A: A service on a Bluetooth device is described in an SDP service record, which is stored in the device's "Service Discovery Database." A service record consists of service attributes, each of which describes some information about the available service.

Q: How does Class of Device (CoD) relate to SDP?

A: The CoD may be retrieved from a Frequency Hop Synchronization (FHS) packet. This information contains, among other things, the Major Service Class of the device. This tells the discovering device what "kind" of device it has discovered (e.g., a printer, an access point, and so on) Using this information, the discovering device can rule out certain devices that are not interesting, and only query the Service Discovery Databases of those devices that are interesting. For many application types, this is likely to result in an efficiency gain.

Q: What's the difference between SDP and SDAP?

A: SDP is a part of the core Bluetooth specification and defines the data representation of SDP data structures as well as the set of transactions used to communicate between SDP clients and servers. The Service Discovery Application Profile (SDAP) is one of the Bluetooth profiles defined by the Bluetooth SIG. It describes usage scenarios for a Service Discovery Application, and suggests primitives for achieving these scenarios that may be implemented in terms of the underlying SDP API.

Chapter 6

Linux Bluetooth Development

Solutions in this chapter:

- **Assessing Linux Bluetooth Protocol Stacks**
- **Understanding the Linux Bluetooth Driver**
- **Using Open Source Development Applications**
- **Connecting to a Bluetooth Device**
- **Controlling a Bluetooth Device**

☑ **Summary**

☑ **Solutions Fast Track**

☑ **Frequently Asked Questions**

Introduction

Bluetooth technology is an open standard while Linux is open source. There's some obvious synergy there: combine low cost devices with free software and you've got a communications technology anybody can afford.

Linux is proving to be the obvious system of choice for students and academics trying to get into Bluetooth technology on tight budgets. But don't think it's just for educational use: Linux is being deployed in real commercial products from local area network (LAN) access points to laptops, and more besides. To give it a real stamp of credibility, Linux Bluetooth development has backing from a Bluetooth Special Interest Group (SIG) promoter with IBM's BlueDrekar middleware, and, of course, a myriad of smaller companies and individuals are contributing to the development of open source, too.

This chapter takes a look at what Linux can do for your Bluetooth applications, and gives you some useful insight from inside the Linux developer's community.

Assessing Linux Bluetooth Protocol Stacks

Until recently, the Linux kernel did not come with a Bluetooth stack among its stock drivers. But shortly after this chapter was originally completed, a new Bluetooth project was released as open source and rapidly accepted into the 2.4.6 kernel. This project is called Bluez (bluez.sourceforge.net), and at the time of this writing, its recent 1.2 release includes stable Host Controller Interface (HCI) and Logical Link Control and Adaptation Layer (L2CAP) drivers, as well as user-space Radio Frequency Communications Port (RFCOMM) and Service Discovery Protocol (SDP) applications leveraged from the OpenBT project (which we'll discuss in short order). Although it has gained acceptance into the mainline Linux kernel, it may not yet be the driver of choice for developers. As of now, it does not support as many features as some of the other available stacks. It does not yet appear to have the developer and user following that OpenBT does, and most importantly, has not been ported back to earlier kernel versions.

Currently, there are two other major Linux Bluetooth protocol stacks: IBM's BlueDrekar and the OpenBT project. Another future contender will be Rappore Technology's stack, which is already ported to Windows and BlueCat embedded Linux.

IBM's BlueDrekar can be downloaded from their project Web site at www.alphaWorks.ibm.com/tech/bluedrekar. This is not an open source stack. What you get for free are the binary modules. If you want the source, you can get it, but according to their documentation and Web site, you must be a SIG member and you must sign a limited license with IBM. You will also need a license to distribute their stack.

SourceForge hosts the OpenBT project. You can find their Web site at www.sourceforge.net/projects/OpenBT. Axis Communications (www.axis.com) originally developed this stack for their embedded Linux product and most of the main developers work there. This is a truly open source stack.

If you're an embedded developer using BlueCat Linux on your target, you can find out more about the status of Rappore's stack at their Web site: www.rappore.com. (This stack is not open source; we won't cover the Rappore stack in detail in this chapter.)

Comparing BlueDrekar with OpenBT by Features

The big factor that distinguishes BlueDrekar from OpenBT is source code availability. Why would you even consider a closed source solution when an open source one is available? For an x86 application developer, BlueDrekar offers more than the OpenBT stack. For embedded developers who need to cross-compile and don't want to license source, OpenBT may be good enough.

Table 6.1 shows a breakdown of the feature differences between the two stacks, which we'll discuss in the following sections.

Table 6.1 Feature Comparison between OpenBT and BlueDrekar

Feature	OpenBT	BlueDrekar
Kernel versions	2.0.x – 2.4.x	2.2.12, 2.2.14
Hardware platforms	X86, ARM, MIPS, PowerPC	X86
Bluetooth protocols	Host Controller Interface (HCI), Logical Link Control and Adaptation Protocol (L2CAP), Service Discovery Protocol (SDP), RFCOMM, HCI-Universal Asynchronous receiver Transmitter (HCI-UART), HCI-USB	HCI, L2CAP, SDP, RFCOMM, Synchronous Connection Oriented (SCO), HCI-UART
SDP server support	Server, XML database	Server, dynamic database
API	Standard Unix device driver	Custom lib Applications Programming Interface (API)
License terms	AXIS OpenBT Stack license	AlphaWorks

The basic Bluetooth host protocols are supported by both stacks. Beginning at the HCI, which links a host to a module, both stacks support the UART transport layer needed for basic serial communications. OpenBT goes on to also support the higher speed Universal Serial Bus (USB). L2CAP, RFCOMM, and SDP are also provided by both protocol stacks.

Kernel Versions

Developers have used the OpenBT source on a wide range of kernel versions, including uCLinux. Because the source is available, people are free to port it to whatever kernel version they require.

The BlueDrekar binaries, on the other hand, are compiled only against certain 2.2.x kernel versions at the time of this writing, so you can't use them with older or newer kernels.

Hardware Platforms

Developers around the world have used OpenBT on a variety of processor types. This author's company has used it on ARM and MIPS, as well as x86 processors, and according to the mailing list archives for OpenBT, some people have used it with PowerPCs as well. Again, because you have the source, if you need to port it or even just cross-compile it for a non-x86 platform, you can do so.

With BlueDrekar, you only get the x86 binaries. You don't have the source unless you apply for a license, so obviously you're limited to just x86 platforms.

Bluetooth Protocols

Here's where BlueDrekar starts to catch up. The OpenBT project does not currently support the Synchronous Connection Oriented (SCO) connections used for voice, which is a major drawback. It does include support for an HCI-USB layer, however.

BlueDrekar does have support for SCO already. For BlueDrekar, you can get the source for their HCI-UART module. This is the one part of their stack, which is open source. IBM released this source under GPL with the hope that others could use it as a basis for developing the other HCI link drivers.

SDP Support

The Service Discovery Protocol (SDP) is used by a client device to find out about the services it can use on a server device. An SDP server maintains a database of services; this can be preconfigured (static), or can be built up dynamically as services register with the database system. Once a database is in place, clients send SDP requests to query its contents, and servers reply with SDP responses giving details of services supported and information needed to connect to those services.

SDP is another area where BlueDrekar is ahead of OpenBT. The OpenBT project does provide an SDP server daemon to handle SDP requests from remote devices. However, it does not yet provide an API for local applications to dynamically register themselves in the SDP database. Another disadvantage is that applications must frame their own SDP request packets and parse the resulting SDP responses.

BlueDrekar is much nicer. It also provides a server daemon, but additionally, it has an API for dynamically registering services in the local database as well as handling a lot of the details of SDP. Applications still need to know the basic components of SDP packets, but they don't have to hand-tool the packets themselves like they do with OpenBT.

API

The OpenBT stack provides a set of device files for applications to use. These are all TTYs (terminals) and follow the standard Linux API for TTY drivers. Stack control is done via blocking *ioctl* calls. Since there's no intervening library layer, all of the control I/O is synchronous. There is no event notification aspect of the API.

The BlueDrekar stack provides a library layer and a daemon (referred to collectively as *middleware*). Although data transfers are handled over standard drivers, control operations are done via library calls. These often employ callback mechanisms for event notification.

License Terms

Licensing is the *big* issue. The OpenBT project is released under the AXIS OpenBT Stack license. You can see the text of this license at http://developer.axis.com/software/bluetooth/OpenBT_license.txt. Basically, it is the GPL with some additional freedoms. If you write applications that use the stack, they will not fall under the GPL and may remain proprietary. But if you write applications that are a derived work of the *applications* in the OpenBT source tree, then they *will* fall under the GPL—unless they have nothing to do with Bluetooth technology. Note that just because the stack is under GPL doesn't mean applications that use the stack must be. However, if you modify or add SCO support to the stack (for example) then these changes would be under GPL.

BlueDrekar is released under IBM's AlphaWorks license. You can download the binaries for free and write applications that use them, but if you want to see the source or distribute the binaries with a product then, you'll need extra permissions. According to their Web site, you must be a Bluetooth SIG member to get this additional permission.

Other Considerations

If you're a PC application developer, then you may not have any control over which Bluetooth stack the user has on his or her PC. The OpenBT and BlueDrekar APIs are not at all similar, so it would be tough to write an application that works on both. It's likely you'd have to pick one particular stack and require users to install it.

If you're an embedded developer, then chances are you're probably not only writing applications, but you're also trying to decide which stack to ship with your device. You have total control over which stack your application will use, because you decide which stack the user gets. Note that at the time of this writing the OpenBT stack produced a somewhat smaller image when compiled for an x86, but probably not enough to make too much of a difference. If size is important, then cross-compile the latest release of OpenBT against your target platform and check it. To compare it with BlueDrekar you'll have to ask IBM about getting this information. The open source nature of OpenBT can be a real bonus for embedded developers because it's easy to check things.

Axis Communications originally designed the OpenBT stack to serve as a LAN Access Profile server on their embedded Linux products. If you need a PPP server over RFCOMM, then once you get the stack running on your platform, you're basically done. However, although it functions well in this regard, developers who want to leverage the stack for other purposes should expect to do some work.

For the rest of this chapter, we're going to discuss using the OpenBT stack. I have to pick one, just like you will. I'm not picking OpenBT because it's a better implementation than BlueDrekar—to be perfectly frank I don't think it is (yet). Instead, I chose it for the following reasons:

- It's freely available.

- I'm under no restrictions to not discuss any aspect of it.

- I have access to the source, so I understand it much better.

- I've used it in the past on several different platforms, for several different kernel versions.

- I've contributed to it in the past.

- I think it has the best chance of making it into the standard Linux kernel tree (eventually).

- If I can encourage you to use it and contribute, then I benefit from your use as you can benefit from mine.

Fair Warning

It's only fair to be perfectly clear on something at this point: the OpenBT stack is a work in progress, and is not feature-complete as a client stack. Here are the big issues, in order of severity:

- There is no way to bind RFCOMM server channels for server applications other than PPP.

- There is no interface for dynamic SDP registration.

- Applications must assemble their own SDP requests and parse the SDP responses.

- There is no SCO support.

- There are no interfaces for supporting other protocols above L2CAP.

- The stack still has many bugs ranging from annoying behavior to full system lockups.

Also, as with any implementation, the stack still has some bugs—especially when supporting client applications. You can get a list of the current known bugs from the OpenBT Web site on SourceForge.

Nonetheless, OpenBT has one major advantage: the source is open. It goes without saying that one of the reasons I know about all these problems is because I can look in the source and see them. I can also look in the source and fix them.

That being said, let's talk about the basics of how the OpenBT stack works. From here on, when I use the term *Bluetooth driver* I'll be referring to the OpenBT stack. Specifically, I will be referring to version 0.0.2, released in March of 2001.

Understanding the Linux Bluetooth Driver

The first thing you should do is go to the OpenBT project Web site, download their latest release, and then follow the instructions for installing and using the driver. Go ahead and play with included applications until you're satisfied that you've got things working on your system. If you don't have Bluetooth hardware, that's okay, because the stack includes several options for simulating hardware connections between two devices. You don't even necessarily need more than one PC to try it out.

Note that the OpenBT stack comes with a lot of options about *user mode*, *kernel mode*, and *real* versus *simulated* hardware connections. In this chapter, I'm going to limit the discussion to using the kernel mode driver with real hardware. In the end, your application will have to work under these conditions anyway.

In this section, we'll first talk about what the Bluetooth driver is, and tour some of its visible properties. Then we'll cover the basics of using the Bluetooth driver interfaces.

Learning about the Kernel Driver

The actual kernel Bluetooth driver is the *bt.o module*. This is built in the linux/drivers/char/bluetooth directory of the OpenBT source tree. This loadable module implements a TTY (terminal) driver and an *ldisc*, the line discipline that affects how the data stream to a terminal is interpreted. I'll explain those terms in more detail after taking a quick look at what happens when you load the Bluetooth driver into the kernel.

Investigating the Kernel Module

To load the Bluetooth driver into the kernel, execute the following command in a terminal window as root:

```
$ insmod bt.o
```

Now let's browse through the proc directory and see what just happened. Enter this:

```
$ cat /proc/devices
```

One of the char driver entries will be listed as *bt*. This is our driver. On the same line, you'll see its *major number*. This major number uniquely identifies the Bluetooth driver in the kernel. Later, when we look at the Bluetooth device files, we'll see that their major number matches up with this, effectively binding them to this driver. This is what tells the Linux kernel which driver to invoke when we make system calls like *open* on those device files.

Now enter this:

```
$ ls /proc/bt_*
```

And you'll see the proc files installed by the driver. Enter this to see some status information on the driver:

```
$ cat /proc/bt_status
```

Finally enter this:

```
$ cat /proc/tty/drivers; cat /proc/tty/ldiscs
```

The first command lists all the TTY drivers currently registered in the kernel. Ours is now one of them. The second lists all the *ldiscs* currently registered in the kernel. Note *bt_ldisc*—that's ours.

What Exactly Is a TTY?

One way to think of a TTY is as a subclass of a character driver. A TTY implements the same interface as a character driver and then some. In fact, you might think of a TTY as a character driver with an attached filter. The filter sits in the kernel between the TTY and an upper layer. This filter is called an *ldisc*, or "line discipline."

So What's an *ldisc*?

A line discipline (*ldisc*) monitors and even modifies the data stream that passes between an upper layer and the TTY. It might do things like look for special control characters in the data stream. It might even reformat the data stream into protocol packets of some kind or other.

Developing & Deploying…

What Exactly Do You Mean by "Character Driver"?

A character driver is one of the basic driver types supported by the Linux kernel (some others are block drivers and network drivers). A character driver represents a connectionless data stream over some type of device. All character drivers must support the following system calls: *open*, *close*, and *write*. Most character drivers also support the *read*, *select*, and *ioctl* system calls. Examples of character drivers you might find on your system are /dev/audio, /dev/ttyS0 (the serial TTY), and /dev/mem.

One really important feature of the relationship between a TTY and its *ldisc* is that you can change the *ldisc* at runtime. In effect, you can swap filters. In the next section, I'll show you how this affects the Bluetooth driver.

Building Driver Stacks in the Linux Kernel

Figure 6.1 is a simplified diagram of the default TTY driver configuration after you load the *bt.o module*. You see how both the bt and serial TTY drivers use the N_TTY *ldisc* as an adapter between themselves and the standard TTY I/O code? The N_TTY *ldisc* is suitable for console TTY drivers. It does things like scan for control characters in the byte stream. But an application can change any TTY driver's line discipline by using a special *ioctl* call. For example, we could have an application change the serial driver's line discipline to be *bt_ldisc* instead of N_TTY.

Figure 6.1 Default TTY Driver Configuration

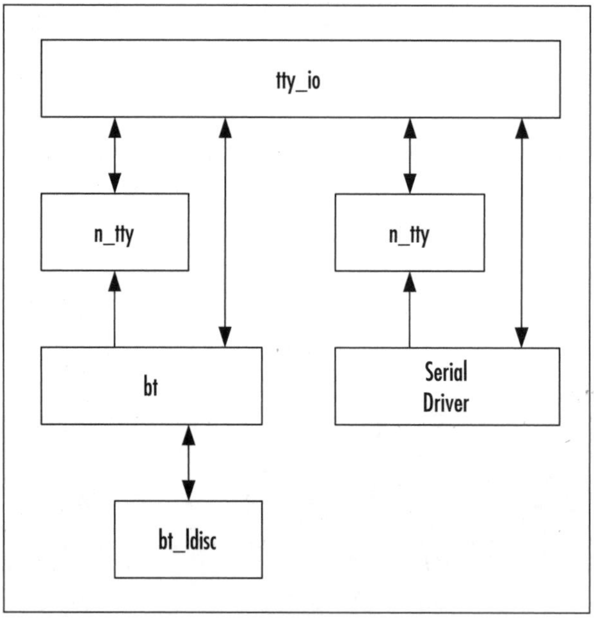

Guess what? That's exactly how we make the Bluetooth driver talk to a Bluetooth card attached by a serial cable. Figure 6.2 shows a picture of this. The *bt_ldisc* in effect will route all data to and from the serial port through the Bluetooth driver. That's where all the parsing and packet assembly will take place.

Figure 6.2 Stacked TTY Driver Configuration

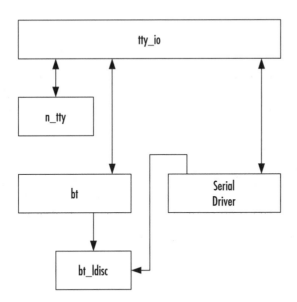

In summary, line disciplines are important because they allow user-space applications to stack TTY drivers in the kernel. Note that this is exactly how PPP works over a TTY—and therefore RFCOMM devices must be TTY drivers.

Understanding the Bluetooth Driver Interface

Now that you understand what the Bluetooth driver is, how exactly do applications use it? They use it by making system calls on the Bluetooth device files.

Investigating the Bluetooth Device Files

You may have noticed during the installation that at one point you had to create some files in the /dev directory. Take a look at them now by entering:

```
$ ls -l /dev/ttyBT*
```

These device files are your application's interface to the Bluetooth driver. Notice that all the devices have the same major number but different minor numbers (if you're not sure how to tell, then check the man page for *ls*). Having the same major number means that the same kernel driver implements them all. The different minor numbers represent different instances of an interface to the kernel driver.

There are two types of Bluetooth device files: *data device* files and *control device* files. Table 6.2 shows the main differences between them.

Table 6.2 Comparison of the Control and Data Device Files

Feature	/dev/ttyBTC	/dev/ttyBT[0-6]
Can open before stack is initialized	YES	NO
Multiple processes can open at the same time	YES	NO
Can transfer data over an RFCOMM connection	NO	YES
Can execute stack control ioctls	YES	NO

Using the RFCOMM TTY Drivers

The data device files are named /dev/ttyBT0 through /dev/ttyBT6. These are all instances of RFCOMM TTYs. Once they're opened and connected, they behave exactly like serial ports, as we'll see later. Only one process at a time can open any individual RFCOMM TTY. All the standard system calls which work over standard character drivers and all of the *ioctls*, which work over standard TTY drivers, also work over the RFCOMM TTY driver.

The minor number for the RFCOMM TTY's has special significance to the Bluetooth driver. Each minor number corresponds to a line number used internally by the driver to index a connection session. Each possible RFCOMM or SDP connection, which the driver can make with a remote peer, is represented internally by a session. Since there are seven RFCOMM TTYs, there are seven session "objects" maintained by the driver.

The only trick to using the RFCOMM TTY device files is in understanding the concept of an RFCOMM session. Within the driver, each RFCOMM session has a state machine. The driver indexes sessions internally by a line number. When opening an RFCOMM device file, the line number comes from the minor number of the device file. When connecting to a remote service, you specify the local line number as one of the connection parameters. Figure 6.3 illustrates the state machine for a single session.

In Figure 6.3, you can see the three parameters that specify the state of a session are: whether or not the device file is open, whether or not the TTY is hung up, and whether or not an RFCOMM connection to a remote peer exists. The important points to take away from this are as follows:

Figure 6.3 The RFCOMM Session State Diagram

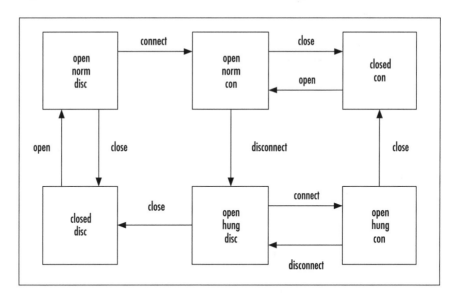

- The driver hangs up the TTY when an existing RFCOMM connection gets disconnected.

- The only way to return a hung-up TTY to normal is to close and reopen the device file.

- Data can only be transferred in the open/normal/connected state.

One very interesting consequence is that one process can establish an RFCOMM connection on a session without opening its device file, and another process can then open the device file and transfer data across the connection.

Multiplexing over RFCOMM

All of the RFCOMM device files operate independently of one another. Each represents a different potential RFCOMM channel. That's all you really need to know about multiplexing! You don't have to worry about it much at the application layer. If you have an application that can handle multiple connections, it should open and listen on multiple RFCOMM device files. Figure 6.4 illustrates this.

When you open an RFCOMM device file, your process gets exclusive access to it. True, other processes can establish RFCOMM connections for it, but yours is the only one that can transfer data through it. None of your data transfers will affect any other RFCOMM session (other than using up some of the link's bandwidth).

Figure 6.4 Multiple Simultaneous RFCOMM Connections

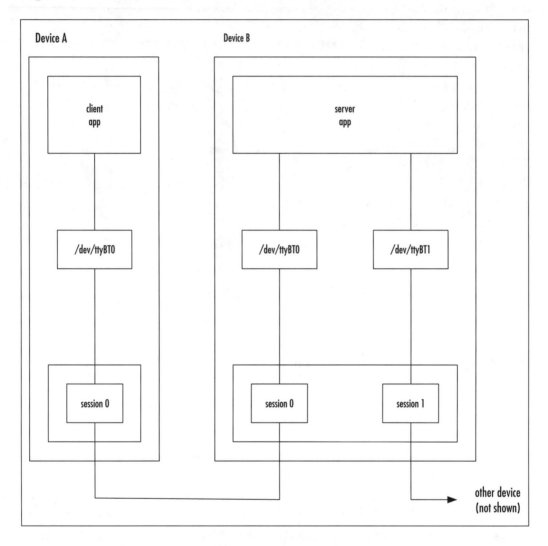

However, there are things your application can do that will affect other processes using the Bluetooth driver. Most of the *ioctl* calls specific to the Bluetooth driver have global affects. For example, if your application decides that it needs to shut down or reinitialize the stack, it could interrupt another application's data transfer.

The OpenBT stack lacks a central stack manager. In other words, there is no single process responsible for running the driver in an orderly fashion. The Bluetooth driver itself does not enforce any policy. For example, it does not decide when to enable Inquiry Scans, or security procedures. All policy is left to the applications. And the OpenBT source tree does not come with a central management application to make sure applications don't conflict with one another. If one application wants Inquiry Scan enabled and another wants it disabled, the winner is whoever issued the *ioctl* call last.

So how can you write applications that cooperate well with others? Short answer: you can't. This is a problem for desktop applications. For embedded developers, odds are you control all the applications that will use Bluetooth and you can design your own cooperation strategy.

The one device file /dev/ttyBTC is a special device, dedicated to controlling the kernel driver as a whole. We'll see later how to use this device to initialize and shut down the Bluetooth stack. Any number of processes can open /dev/ttyBTC at the same time.

Note that there are no device files for SDP, L2CAP, or any of the other Bluetooth protocols implemented by the driver. We'll see that we can access SDP and HCI using *ioctl* calls on any of the devices. And there simply is no interface to L2CAP—it's completely internal to the driver.

Can you add your own device files to implement other protocol layers above L2CAP? That's a pretty frequent question to the bluetooth-dev mailing list. And the disappointing answer is no, not without modifying the stack itself—but remember, you do have the source.

Although the Bluetooth driver is "just another TTY driver," there are some specific things you need to understand about its interface. You need to be familiar with some of the more important *ioctl* calls used to control Bluetooth-specific features, and you need to know the difference between the control device file and the other device files.

Installing a Line Discipline over an RFCOMM TTY

Because the RFCOMM device files are TTYs, you can set up line disciplines above the RFCOMM layer. This is exactly the way PPP works. In the same way that the Bluetooth driver sets up a line discipline above the serial driver, PPP sets up a line discipline above the Bluetooth driver. The whole key to using RFCOMM comes from understanding this principle. Any application that works over a TTY will work over an RFCOMM TTY, once the underlying RFCOMM connection has been established. *Any* process can establish that connection—it doesn't have to be the process that will use the TTY to transfer data.

Using the Control Driver

The control device file is /dev/ttyBTC. Unlike the other Bluetooth device files, this one isn't used to transfer data between different devices. This one is only used to control the local Bluetooth driver. Whenever you need to issue a stack control *ioctl*, you should do it using this device file. This includes the *ioctl* calls for initialization, shutdown, security, connection, hardware control, and so on.

The most important role of this device file is to initialize the driver. Until the driver is initialized, you cannot open any of the other device files. You can only open the control device file. However, once the stack is initialized, you cannot only open the other devices' files, but you can use them to execute all of the stack control *ioctls* which can be used on /dev/ttyBTC. In a way, the only purpose of the control device file is to initialize the stack.

Using Open Source Development Applications

The OpenBT source tree comes with several applications. You can use these applications to:

- Provide your SDP server.

- Manually establish PPP connections between devices.

- Manually establish RFCOMM connections between devices.

- Browse the SDP database on a target device.

- Provide examples to learn how to write applications for the stack.

- Provide a starting point for your own application.

Depending on what you want to use the Bluetooth stack for, you may not need to write any code at all. For instance, once you establish a PPP connection over RFCOMM, all the power of the standard GNU network applications is at your disposal—the Bluetooth connection is just like any other network connection. All existing applications that use a socket interface are instantly ported to use Bluetooth: Web browsers, Web servers, FTP, Telnet, and so on.

Investigating the OpenBT Applications

The OpenBT source tree comes with some applications. Table 6.3 summarizes their features.

Table 6.3 Summary of Features in OpenBT Applications

Application	Features
btd/btduser	Initialize the stack
	Do HCI Inquiries
	Establish RFCOMM connections
	Spawn PPP over RFCOMM
	Send test data over an RFCOMM link
sdp_server	Query an XML database
	Receive and parse SDP requests
	Compose and send SDP replies
BluetoothPN	Browse a remote device's SDP database

Understanding the btd and btduser Applications

The btd application will probably be the most useful for you. The difference between btd and btduser is that btd is meant to work with the kernel mode Bluetooth driver, while btduser works with the user mode Bluetooth driver. Many people prefer btduser since it is less prone to lock up your system if things go badly. However, the OpenBT developers do not support it as well as btd.

For btd you have to install the Bluetooth kernel driver (i.e., *insmod bt.o*). For btduser, you don't. Other than that, their usage is basically the same.

The btd application can take a number of different arguments on startup. An example follows. If you're curious about other arguments besides the one I mention, then look in the sdp.c source file. At the top of the *main()* routine, you'll see the argument parsing. From that, you can figure out what the other arguments to btd are. The README that comes with OpenBT talks about starting btd, but it is not always up-to-date. Remember, OpenBT is still early in its development, and often the source code is the best documentation.

Understanding the sdp_Server Application

The sdp_server application provides you with an SDP database server daemon. Once you've installed the Bluetooth driver, you can start this daemon and it will automatically receive and respond to SDP queries from remote devices.

If you start the daemon with no arguments, it will automatically use /etc/sdp.xml as the SDP database file and /tmp/sdp_sock as the source of SDP requests. The /tmp/sdp_sock file is a Unix socket created by the btduser application. You can specify a different XML file as the first argument to sdp_server and a different source device as the second argument. Note that if you provide one argument, you must provide the other as well. If you want to use the SDP server when the Bluetooth driver is in kernel mode, then you should specify /proc/sdp_srv as the source of SDP requests.

The following is an example of starting the sdp_daemon with command-line arguments:

```
$ sdp_daemon /tmp/my_sdp_database.xml /proc/sdp_srv &
```

SECURITY ALERT

Warning! Never remove the Bluetooth driver while the sdp_server daemon is using /proc/sdp_srv. If you do so in the current release version of the stack (0.0.2 at the time of this writing), you will get a kernel panic when you stop the daemon. Future versions of the stack will probably not allow you to remove the driver while the sdp_server daemon is using it.

Understanding the BluetoothPN Application

This application provides a GUI that displays the SDP database on a remote device. It provides some examples of how to make SDP requests and process their results.

Establishing a PPP Connection Using the btd Application

The quickest, most useful way to establish and exploit a Bluetooth connection from Linux is to use the standard GNU network applications over PPP. And the easiest way to do that is with the btd application. Let's look at an example.

It assumes the following setup:

- Two Linux PCs configured to use PPP; one will be the server and one the client.

- Both PCs are connected to Ericsson Bluetooth Developer kits via RS232 to /dev/ttyS0.

- The OpenBT Bluetooth driver is installed in both PCs kernels.

- There us an open terminal window with root permissions on each PC.

- The server should have the "local" and "nodetach" options specified in its /etc/ppp/options file (see man(8) pppd).

- The client should have the "local," "nodetach," and "noauth" options specified in its /etc/ppp/options file.

Here are the steps:

1. On the server:
   ```
   $ btd --server --physdev=/dev/ttyS0 --speed=57600 --modem=0
   ```

2. On the client:
   ```
   $ btd --client --physdev=/dev/ttyS0 --speed=57600 -modem=0
   ```

3. On the client, you will now see a menu of options. Select an HCI Inquiry for one device, with a maximum timeout of about five seconds:
   ```
   > inq 1 5
   ```

4. If the inquiry succeeds, the program will report the Bluetooth Device Address (BD ADDR) of the server's Bluetooth card on the terminal. For example, it might return 11:22:33:44:55:66 (it's unlikely, but this is just an example). Next, create an RFCOMM connection to server channel 2 of that device, using line 0. When the server btd application detects the connection, it will spawn PPP and pass in /dev/ttyBT0 on the command line as the TTY. The line 0 argument maps to /dev/ttyBT0 on the local device. When the client btd application spawns PPP, it will also pass /dev/ttyBT0 to the local PPP as the TTY. Here's the command:
   ```
   > rf_conn 11:22:33:44:55:66 2 0
   ```

5. If the command succeeds, then after a few seconds you will see the connected message on the client's terminal window. On the server, you should see PPP start up and wait for an incoming PPP connection. At this point, we're ready to start PPP on the client. Here's the command:
   ```
   > ppp
   ```

6. If the PPP connection succeeds, you should see a message like this on both the client and server side:
   ```
   local IP address 192.168.1.249
   remote IP address 192.168.1.17
   ```

7. At this point, you can test the connection. First, on either the client or server, open a terminal window and use **ifconfig** to determine the IP address of the remote PPP connection. It should report the **ppp** connection similar to this:

```
> ifconfig

ppp0   Link encap:Point-to-Point Protocol
       inet addr:192.168.1.249 P-t-P:192.168.1.17
```

8. Now, open another terminal window on the client and ping the remote IP.

```
> ping 192.168.1.17
```

Those ping responses are coming back across the Bluetooth link! Pretty exciting, eh? Well, the first time anyway. You can also go ahead and try some other network commands like Telnet and FTP. Have some fun.

Debugging...

Watching Driver Debug Messages

If you want to watch exchanges between the stack and the card (a good idea for debugging problems) then you can turn on some of the debug messages before you compile the stack. Edit the btdebug.h file in the OpenBT source tree. My favorite macro to turn on is BT_DATAFLOW_DEBUG. Change its #define from 0 to 1 and then recompile and insert the OpenBT module. Then, when you're running your application, open another terminal and execute this command to see the running transactions between the host and the card (on most systems you must be root to do this):

```
$ tail -f /var/log/messages
```

If you see a lot of messages to the effect of "HCI timeout" in this debug, then chances are your card is not responding to HCI commands from the host. You should make sure your serial port is set up right and you are using the right type of cable (null modem for Ericsson Bluetooth Developer Kits; other hardware may vary). A good way to double-check your serial port settings is to do this:

```
$ cat /proc/tty/driver/serial
```

The btd application provides the quickest way to get started, but it assumes that:

- You know the remote server channel number without doing an SDP discovery.

- You want to use PPP over RFCOMM, and not some other application.

If you have other requirements, then you'll need to produce your own application. If you're willing to accept a GPL-like license on your application, then you can use btd.c as a starting point to make a derived work.

Writing Your Own Minimal Application

Admittedly, btd.c has grown to become rather large and complicated. You're probably wondering, "What's the bare minimum I need to establish a connection?" The following source will give you a starting point. This program does essentially the same thing as btd, and makes the same assumptions. But it boils down btd.c into the absolute minimum amount of code needed to establish an RFCOMM connection.

```
#include <linux/bluetooth/btcommon.h>

#include <linux/bluetooth/l2cap.h>

#include <linux/bluetooth/rfcomm.h>

#include <sys/ioctl.h>

#include <sys/time.h>

#include <fcntl.h>

#include <termios.h>

#include <errno.h>

#define SYSCALL(v,x,s) if ((v) = (x)) < 0) { perror(s); exit(errno); }

void tty_init(int fd)

{

        int ret;

        struct termios t;

        SYSCALL(ret, ioctl(fd, TCGETS, &t), "TCGETS");

        cfmakeraw(&t);

        t.c_cflag &= ~CBAUD;

        t.c_cflag |= B57600 | CS8 | CLOCAL;

        t.c_oflag = 0;

        t.c_lflag = 0;
```

```
        t.c_cflag &= ~CRTSCTS;

        SYSCALL(ret, ioctl(fd, TCSETS, &t), "TCSETS");
}
int main(int argc, char **argv)
{
        int phys_fd, bt_cfd, bt_ldisc = N_BT, ret, wrscan = 0x03;
        bt_connection_con = {
                { 0x00, 0xd0, 0xb7, 0x03, 0x48, 0x9a }, /* BD ADDR */
                CREATE_RFCOMM_ID(0, 2)
        }
        SYSCALL(phys_fd, open("/dev/ttyS0", O_RDWR, 0), "/dev/ttyS0");
        tty_init(phys_fd);
        SYSCALL(ret, ioctl(phys_fd, TIOCSETD, &bt_ldisc), "TIOCSETD");
        SYSCALL(bt_cfd, open("/dev/ttyBTC", O_RDWR, 0), "/dev/ttyBTC");
        SYSCALL(ret, ioctl(bt_cfd, BTINITSTACK), "BTINITSTACK");
        SYSCALL(ret, ioctl(bt_cfd, HCIWRITESCANENABLE, &wrscan),
                "HCIWRITESCANENABLE");
#ifdef CLIENT
        SYSCALL(ret, ioctl(bt_cfd, BTCONNECT, &con), "BTCONNECT");
#endif
        for(;;) sleep(10);
}
```

I'll explain most of the things this application is doing in the next section, "Connecting to a Bluetooth Device," but first I'll show you how to use the application.

I defined the SYSCALL macro so that I could show a real example of checking system call returns while conserving space in the text. It does a primitive form of exception handling (if you can call exiting the application exception handling) that shows the user what the error is.

The *tty_init* routine is based on the *fd_setup* routine in btd.c. It sets up the serial port TTY to work in raw mode, sets the baud rate, hardware flow control, and so on.

Note that this program has the server device's BD ADDR hard-coded into the declaration of the *bt_connection* struct! Yours will differ, so change this before trying it. A real-world application wouldn't do this, of course.

To build the program, put the following Makefile in the same directory:

```
bt_mod_inc_dir := /home/gmcnutt/OpenBT/linux/include
OCFLAGS += -g -MD -I$(bt_mod_inc_dir) $(EXTRA_FLAGS)
```

Change the *bt_mod_inc_dir* variable to match the location where you installed the OpenBT source tree. Assuming you saved this file as *simple.c*, to make the server, type:

```
$ make simple
```

And to make the client, type:

```
$ make EXTRA_CFLAGS=-DCLIENT simple
```

First run the program on the server, and then run it on the client. Next, open new terminal windows on the server and client. On one, type:

```
$ cat /dev/ttyBT0
```

And on the other one, type:

```
$ echo hello > /dev/ttyBT0
```

You should see "hello" appear on the opposite side. Any program that works over a character device or a TTY should work over this connection. Go ahead and try some others. Try catting a binary file, too, just to see why we need to make TTYs raw before we can safely transmit binary data.

Connecting to a Bluetooth Device

At this point, you're probably impatient to start writing some code. I know I would be. In fact, if you're like me, this is probably the first section you jumped to. In this section, I'll give you some examples to start with and talk through some of the issues. I'll show you how to get the stack up and talking to the hardware, how to discover other Bluetooth devices, and how to find and connect to applications on those devices.

For all of these examples, I used the following setup:

- The OpenBT Bluetooth driver version 0.0.2
- Ericsson Bluetooth development h/w, ROK 101, firmware revision P9A
- RS-232 connection between the host and the Ericsson card
- Red Hat 6.2
- Linux 2.2.18 kernel

In the rest of this section, we'll see how to initialize the stack, look for remote devices, do SDP queries and initiate and shut down connections. I'll also show an example of adding a new service to the XML database.

Initializing the Bluetooth Stack

Figure 6.1 illustrated what your system is like after you load the Bluetooth module and connect the serial cable between the host and the card. At this point, the Bluetooth driver and the serial port driver are both registered as TTY drivers in the kernel, but both are idle. Both are using the default N_TTY line discipline and standard termios settings. The Bluetooth line discipline is registered in the kernel, but nothing is using it. No data is moving between the host and the card.

The Bluetooth driver must use the serial driver to talk to the card. In order to do this, we need to "hook up" the Bluetooth driver on top of the serial driver so that when it sends data, it sends it through the serial driver to the serial port; and when the serial driver receives data from the serial port, it pushes it up to the Bluetooth driver.

We also must change the default settings of the serial driver. For one thing, the default settings are not compatible with binary data. That's because TTYs are commonly used for things which require some control character processing, like consoles. That won't work for us because this processing might change, replace, or insert certain values in the data passing through the TTY. We just want the TTY to pass the data exactly as we tell it to.

Also, the default baud rate for serial ports is typically 9600. But the Ericsson Bluetooth Developer's kit will expect us to talk to it at 57600—at least until we can tell it to switch to a different baud rate. This default baud rate is vendor-specific. Unfortunately, it is not part of the HCI UART spec.

Of course, if you're using USB instead of serial, then you don't have to worry about any of this. The USB Bluetooth driver provides a TTY interface, but the baud rate is meaningless.

Preparing the Serial Driver

The following example shows how to open the serial port and make it a raw TTY. When it's raw, that means it won't mess with our data as it moves between the Bluetooth driver and the serial port. If you don't make it raw, it will try to filter the data stream looking for special characters. If you think *this* is confusing, just try using **cat** on /dev/ttyS0. It works great… for text files. Try it with a binary and you'll probably hose your terminal settings. But we can fix this by using a raw TTY. The following code shows how to do this:

```
int fd;
struct termios t;
/* open the device for reading and writing */
fd = open("/dev/ttyS0", O_RDWR, 0);
/* get a copy of the driver settings */
ioctl(fd, TCGETS, &t);
/* raw mode settings */
cfmakeraw(&t);
/* set the baud rate to 57600 baud, 8 data bits,
   1 stop bit */
t.c_cflag &= ~CBAUD;
t.c_cflag |= B57600 | CS8;
t.c_cflag |= CLOCAL;
t.c_oflag = t.c_lflag = 0;
/* hardware flow control */
t.c_cflag &= ~CRTSCTS;
/* put the setting into effect */
ioctl(fd, TCSETS, &t);
```

Whether or not you need hardware flow control depends on the Bluetooth hardware you're using. Some products are okay with it, while some specifically tell you not to use it. The Ericsson hardware seems to work okay either way. Note that many embedded devices have custom UART hardware. Sometimes these don't support the hardware lines necessary for hardware flow control. If you have trouble getting the Bluetooth driver to talk to the card, then find out whether or not this setting is correct for your hardware.

Observant readers will wonder if we need to fix the termios setting for the Bluetooth driver itself. After all, it's a TTY driver. Won't we have the same problem with binary data? Yes—once we start trying to read or write from it. But that's fine at this point. It won't affect any of the *ioctl* calls we'll be doing. Later, when we want to transfer binary data, we'll address this. If we just set the driver up for another application like PPP, then that application should be responsible for dealing with this (PPP does).

Stacking the Drivers

Now that the serial driver is ready, we can connect it to the Bluetooth driver. Remember that the Bluetooth stack registered its own line discipline with the kernel when we loaded the module. The way we stack the drivers is by telling

the serial port to switch from using the N_TTY line discipline to the Bluetooth
line discipline. That way, when the serial driver receives data, it will push it up
into the Bluetooth stack, and when the Bluetooth stack wants to send data, it has
a handle to the serial driver.

```
/* hookup serial driver and Bluetooth driver */
int bt_ldisc = N_BT;
ioctl(fd, TIOCSETD, &bt_ldisc);
```

The N_BT constant uniquely identifies the Bluetooth line discipline among
all other line disciplines registered in the kernel. This identifier is what tells the
serial TTY to use the Bluetooth stack as its upper layer interface. It's defined in
btcommon.h—part of the OpenBT source tree.

The TIOCSETD *ioctl* replaces the serial port's current line discipline with the
one specified. It also causes the Bluetooth line discipline's *open()* routine to be
called, passing in the serial port's TTY. This gives the Bluetooth stack a handle to
the serial TTY driver so it can use it as the lower layer. At this point, Figure 6.2
shows our driver configuration in the kernel.

Starting Communication between the PC and the Card

Once the drivers are stacked, the host can start talking to the hardware. There are
some specific things the Bluetooth stack needs to find out from the card before it
does anything else. It also needs to do some internal initialization as well.

```
/* open the bt control channel */
bt_cfd = open("/dev/ttyBTC", O_RDWR, 0);
/* initialize the stack */
        ioctl(bt_cfd, BTINITSTACK);
```

If you're going to initialize the stack, you have to use the /dev/ttyBTC device
(the control device). The Bluetooth data devices (for example, /dev/ttyBT0) won't
work. In fact, you can't even open these other devices until the stack is initialized.
This, and the fact that multiple processes can open /dev/ttyBTC at the same time,
makes it unique. Note that closing /dev/ttyBTC is safe. The stack will remain initialized. To shut it down, we'll use the BTSHUTDOWN *ioctl*. You'll learn more
on that in the section "Disconnecting" later in the chapter.

The BTINITSTACK *ioctl* tells the Bluetooth driver to initialize itself and begin
talking to the Bluetooth hardware. It will query the hardware for things like buffer
sizes and numbers, read the local BD_ADDR, and so forth. As an application
writer, you don't really need to worry about the details. There is one thing you

should know, however: this *ioctl* call can return before initialization is complete. For this reason, it's sometimes a good idea to pause your application before continuing.

Debugging…

Detecting UART Overruns

A common problem people have (especially on embedded devices) is UART overruns. A UART overrun is what happens when data is coming in on the serial port too fast for the serial driver to read it. Embedded devices with slow CPUs, bad IRQ latency, and/or cheap UART hardware sometimes see this problem.

```
$ cat /proc/tty/driver/serial
```

The preceding command will show you if your UART is getting receive overruns. If an "oe" field appears in the report, then this gives a count of the number of UART overruns detected by the serial driver. If you are having problems with data corruption, then definitely check for this.

Switching to a Higher Baud Rate

If we want the Bluetooth driver and the hardware to use a higher baud rate we can tell it to do so now. At 57600 baud, the bottleneck will be the serial connection between the host and the card. This doesn't mean we'll lose data. We just won't be taking full advantage of what the radio can do. If we jack it up to 115200 baud, then we're more in line with the maximum radio data rate of 723.2 Kbps, which is already pretty slow compared to currently extant wired media. Keep in mind that this only affects the baud rate between the host and the Bluetooth hardware. In other words, we're not changing the radio characteristics of the card in any way.

NOTE

Keep in mind that if you change the baud rate from the power on default, if you ever shut down the stack, you'll need to physically reset the hardware before starting it up again. Both the stack and the hardware have to start up at the same baud rate or they won't talk to each other.

```
/* tell the card to switch baud rates */
int final_baud_rate = 115200;
if (ioctl(bt_cfd, HCISETBAUDRATE,
          &final_baud_rate) == 0) {
      /* switch the serial port baud rate */
      struct termios t;
      ioctl(fd, TCGETS, &t);
      t.c_cflag &= ~CBAUD;
      t.c_cflag |= B115200;
      ioctl(fd, TCSETS, &t);
}
```

The HCISETBAUDRATE *ioctl* will try to send a vendor-specific command to tell the hardware to change the baud rate. Keep in mind that the command to switch baud rates is vendor-specific. Some vendors might not provide this feature. This is an example of why it's important for your application to check the return results of system calls. In this case, if the *ioctl* call fails, then presumably the card won't change its baud rate. This could be because it has a fixed baud rate, or because it uses a different vendor-specific command, either way we'd better just leave the serial port baud rate alone or the Bluetooth driver will lose communication with the card.

Developing & Deploying…

Avoiding Race Conditions When Changing Baud Rates

Incidentally, there's something of a race condition here between when the card switches baud rates and when the serial port switches baud rates. What happens if the card sends us data at the higher baud rate before we manage to change the serial port settings? If this happens, it is usually not fatal, but it's essential to change the serial port immediately after changing the card's baud rate. You should also stop any data streams before changing baud rates.

Finding Neighboring Devices

Now that the Bluetooth driver is talking to the hardware we can engage in some Bluetooth traffic. Of course, we'll need somebody to talk to. In order to find other Bluetooth devices in range, we'll do an HCI Inquiry. Also, we probably want to let other devices find us, too, so we'll see how to tell the hardware to respond to other device's Inquiries.

Letting Other Bluetooth Devices Discover Us

By default, the Ericsson Bluetooth Development Kit hardware doesn't respond to other device's inquiries. This is okay, because we don't really want other people trying to connect with us until we're ready. The following example shows how to enable both scan and inquiry responses:

```
/* enable page scan & inquiry scan */
#define PAGE_SCAN_ENABLE     0x01
#define INQUIRY_SCAN_ENABLE 0x02
int wrscan = (PAGE_SCAN_ENABLE |
              INQUIRY_SCAN_ENABLE);
ioctl(bt_cfd, HCIWRITESCANENABLE, &wrscan);
```

The HCIWRITESCANENABLE *ioctl* takes a bit mask parameter. Only the first two bits have meaning. Bit 0 corresponds to Page Scan, and bit 1 corresponds to Inquiry Scan. You set the bit to enable the corresponding scan type. To find out more about Page Scan and Inquiry Scan, consult the Bluetooth Core Specification. For now, just realize that other devices won't see you if you don't turn on *scan enable*.

Sending an HCI Inquiry

To find other neighboring devices use the HCIINQUIRY *ioctl*. This *ioctl* takes a parameter of type *inquiry_results*, which serves both as an in-param and an out-param. The *btcommon.h* header defines this structure.

```
typedef struct inquiry_results {
        u32 nbr_of_units;
        u32 inq_time;
        u8 bd_addr[0];
}
```

The *nbr_of_units* field specifies the maximum number of responses, which the hardware should listen for before ending the Inquiry procedure. The valid range for this value is 0 through 255. But 0 means an unlimited number of responses! Not a good idea since you've only allocated a finite amount of space in which to receive responses.

The *inq_time* field specifies the time, in units of 1.28 seconds, which the hardware should allow for the Inquiry to finish. The hardware will terminate the Inquiry procedure if either it receives the maximum number of responses, or the said amount of time expires—whichever comes first. The valid range for this value is 0x01–0x30, or 1.28–61.44 seconds.

The *bd_addr* field marks the start of a block of memory set aside for the Inquiry responses. By default, there isn't any space for responses. One way to make space is to allocate enough memory for the inquiry_results structure, plus some extra for the responses. It turns out that the driver will only store the BD ADDR from each response, so you'll need to set aside 6 bytes per response. One way to do this is to wrap it with your own structure that has a static buffer, like this:

```
typedef struct my_inq_result {
        inquiry_results hdr;
        unsigned char buf[MAX_RESPONSES * 6];
} my_inq_result_t;
/* issue the inquiry and block */
my_inq_result_t inq;
inq.hdr.inq_time = 5;
inq.hdr.nbr_of_units = MAX_RESPONSES;
ioctl(bt_cfd, HCIINQUIRY, &inq);
/* parse the results */
for (i = 0; i < inq.hdr.nbr_of_units; i++) {
        unsigned char *bd_addr = inq.buf + i * 6;
        printf("%x:%x:%x:%x:%x:%x\n", bd_addr[0],
                bd_addr[1], bd_addr[2], bd_addr[3],
                bd_addr[4], bd_addr[5]);
}
```

The inquiry response actually carries extra information, like the class of device responding. This information is not passed up the stack at the moment, but it's worth being aware that it's there, as in the future the driver may change to store more information. If that happens, of course, more memory would have to be allocated for responses.

The *ioctl* call will block until either the Inquiry completes, or an error occurs. Possible errors include timeouts waiting for the hardware to send the expected HCI commands. If the call is successful, then the *inq* argument contains information from any inquiry responses received. Note that the *ioctl* returns success even if no remote devices responded.

Upon successful return, the *nbr_of_units* field now indicates the actual number of responses received (this is less than or equal to the number you specified) and the *bd_addr* field contains the received BD ADDRs of remote devices.

Using Service Discovery

Once you've discovered another device, you're ready to find out what services it offers. Likewise, you may want other devices to discover the services your application provides. By now, you probably know that this is where the Service Discovery Protocol comes into play.

Let me reiterate some of the caveats regarding SDP on OpenBT:

■ You cannot dynamically register a new service in the SDP database.

■ Your application must know how to assemble SDP requests and parse SDP responses.

■ Services cannot register themselves with the RFCOMM layer.

With these limitations, you may well wonder what's the point of even discussing SDP. Well, there are some benefits:

■ You can statically add services to the SDP database, and for embedded developers this may work well enough.

■ Your client applications will know how to discover and connect to services running on a stack, which correctly supports RFCOMM registration.

■ OpenBT

In the rest of this section, we'll talk about how to connect to a remote SDP server, how to send requests, and how to process responses. This will cover the client side of things and should be useful even with the current state of the OpenBT stack. After that, we'll look at an example regarding how to add a service to the SDP database.

Connecting to a Remote SDP Server

Before you can do a query, you need to establish an SDP connection with the remote device. Anytime we need to establish a connection, we'll use the BTCONNECT *ioctl* call. This call takes a parameter of type *bt_connection*. The *btcommon.h* header defines this structure.

```
struct bt_connection {
        u8 bd[6];
```

```
            u32 id;
};
```

The *bd* field is the BD ADDR of the remote device you want to connect to. For instance, you can use one of the BD ADDRs discovered in your inquiry.

L2CAP uses a Protocol Service Multiplexor field (PSM) to uniquely identify an instance of a higher layer protocol using an L2CAP connection. For some protocols, this value is well-known (i.e., in the Core Specification), and for others you have to discover it. The Bluetooth Core Specification defines the PSM for SDP to be 1.

The *id* field is a combination of the PSM for the protocol instance you want to connect to: the line number and the SDP ID. The high 16 bits of the *id* field indicate the PSM. The next 8 bits of the *id* field specify the line or session number. Remember the session state machine in Figure 6.2? This value identifies one of those sessions. It also maps to the minor number of a Bluetooth TTY (/dev/ttyBT0, and so on). When we specify a line here, we're telling the Bluetooth driver to use the session associated with one of the Bluetooth TTYs.

The lowest 8 bits represent the SDP connection ID. For the BTCONNECT call, these are not important. Later, when we look at the BT_SDP_REQUEST *ioctl*, we will see how these bits are used.

To make things easier on yourself, you should include the *sdp.h* header so you can use the CREATE_SDP_ID macro. This macro automatically fills in the PSM. The following example shows its usage:

```
/* set remote BD ADDR from the inquiry results */
bt_connection con;
memcpy(con.bd, inq.hdr.bd_addr, 6);
con.id = CREATE_SDP_ID(SDP_LINE, 0);
sdp_con_id = ioctl(bt_fd, BTCONNECT, &con);
```

The BTCONNECT *ioctl* blocks until the connection completes or an error occurs. It returns an SDP connection ID on success. This is a little out of the ordinary for a system call, which should normally return 0 on success!

Sending an SDP Request

After a successful BTCONNECT call, we can start sending SDP requests to a remote device. We'll send SDP requests (and receive responses) by using the BT_SDP_REQUEST *ioctl*. This call takes a parameter of type *bt_sdp_request*. The header *btcommon.h* defines this structure.

```
typedef struct bt_sdp_request {
    u32 conID;
    u8 sdpCommand;
    u8 pduPayload[256];
    int pduLength;
    u8 requestResponse[256];
    int responseLength;
    } bt_sdp_request;
```

Developing & Deploying…

Picking an SDP Line Number

When you specify a line number for an SDP connection, you must specify the line number of a session that is in the closed/disconnected state. Unfortunately, there is no way for your application to know *a priori* which sessions are in this state. Until the OpenBT developers introduce a fix for this problem, your application will have to use a trial-and-error algorithm. If a BTCONNECT *ioctl* fails, this means the session state is not suitable for SDP, and your application can try another one. This problem is not specific to the Bluetooth stack—it applies to any device file.

The *conID* field has the same format as the *id* field of the *bt_connect* structure. Again, we'll use the CREATE_SDP_ID macro, but this time, when we pass in the SDP index, it will be the value returned by the BTCONNECT *ioctl*.

The *sdpCommand* field is the actual SDP command. For example, the *ServiceSearchRequest* command is 0x02. See the SDP chapter of the Bluetooth Core Specification for other commands.

The *pduPayload* field is a buffer where we have to put the raw SDP protocol, which comprises our request. The driver will build the SDP packet header for us, but we have to provide the payload of the request in this buffer. Unfortunately, nobody has provided a nice library to build these requests for us. Yet. You can consult the Core Specification or other references to learn more about constructing your own payloads. But one thing you need to note: the SDP specification defines multibyte fields to be "big endian." So, when you define these fields in your payload, you need to put the high bytes first.

The *pduLength* field indicates the number of bytes in our payload buffer. Note that we're limited to 256 bytes.

The *requestResponse* field is a buffer where we'll find the response to our request when the *ioctl* call returns (assuming we received a response).

The *responseLength* field tells us how many bytes we received in our response when the *ioctl* call returns. If this is zero, then it's safe to assume we didn't get the response.

Let's look at an example of a service search request for our custom echo service:

```
bt_sdp_request sdp_req;
int i = 0;
memset(&sdp_req, 0, sizeof(sdp_req));
sdp_req.conID = CREATE_SDP_ID(0, 0);
sdp_req.Command = 0x02; /* service search req */
sdp_req.pduPayload[i++] = 0x35; /* des hdr */
sdp_req.pduPayload[i++] = 0x03; /* des sz */
sdp_req.pduPayload[i++] = 0x19; /* uuid hdr */
sdp_req.pduPayload[i++] = 0x13; /* uuid[1] */
sdp_req.pduPayload[i++] = 0x02; /* uuid[0] */
sdp_req.pduPayload[i++] = 0x00; /* count[1] */
sdp_req.pduPayload[i++] = 0x03; /* count[0] */
sdp_req.pduPayload[i++] = 0x00; /* continuation */
sdp_req.pduLength = i;
ioctl(bt_fd, BT_SDP_REQUEST, &sdp_req);
```

Remember my warning about multibyte fields and endianness? Look at the Service Class UUID field in our example. We put the high byte before the low byte in our buffer. Likewise for the *MaxServiceRecordCount* field. Sometimes developers are tempted to define structs, which correspond to protocol packets so that they can fill out the struct and then copy it to the buffer (or cast the buffer to a struct of that type). Beware of doing this! If your application is running on a little-endian processor, then this will not work correctly for SDP. You will get the bytes backwards. The ugly but reliable technique in the previous example will work correctly regardless of the endianness of your host processor. Another alternative is to define or use existing macros that do safe byte-swapping conversions.

Processing an SDP Response

The BT_SDP_REQUEST *ioctl* call will block while the Bluetooth driver sends the request and waits for the response. If the *ioctl* succeeded, then the response will appear in the *bt_sdp_request* struct, which you passed in.

The *responseLength* field tells you how many bytes are in the *requestResponse* buffer. If this field is zero, then the Bluetooth driver did not receive any response before timing out.

The first byte of a well-formed response indicates the SDP status of the response. Zero means success; non-zero indicates an SDP error. Consult the SDP spec if you want your application to decode the error type. Remember: the *ioctl* call can succeed even when the SDP request fails.

```
/* any response? */
if (sdp_req.responseLength == 0) {
        printf("SDP response length zero\n");
        exit(0);
}
/* was it an error? */
if (sdp_req.requestResponse[0] == 0x01) {
        printf("SDP Error Code 0x%x\n",
                sdp_req.requestResponse[5] << 8 |
                sdp_req.requestResponse[6]);
        exit(0);
}
/* any matching service records? */
if (!sdp_req.requestResponse[8]) {
        printf("No remote service!\n");
        exit(0);
}
/* get the first service handle */
server_hdl = sdp_req.requestResponse[9]  << 24 |
                sdp_req.requestResponse[10] << 16 |
                sdp_req.requestResponse[11] << 8  |
                sdp_req.requestResponse[12];
```

If the number of ServiceRecords is zero, then the remote device does not support the service we were looking for. Otherwise, using the service handle, we can send more SDP requests to fetch back attributes of the matching ServiceRecords. The ultimate goal is to establish a connection, so we should send an AttributeRequest for the ProtocolDescriptorList next and parse the RFCOMM server channel out of the response. The purpose of this chapter is not to teach you how to parse SDP, so I'll leave that as an exercise for the reader.

When your application is finished making requests, it should close the SDP connection by using the BTDISCONNECT *ioctl* call. That way, the remote server can free up any resources it has committed to servicing your connection. However, the current release of OpenBT appears to have a bug in it such that BTDISCONNECT does not work for SDP connections.

Adding a Service to the Local Database

The SDP service database is an XML file. Remember that we can use the sdp_server daemon to handle SDP queries from remote devices to our local database. To add a service, we edit an XML file and pass it as an argument when we start the sdp_server daemon.

Example: Adding an Echo Service

Here's an example of adding an echo service. It uses RFCOMM over L2CAP as its protocol stack. We place it within the <bluetoothSDP></bluetoothSDP> tags of the XML file:

```
<EchoServerServiceClassID ServiceRecordHandle = "0x0111ffff">
    <ServiceRecordHandle Parameter0 = "0x0a0111ffff">
    </ServiceRecordHandle>
    <ServiceClassIDList NbrOfEntities = "1">
      <EchoServerServiceClassID>
      </EchoServerServiceClassID>
    </ServiceClassIDList>
    <ProtocolDescriptorList NbrOfEntities = "2">
      <L2CAP type = "DES" Parameter0 = "0x0003">
      </L2CAP>
      <RFCOMM type = "DES" Parameter0 = "0x0802">
      </RFCOMM>
    </ProtocolDescriptorList>
    <BrowseGroupList NbrOfEntities = "1">
      <PublicBrowseGroup>
      </PublicBrowseGroup>
    </BrowseGroupList>
    <BluetoothProfileDescriptorList NbrOfEntities = "1">
      <EchoServerServiceClassID type = "DES" Parameter0 = "0x090100">
      </EchoServerServiceClassID>
```

```
    </BluetoothProfileDescriptorList>
    <ServiceName>Echo Server</ServiceName>
    <ServiceDescription>Echo Server</ServiceDescription>
    <ServiceAvailability Parameter0 = "0x0815">
    </ServiceAvailability>
  </EchoServerServiceClassID>
```

In the <ServiceClasses> tag, add this:

```
EchoServerServiceClassID = "0x1302"
```

I pulled the EchoServerServiceClassID out of thin air (there is no echo server in the Bluetooth specification), so for all I know it conflicts with an existing class ID! Just another reason why OpenBT needs an SDP interface before armies of irresponsible hackers like myself start filling the world with pirate IDs. I did make sure that the ServiceRecordHandle didn't conflict with any of the other ones in the file, however.

The "Bluetooth assigned numbers" part of the Bluetooth specification lists the numbers that have been allocated. You can use Universally Unique IDs (UUIDs) to safely allocate your own numbers.

Querying the Local Database

Currently there is no interface to query the local SDP database from within your application. If you want to do this, then you can look at how the sdp_server code invokes the XML parser and processes queries from remote devices.

Connecting to a Bluetooth Service

Usually the purpose of making SDP requests is to discover if a remote device supports a particular service, and if so, what the pertinent connection parameters are. Once this discovery phase is over, your application needs to connect to the actual service. Connecting involves two steps: opening a data device and connecting its associated line.

Using a Data Device

So far, all of the examples have used /dev/ttyBTC as the device. Once we're ready to actually begin transferring data across a session, we'll need to open one of the data TTYs. Recall from our session state machine that a session must be in the opened/normal/connected state to transfer data. If you look back at Figure 6.2, you'll see that it really doesn't matter whether we establish the RFCOMM connection first or open the TTY first.

Opening a data device is trivial, but here's the code in case you have any doubts about how to do it:

```
int bt_fd = open("/dev/ttyBT0", O_RDWR);
```

On success, the device is all yours. If the *open* fails and *errno* is EBUSY, then some other process already has it. In this case, you can just keep trying the other devices (e.g., /dev/ttyBT1) until you find one that's available. Unfortunately, there isn't really a cleaner way to tell if a device is already being used.

If the *open* fails and *errno* is EPERM, then the stack is not initialized. In this case, you can open the control device and use the INITSTACK *ioctl* call (see earlier) to initialize it and then try again.

Creating a Connection

The SDP transactions give you the parameters you need to know to establish a connection to a remote service. And, in fact, you've already seen the command to establish a connection: the BTCONNECT *ioctl*. We used it to establish an SDP connection. But this time, you'll be connecting to a different protocol to access the service—which protocol depends on the particular service and what it's ProtocolDescriptorList indicated.

Here's an example of establishing an RFCOMM connection.

```
bt_connection con;
int server_channel;
/* do the SDP queries, assign 'server_channel'
   a value based on the results */
/* connect via RFCOMM */
memcpy(con.bd, inq.hdr.bd_addr, 6);
con.id = CREATE_RFCOMM_ID(line, server_channel);
sdp_con_id = ioctl(bt_fd, BTCONNECT, &con);
```

The CREATE_RFCOMM_ID macro is similar to the CREATE_SDP_ID macro. You can find it in the *rfcomm.h* header.

The *line* parameter should match the minor number of the TTY you intend to use for data transfers.

The *server_channel* parameter should match the value obtained from the ProtocolDescriptorList you get during the SDP session. See the SDP chapter of the Bluetooth Core Specification for an explanation.

Accepting a Connection

Remember the caveat about not being able to register services with RFCOMM? Well, that makes accepting a connection random luck. It could be done better, and maybe in the future it will be, so I'll start by explaining how I believe connection acceptance should work. At the moment, the protocol stack has many compromises, and you'll have to use it as is, so I'll go on to explain how connection acceptance works now.

Understanding the Way It Should Work

When you register a service with SDP, and you provide a parameter in the RFCOMM Protocol Descriptor, that parameter is supposed to identify the server channel your application will be listening on. The remote client gets this value and uses it to request a connection to your service. When the RFCOMM driver sees a connection come in on that channel number, it should make sure that the correct server application gets it.

Understanding the Way It Does Work

The problem with the OpenBT stack is that there is no way for the RFCOMM driver to map a server channel to a session on the side receiving the connection request (everything works fine on the side initiating the connection—it just associates the connection with the session indexed by the line number).

Instead, when the RFCOMM driver gets a connection request it looks for the first available TTY, starting with minor number 0, and associates the connection with that session.

This is why btd works. It doesn't really matter which server channel the client requests as long as it is a legal value (even numbers 2 through 60). The first connection on the server side will go to the session for ttyBT0—which is what btd, by default, passes to PPP when it spawns it.

In other words, the only way to make sure the correct server accepts the connection is to carefully control the order in which connections are made. For a shipping product with more than one server application, this would be totally unacceptable. On the other hand, the client side works fine. So, if a product is shipping with only client applications, then this problem won't be an issue.

Transferring Data

Since the Bluetooth driver is just another TTY driver, transferring data is as simple as reading and writing from a file or any other device. You can find any number of books discussing I/O in C for Unix clones, so I'll just provide an example showing an *echo* application.

Don't forget that Bluetooth devices are TTY devices and by default they are not raw. Remember how we had to set up the serial device so that it wouldn't interfere with a binary data stream? The same thing applies to the Bluetooth data devices. If your application is going to use read and write calls on a Bluetooth device to transfer binary data, then follow the earlier examples used on the serial device to make it raw.

```
/* declare a buffer to fetch & hold the data */
char buf[BUF_SZ];
/* while we can read more data… */
while ((n = read(data_fd, buf, sizeof(buf))) > 0) {
        /* echo the data back out the same
          channel */
        write(data_fd, buf, n);
}
```

This loop will read and echo data from our RFCOMM channel as long as it remains open. The call to read will block until data becomes available, the channel closes, or an error occurs. If, and only if, some data becomes available, then read will copy as much as it has or will fit into the buffer and return the number of bytes it put in the buffer. If the channel closes, read returns 0. If an error occurs, read returns a negative error number.

The write will queue up the data for transmission. Its semantics are similar to read. Note that this is not a perfectly reliable echo routine since it just assumes that all the bytes went out okay, but it shows the basics of I/O.

Disconnecting

Disconnecting always takes two steps: a Bluetooth disconnect and a system call to close. At most, only one side of the connection needs to execute a disconnect, and in cases where two devices go out of range, the Bluetooth stack cleans up the connection automatically. But your application will always need to do a close after a disconnection occurs. Refer to Figure 6.3 to see the state machine.

If your client application succeeds in making a connection, then it's important to disconnect before exiting. If you don't, then the Bluetooth driver won't let anyone else use the line associated with the connection until someone reinitializes the stack with a BTSHUTDOWN or BTINIT *ioctl* call. Note that the Bluetooth driver will not automatically disconnect a line if the application closes the file descriptor or exits. You have to explicitly tell it to disconnect.

You close a connection with the BTDISCONNECT *ioctl* call. This call takes a parameter of type *bt_connect*. If you like, you can use the same one you passed in to the BTCONNECT *ioctl*.

```
ioctl(bt_fd, BTDISCONNECT, &con);
```

Even after doing a BTDISCONNECT, no other process can use the line associated with your device file until your application either explicitly calls *close* or exits. So, if you disconnect the line but don't close the file descriptor, other applications will get EBUSY if they try to open that device file.

An application can always tell when the session disconnects from below. An RFCOMM link can disconnect if it or any layer below it disconnects, or if the remote peer goes out of range. In all these cases, the Bluetooth driver will do a hang-up on the upper TTY. This means that any time your application does a *select*, *read*, or *write* on the file descriptor, these system calls will return a negative value. If it is blocked on one of these calls, it will return immediately.

When this happens, your device file descriptor is pretty much out of commission. You won't be able to do anything else with it until you close and reopen it. In this case, there's no need to do a BTDISCONNECT *ioctl* call. It will just return an error since the connection doesn't exist any more.

To summarize, when an application wants to end a session, it should call BTDISCONNECT followed by *close*. If an application detects a disconnection during a session, it should only call *close*.

Controlling a Bluetooth Device

The following list covers everything a Bluetooth application can do:

- Transferring data
- Establishing connections
- Controlling Bluetooth features

Not all applications will do all three things. For example, PPP transfers data over an RFCOMM TTY, but it knows nothing about establishing the connection it uses. In the previous section, we covered the first two items on this list. In this section, we'll talk about controlling features of the Bluetooth device itself. We'll see the differences between applications that use the stack and applications that control the stack, we'll learn what things an application can control, and we'll cover the basic scenarios that a controlling application must be able to deal with.

Distinguishing between Control and Data Applications

PPP uses the Bluetooth stack without knowing it. It requires a TTY interface. It relies on another application to set up the connection for it. For example, we saw how to use the btd application to set up the connection and then spawn PPP. Of the three items on our list, an application can do any combination of one or more of those things by itself, and cooperate with other applications to provide any capabilities it doesn't do.

We already saw that the OpenBT project does not come with a stack manager. The btd application provides some features of a stack manager, but you'll probably need to either extend it or write your own application that gives you a broader set of features.

In this section, let's talk about designing our own hypothetical stack manager. On a desktop PC, this application might provide an interface for the user to monitor and control the Bluetooth device. In an embedded device, this application may provide hooks for other applications like power management, or a control panel driver to affect the Bluetooth driver.

Using *ioctls* to Control the Device

The first thing we should consider is what exactly an application can monitor and control. As with any other device driver, an application uses *ioctl* calls to perform control of the Bluetooth driver. Some *ioctl* calls are strictly informational and provide a way to monitor certain parameters of the Bluetooth driver.

Table 6.4 provides a summary of the *ioctl* calls currently supported by the OpenBT Bluetooth driver. Although you should always program to an interface and not an implementation, this advice assumes that the interfaces are stable and well documented! Currently, the only documentation on these *ioctls* is the source code. You can find the implementation for all of these calls in the linux/drivers/char/bluetooth/bluetooth.c file in the OpenBT source tree. Some of these are *ioctls* we've already seen in previous sections. I include them here just to give you a complete reference.

Table 6.4 Summary of OpenBT *ioctls*

Name	Description
BT_SDP_REQUEST	Sends an SDP request and blocks (with no timeout) until the response returns.
BTCONNECT	Requests an SDP or RFCOMM connection with a remote device. Blocks until the connection operation completes or, in the case of RFCOMM, a timeout occurs.
BTDISCONNECT	Disconnects an existing RFCOMM connection. Blocks until the disconnect operation completes or a timeout occurs.
BTWAITFORCONNECTION	Checks if a connection exists on the specified line and, if not, blocks until one appears on that line. Does not return on stack shutdown.
BTWAITNEWCONNECTIONS	Blocks until a new connection appears on any line. Does not return on stack shutdown.
BTISLOWERCONNECTED	Checks if a connection exists on the specified line and returns the result in the out-parameter.
BTINITSTACK	Initializes the driver. If the driver is already initialized, it implicitly performs the equivalent of BTSHUTDOWN first.
BTSHUTDOWN	Shuts down the driver, disconnecting all active connections and hanging up their associated TTYs.
BTREADREMOTEBDADDR	Returns the BD ADDR of the last remote device to establish a link-level connection in the out-parameter.
BTISINITIATED	Checks if the driver has been initialized yet and returns the Boolean result in the out-parameter.

Continued

Table 6.4 (continued)

Name	Description
BTHWVENDOR	Returns a string describing the name of the hardware, which the stack was compiled to support. Warning: currently, this does not limit the size of the string being copied into the user's buffer.

HCIINQUIRY
HCILINKKEYREPLY
HCILINKKEYNEGATIVEREPLY
HCIPINCODEREPLY
HCIPINCODENEGATIVEREPLY
HCISWITCHROLE
HCISETLOCALNAME
HCIAUTHENTICATION_
   REQUESTED
HCISETCONNECTION_
   ENCRYPTION
HCIRESET
HCICREATE_NEW_UNIT_KEY
HCIREADSTOREDLINKKEY
HCIWRITESTOREDLINKKEY
HCIDELETESTOREDLINKKEY
HCIREADSCANENABLE
HCIWRITESCANENABLE
HCIWRITEPAGESCANACTIVITY
HCIWRITECLASSOFDEVICE
HCIREAD_AUTHENTICATION_
   ENABLE
HCIWRITE_AUTHENTICATION_
   ENABLE
HCIREAD_ENCRYPTION_MODE
HCIWRITE_ENCRYPTION_MODE
HCISET_EVENT_FILTER
HCIREADLOCALBDADDR
HCIENABLEDUT
HCISETBAUDRATE
HCIWRITEBDADDR
HCISENDRAWDATA
BTSETMSSWITCH
	These ioctls all provide access to the HCI Protocol commands. See the HCI chapter of the Bluetooth Core Specification for a description of what these commands are used for.
	
	If a command does not provide any status information back to the Host, it returns immediately.
	
	If a command expects a Command Complete event, it blocks until either the Host Controller sends this event or a timeout occurs.

Covering Basic Scenarios

Now that we know what our stack manager *can* do, what *should* it do? What features should it provide? Let's consider the bare minimum. You can always add more to fit your needs. One basic assumption of our design is that the stack manager is responsible for the parameters that affect the entire driver or the hardware. In other words, a bare-bones stack manager won't concern itself with establishing RFCOMM connections or transferring data.

As a bare minimum, the stack manager should initialize and shut down the stack at the proper times. It should detect link loss and cleanup if necessary. It would also be helpful if it kept tabs on remote devices coming in and out of the vicinity.

Example: Startup

In previous sections, we've seen examples of how to initialize the stack and to set it up over a lower TTY like the serial driver so that it can talk to hardware. These steps will always be necessary at some point. For an embedded solution, the Bluetooth hardware might be on board, interfacing with the CPU via a UART or some other bus. In these cases, you might have to provide your own TTY driver over a custom hardware interface. Remember, the Bluetooth driver relies on the ability to use a line discipline in order to communicate with the hardware driver. Only TTY driver's use line disciplines, so the hardware driver must be a TTY.

But when should your stack manager start up the driver? It depends on the application. You can start it automatically when the application runs, or you can wait for a command from a User Interface (UI), or a signal from another process, and so on.

Probably the simplest thing to do on an embedded device is to start the stack on system bootup. You can do this by having the *init* process automatically start your stack manager from /etc/rc.local or whatever startup script you use for your configuration.

Example: Link Loss

There really isn't any way for a central stack manager to detect a link loss. When a link with another device goes down, the Host Controller sends the host a sequence of disconnection event notices for each handle on the link. The Bluetooth driver processes these events by disconnecting all sessions on that link. Any processes using the TTYs for these sessions can detect a hang-up. But a central stack manager won't necessarily get any kind of notification if it's not using one of those TTYs.

Is this important? It could be if the stack manager kept local cached data about link status or peers. In that case, it would be nice to get notification so that it could clean the caches. But as it is, any active processes using the links for data will be notified. If a stack manager worked in the mode of establishing connections and then spawning applications to use them (this is how btd works with PPP), then it can determine when the process terminates on a hang-up using normal Linux process handling.

The following example illustrates this model.

```
for (;;) {

retry:

        if (!do_hci_inquiry()) goto retry;

        if (!do_sdp_request()) goto retry;

        if (!do_connect()) goto retry;

        if ((pid = fork()) == 0) {

                execvp(APP, APP, APPARGS);

        } else {

                wait(pid);

                do_disconnect();

        }

}
```

The *do_hci_inquiry()* function and its friends would do what their names imply (the previous section illustrated code for implementing these kinds of functions). Once a connection is ready, the stack manager spawns a child process to use the connected TTY, then it waits for the child to exit. When the child exits, the stack manager makes sure the session is disconnected and then repeats the process.

If the link goes down at any point prior to the connection being made, one of the functions will fail and we'll go back to try again. If the link goes down after the connection is made, the child process will exit when it detects the hung-up condition of the TTY (actually, this depends on the behavior of the child application, but most legacy applications that use TTYs will exit by default when they can't use the TTY anymore). The *do_disconnect* is benign if the connection was already severed, but it makes sure the connection is cleaned up in case the child exited for a reason other than a TTY hang-up.

Note that a stack manager could handle a whole set of child applications like this, where each application is kept in a structure associating it with the relevant info needed to do SDP queries for the services it likes.

Example: User-Initiated and Automated Shutdown

If your stack management application has a user interface, then it can give the user the option of starting up or shutting down the driver. Alternatively, it might provide a means for other processes (like a power management service) to initiate a shutdown or startup via an IPC (InterProcess Communication) mechanism.

This example shows how a stack manager might install a signal handler to shut down or start up the stack based on requests from other processes.

```
static int stack_init = 0, bt_cfd;
void handler(int sig)
{
        if (stack_init) {
                ioctl(bt_cfd, BTSHUTDOWN);
                stack_init = 0;
        } else {
                ioctl(bt_cfd, BTINITSTACK);
                stack_init = 1;
        }
}
int main(int argc, char **argv) {
        do_init_stack();
        stack_init = 1;
        signal(SIGUSR1, handler);
        for (;;) do_whatever();
}
```

This example assumes that if a user or another process wants to shut down the stack or bring it back up, then they will send the stack manager a SIGUSR1. Other forms of IPC might be more pertinent in different cases. The BTSHUT-DOWN and BTINITSTACK *ioctls* take care of all the gritty details, shutting down connections, hanging up TTYs, flushing buffers, and so on.

Example: Idle Operation

Stack management applications can keep tabs on what other remote devices are in the area by doing periodic inquiries and keeping the results cached locally. You could provide an API for other applications to access this cache so that they don't have to do their own inquiries. You could even keep a cache of remote SDP databases for devices in range.

This example shows how a stack manager might maintain a remote BD ADDR cache. You could extend this example to keep other information about remote devices in the local cache. It polls a local socket for requests from local processes to retrieve the cache. You extend this by providing a functional API to handle IPC with the stack manager daemon.

```
typedef char BD_ADDR[6];

BD_ADDR cache[MAX_ADDRS];

for (;;) {
        ioctl(bt_cfd, HCIINQUIRY, &inq);

        memcpy(cache, inq.buf, inq.hdr.nbr_of_units);

        do_listen_for_cache_requests_with_timeout();
}
```

This is just a simple example. It uses the HCIINQUIRY command (see previous sections) with one of our wrapper structs for the inquiry results. It also has a buffer for keeping the results of HCI inquiries. Every so often it executes an HCI inquiry request to see what remote units are in the vicinity and puts their BD ADDRs in the cache.

The *do_listen_for_cache_requests_with_timeout()* could implement any form of IPC you like to field requests from other processes for the latest inquiry results. Every once in a while the process stops listening for requests and refreshes the cache.

The usefulness of something like this depends on how many processes are potentially doing their own HCI inquiries. But you could extend the idea to cover more expensive operations like searching remote SDP databases. Also, since we won't automatically receive notice when another device modifies its SDP database, the process could periodically update its cache of another device's SDP database.

Summary

The publicly available Bluetooth stacks for Linux are limited in number. As of this writing, the only two released implementations are IBM's BlueDrekar and the OpenBT project. BlueDrekar has some nice features, looks pretty complete, and is freely available for download in binary form for x86 platforms running 2.2.x kernels. OpenBT is an open source project with support for most stack protocols and features and may work well enough for embedded devices. It has been ported to a variety of processors and can be cross-compiled, but it is still early in its development and not a fully-featured implementation. The focus of the discussion and examples is on OpenBT in this chapter because it is open source and may someday be a part of the standard Linux distribution as a stable implementation.

The OpenBT stack provides a loadable kernel module, which implements a TTY driver. It currently supports six data TTYs for RFCOMM connections and one control TTY for managing the driver. The driver internally manages RFCOMM connections with a session state machine. Applications use *ioctl* calls to establish the RFCOMM connection. Once an RFCOMM connection exists on a session, any application can use the TTY for that session, just like any other TTY device.

The OpenBT source tree comes with some applications that you can use as examples or starting points for derived works. The entire source is released under a modified form of the GPL, so if you create derived works that are used to implement Bluetooth operations, then these derived works will fall under the same license. The btd application provides a quick way to get network connections working over a Bluetooth link via PPP over RFCOMM. The sdp_server daemon will handle SDP requests from other devices.

Connecting to a Bluetooth device takes several steps. If your application functions as a stack manager, then it must first stack the Bluetooth driver over an underlying hardware TTY driver like serial or USB. Next, it must use a sequence of *ioctl* calls to initialize the stack, discover remote devices, and browse remote SDP databases to find services and connection parameters. Once an application has identified a remote service to connect to, it uses an *ioctl* call to establish an RFCOMM connection session. At that point, it or any other application may use the corresponding data TTY for data transfers. When the RFCOMM session disconnects, the driver performs a hang-up on the data TTY, thus signaling the end of the session.

Applications can do three things with the Bluetooth driver: transfer data, manage individual connections, and manage the overall driver. Not all applications need to do all three. Legacy applications (like PPP) that just use a TTY require another application to set up the connection and perform stack management for them. Developers may want to provide a stack management process for their system, which handles scenarios like link loss, system shutdown requests, and caching remote device data.

Solutions Fast Track

Assessing Linux Bluetooth Protocol Stacks

☑ The standard kernel source tree only recently accepted the Bluez Bluetooth stack, but it may not yet possess all the features some application developers require. It requires Linux 2.4.4 or greater.

☑ IBM's BlueDrekar is a nice-looking implementation distributed in binary form for x86 platforms running 2.2.x. Source is not freely available to the general public.

☑ The OpenBT project is a not-as-nice open source project that works for most things an embedded developer would want. Source is available and has been used on x86, ARM9, ARM7, MIPS, and PowerPCs.

Understanding the Linux Bluetooth Driver

☑ The OpenBT stack implements TTY drivers for RFCOMM, SDP, and stack control.

☑ The Bluetooth driver must be stacked over a lower-layer hardware driver that implements a TTY.

☑ Any legacy application that uses a TTY can use RFCOMM once another application sets up the underlying RFCOMM connection.

☑ SDP, connection setup, and stack control are accomplished with *ioctl* calls.

☑ No interface exists for SCO, or L2CAP, although *ioctls* are available to support most HCI commands.

Using Open Source Development Applications

☑ The OpenBT source tree comes with some applications: btd/btduser, sdp_server, and BluetoothPN.

☑ The difference between btd and btduser is that btd is meant to work with the kernel mode Bluetooth driver while btduser works with the user mode Bluetooth driver. Many people prefer btduser since it is less prone to lock up your system if things go badly. However, the OpenBT developers do not support it as well as btd.

☑ The sdp_server application provides you with an SDP database server daemon. Once you've installed the Bluetooth driver, you can start this daemon and it will automatically receive and respond to SDP queries from remote devices.

☑ This application provides a GUI that displays the SDP database on a remote device. It provides some examples of how to make SDP requests and process their results.

☑ The quickest, most useful way to establish and exploit a Bluetooth connection from Linux is to use the standard GNU network applications over PPP. And the easiest way to do that is with the btd application.

Connecting to a Bluetooth Device

☑ An application manager must set up the driver stack over the hardware TTY and initialize the Bluetooth driver. This can be any application; the OpenBT source tree does not provide a general stack manager.

☑ Client applications must obtain the Bluetooth Device address of the remote device and—for RFCOMM connections—the channel number of the remote service in order to establish a connection.

☑ Once a connection is established, any application can use the TTY associated with the connection for data transfer.

☑ The driver indicates a disconnection event with a hang-up of the associated TTY.

Controlling a Bluetooth Device

☑ Use *ioctl* calls to control the device and get information about device status.

☑ Use /proc/bt_status to get information about device status.

☑ A stack manager must be able to deal with link loss and system shutdown requests. It should provide an interface for users as well as other processes like power management to signal shutdown requests.

Frequently Asked Questions

The following Frequently Asked Questions, answered by the authors of this book, are designed to both measure your understanding of the concepts presented in this chapter and to assist you with real-life implementation of these concepts. To have your questions about this chapter answered by the author, browse to **www.syngress.com/solutions** and click on the **"Ask the Author"** form.

Q: Is the OpenBT stack really ready for prime time on an embedded Linux device?

A: It's the closest thing to it that has freely available source. You can ask IBM about licensing and distribution costs for BlueDrekar, but it's hard to beat the price/performance ratio of OpenBT. If you're faced with the prospect of leveraging OpenBT or developing your own Bluetooth stack… well, you know your project schedule better than I do!

Q: How can I get the latest source for OpenBT?

A: Go to the OpenBT Web site (www.sourceforge.net/projects/OpenBT) and look for the instructions on accessing the CVS repository. This will give you the very latest, bleeding-edge code. Occasionally new tarballs appear for download on this site as well. You might also want to subscribe to the mailing list to keep in touch with progress on this front.

Q: Can a Java application use the Linux Bluetooth stack?

A: Any language that provides some kind of access to the standard I/O system calls (*read*, *write*, and *ioctl*) can use the OpenBT.

Q: When I try to "insmod bt.o" I get an error about missing kernel symbols. What is this and how do I fix it?

A: This happens because the kernel which *bt.o* was compiled against does not match the kernel you are trying to load it into. When you build bt.o, make sure you provide the *INCLUDE_DIR=<path>* argument to make, indicating the path to your target kernel's include files. Also, if your kernel has symbol versioning configured, then make sure *linux/include/modversions.h* is being included in the build process.

Q: I just want to use L2CAP and HCI, not RFCOMM. Is there an interface I can use to access these layers?

A: Not with OpenBT. However, if you aren't limited to using a Linux kernel version earlier than 2.4.4 then Bluez is probably what you want. The Bluez Bluetooth stack has been distributed with kernel source since kernel version 2.4.6; the latest is available from bluez.sourceforge.net.

Chapter 7

Embedding Bluetooth Applications

Solutions in this chapter:

- **Understanding Embedded Systems**
- **Getting Started**
- **Running an Application under the Debugger**
- **Running an Application on BlueCore**
- **Using the BlueLab Libraries**
- **Deploying Applications**

- ☑ **Summary**
- ☑ **Solutions Fast Track**
- ☑ **Frequently Asked Questions**

Introduction

Bluetooth wireless technology is proving popular for handheld mobile devices such as mobile phones and headsets, which have very limited space and power. Using an extra host processor to run applications takes up extra space, uses extra power, and adds cost, too. For the ultimate in compact design, low cost, and energy efficiency Bluetooth applications can be run directly on the same processor that drives the Bluetooth baseband.

Vendors who supply designs for Bluetooth Application-Specific Integrated Circuits (ASICs) also provide interfaces which allow custom applications to run on the same microprocessor which drives the Bluetooth baseband. It is also possible to run applications on commercially available chips. This chapter looks at embedded applications using as an example Cambridge Silicon Radio (CSR)'s BlueLab system for programming embedded applications on BlueCore chips.

Not every application is suitable for embedding on a BlueCore chip. Small simple applications such as the Headset and Audio Gateway profiles, as well as things like central heating controllers or TV remote controllers, are suitable for embedding on a single chip. High-bandwidth or complex applications such as a local area network (LAN) access point are better suited to implementation using a separate host processor.

This is because when an application is running on the same chip as a Bluetooth protocol stack, the application and firmware stack must share the available RAM on the chip. For a single channel RFCOMM-based application, the available RAM is several hundred words. The application code and its constant data must fit into just under 32K words.

Embedded applications running on a BlueCore chip are run under an interpreter called the Virtual Machine (VM). Interpreting application opcodes confers a significant performance penalty which limits suitable applications. For devices such as headsets, most of the time all that is happening is audio input/output (I/O). Control operations are comparatively infrequent, and involve simpler operations than would happen on devices such as LAN access points.

In this chapter, we'll look at some of the implications of these limitations and give some examples of how much can still be done in embedded applications. We'll take you through how to build applications which can be run on BlueCore, and explain how to build run and debug both on PCs and on the BlueCore chip itself.

What you need to know before reading this chapter is:

- The C programming language
- The basics of embedded programming: tasks and message queues

Understanding Embedded Systems

This section assumes that you've done some programming, but you don't have embedded experience. If you've worked with embedded systems before, you might want to skip straight to the "Getting Started" section. For the rest of you, we'll go over tasks, queues, stacks, interrupts, and the difference between running code on a PC and code embedded on hardware.

Understanding Tasks, Timers, and Schedulers

In a Bluetooth system, there are many different tasks to take care of: Link Management messages must be processed; incoming data must be dealt with as it arrives; outgoing data has to be sent to the baseband and radio; if there is a separate host communications through the host controller interface this must be addressed; all this and more must be handled simultaneously. Having a microprocessor for each task would be far too expensive, so the solution is for one microprocessor to swap between tasks, spending a little time on each in turn. This is called *multi-tasking*.

Each task has its own call stack, its own I/O queues, and each task gets a turn at the processor. There is one task which coordinates the rest. This is usually called a *kernel*, but is also referred to as a *scheduler*. Different embedded systems handle swapping between tasks in different ways, some assign priorities to tasks, so that a low-priority task does not stop a high-priority task from running. The BlueCore01 system has a simple round-robin scheduler, which runs each task in turn.

The scheduler stops running a task when the task blocks. A task blocks by making a system call which waits for an event. This behavior means that the scheduler is vulnerable to a task putting itself into an infinite loop. Since the task would never block, no other task would ever get a chance to run.

On the face of it, this means you could disable the whole Bluetooth system if your application didn't block often enough. Since there are many time-critical operations within the Bluetooth stack, you could easily stop the stack from working properly. To solve this problem BlueCore provides an environment called the Virtual Machine which protects the stack code from applications which try to take too much processor time. Instead of your application code being called directly, the scheduler calls the Virtual Machine. The Virtual Machine then runs a number of operations through its interpreter, and afterward blocks so the scheduler can call another task. It doesn't matter if your code is in an infinite loop, the Virtual Machine will still only run a preset number of your application's instructions, so your endless loop can't run endlessly!

The processor time used by other tasks in the system will vary. For instance, when the Link Manager task is in the middle of negotiating link configuration, it will require more processor time than when no Link Management messages are being received. This means that the time between calls to the Virtual Machine will vary. The impact on your application is that BlueCore does not provide Real Time Operating System (RTOS) capabilities because it makes no guarantees regarding how often it will call your application code.

Understanding Messaging and Queues

The tasks in a system need some way of passing information to one another. One task may not be ready to receive a message at the same time another task wants to send it, so some way is needed to store messages for a task until it is ready to deal with them.

Each task has a queue where messages can wait to be picked up. A queue is a first-in first-out (FIFO) data structure. That is to say, the first message to be put into the queue is the first message to come out: messages are received in the same order that they were sent. Several different tasks can send messages to one task by putting messages onto that task's queue.

When a message is created and sent, some memory is temporarily allocated to store the message. It then waits on the queue until it can be processed by the receiving task. After processing, the message is destroyed and its memory is returned to the free pool.

The message queues allow tasks to send one another messages asynchronously: it doesn't matter if tasks run at different speeds, the queues buffer messages so that they can still communicate. The exact mechanisms for sending and receiving messages are explained in more detail in the following section on the message library: "Using Tasks and Messages."

Using Interrupts

Embedded systems need to react to the outside world. A typical embedded system will be connected to some electronic hardware and must react to signals from it, and send signals to control it. Interrupts provide the means for hardware to interact with software. An interrupt is a signal which makes the CPU stop running its current program and jump to a special interrupt routine. The interrupt routine is essentially just another subroutine—you just get to the interrupt routine because of an interrupt signal, rather than because you were called by another function.

Hardware is connected to pins which cause interrupts—commonly called *interrupt lines*. BlueCore01 has two interrupt lines available for connecting up to custom hardware. But keep in mind, the number of interrupt lines available will vary from system to system.

On BlueCore01 the interrupt routines are already written. If the interrupt lines change state, the interrupt routine will cause an event to be generated. The event is VM_EVENT _PIOINT, which stands for Virtual Machine Event Parallel Input Output Interrupt.

Interrupts usually have to be enabled before they can be used. This stops lines which are not currently in use, causing undesirable effects. BlueLab works just the same: by default, no events will happen, if you want your application to respond to an event, you must enable that event using the following call:

```
uint16 EventEnable(vm_event_source source, int enable);
```

So, to enable the PIO interrupt event you use:

```
EventEnable(VM_EVENT_PIOINT, 1);
```

A common use for an interrupt line is to connect a push button switch so that software can react to a user pressing a button. One problem, which is not immediately obvious, is that switches don't just move straight from one state to another. As the contacts close, there is usually a "bounce," which causes the switch to rapidly open and close several times (see Figure 7.1). Software can run fast enough for one push on a button to trigger several interrupts.

The solution to the problem is to debounce interrupt lines which are connected to pushbuttons, keyboards, or any other hardware which might oscillate before settling to a stable value. On many embedded systems, you will have to write a debounce function which catches the first interrupt from a line, disables interrupts, and then samples the line state periodically until it is stable. System code on BlueCore01 includes a debounce engine, and BlueLab provides a function for you which accesses it. All you need to do is call:

```
Void Debouncesetup(uint16 mask, uint16 count, uint16 period);
```

This sets up the debounce engine so that when the interrupt line specified by the *mask* parameter changes, the engine begins reading the pin at the interval specified by the *period* parameter (in milliseconds), until it has seen the same value *count* times. Once the line is stable, the engine sends the VM_EVENT_PIOINT event to application code. The application code can then get the stable value of the interrupt line using the call:

```
uint16 DebounceGet (void);
```

So, for instance, to sample PIO line 5 at 2 millisecond (ms) intervals and wait until it has been stable four times in a row, you would use:

```
Void Debouncesetup(1 << 5, 4, 2);
```

Setting the sampling period to zero switches off debouncing, so you then get an event for every single transition of the line. To switch off debounce on PIO line 5, you would use:

```
DebounceSetup(1<<5, 1, 0);
```

Figure 7.1 Switch Bounce

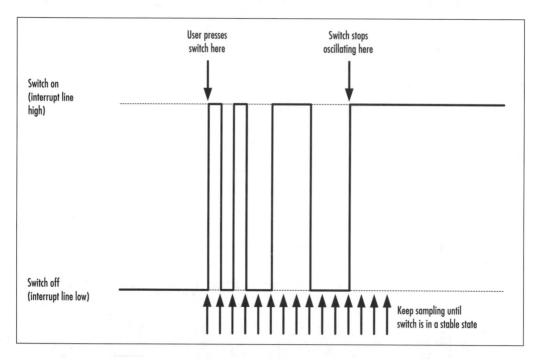

Hardware interrupts aren't the only type of interrupt. Many systems also allow software to generate interrupts. This is done when errors happen, such as a divide by zero operation, or an attempt to access memory that doesn't exist. Software interrupts are usually irrecoverable and result in a system reset. To prevent this from happening, the Virtual Machine interpreter checks user application code on BlueCore for illegal accesses.

Getting Started

BlueLab builds code for CSR's BlueCore chips. So, in addition to BlueLab, you will need a Casira development system. The development tools run on a Win32 PC—therefore, you will need administrator rights on the PC to install the tools.

The BlueCore module is supplied on a carrier board which slots into a blue plastic carrier in the center of the Casira (see Figure 7.2). The circuitry on this board is what would be used in most end-user products. The rest of the Casira development kit provides extra facilities to allow you to develop and debug applications, providing a variety of useful interfaces:

- **SPI interface** Connects to a PC parallel port, and allows you to reconfigure the Casira using the PSTool utility. Images can also be downloaded to the Casira using the Serial Peripheral Interface (SPI).

- **Serial interface** Connects to a PC serial port. BlueLab uses BlueCore Serial Protocol (BCSP), so you must ensure your Casira is configured to use BCSP. (Casiras are sold ready to use BCSP.)

- **USB port** Connects to a PC USB port, and supports the Bluetooth Specification's USB protocol (H2) when correctly configured.

Figure 7.2 Casira Development Kit

- **Audio I/O**   An audio jack which connects to the headsets supplied with the Casira.

- **LEDs**   These can be used to monitor applications running on the BlueCore chip.

- **PIO lines**   Parallel Input–Output lines; useful for connecting custom hardware.

Developing & Deploying…

BCSP and H4

The 1.1 Bluetooth Specification provided two serial interfaces: UART (H4) and RS232 (H3). Casiras can be configured to use the UART (H4) protocol across its serial port interface, but they are sold configured to use BlueCore Serial Protocol (BCSP). BCSP provides extra error checking on the serial interface, so it is more reliable in situations where errors can happen on the serial interface. BCSP also provides separate channels for voice, control, and data. This allows data to be flow-controlled while voice traffic flows remain uninterrupted.

Some stack vendors support BCSP, but not all do. To compensate, Casiras may be reconfigured to support the 1.1 Specification's UART (H4) interface.

The serial port settings are stored in the BlueCore persistent store (flash). A Persistent Store tool (PSTool) utility is available to change these settings.

The procedure for changing the serial port settings to BSCP is as follows:

- Connect the SPI cable between the Casira and a PC parallel port.
- Give the PSTool utility low-level access to the parallel port by installing a device driver. To do this, run the batch file BlueLab20\bin\InstParSPI.bat (this requires administrator rights).
- Register the PSTool user interface in the Windows registry by running BlueLab20\bin\RegPSToolocx.bat.
- Run the PSTool utility, selecting **SPI interface**.
- Access the developer list of tools by pressing **Ctrl+Alt+D**.
- Set the key **Host Interface** to **UART link running BCSP**.

Continued

> - Set the key **UART Configuration Bitfields** to **6**.
>
> To set a Casira to use the 1.1 Specifications UART protocol (H4), the following settings are used:
>
> - Set the key **Host Interface** to **UART link running H4**.
> - Set the key **UART Configuration Bitfields** to **168**.
>
> Note that to set a PS key, the Set button in the PSTool application *must* be pressed. Simply typing in the new value will not work. To be absolutely sure you have successfully set the new value, you can use the **Read** button to read back the current value.

Installing the Tool Set

BlueLab uses Cygwin, a Unix- like environment run under Windows. Cygwin is installed by running setup.exe from the Cygwin directory on the BlueLab CD. When prompted, choose to **Install from local directory**, and press **Next** twice. Now choose your installation directory, **Unix** text file type, and install for **All**. This installs all the tools which BlueLab needs.

The debugger from BlueLab is written in Java and requires version 1.3 or later of the Java2 runtime environment. To install the Java2 runtime environment, run the file setup.exe from the Java directory on the CD and follow the instructions. Finally, install BlueLab by running BlueLab.exe from the main directory on the CD.

Building a Sample Application

To test the installation, it is a good idea to compile a sample application. Starting Cygwin, go to the relevant directory and run **make**.

```
$ cd /cygdrive/c/BlueLab20/apps/hello
```

```
$ make
```

The main compiler xap-local-xap-gcc is derived from the GNU C compiler. This compiles the C code and produces an object file hello.o. The linker then works with the assembler xap2asm to analyze the object file, link in libraries and produce the application files hello.app, hello.dbg, hello.sym, and hello.xap. (See Figure 7.3.)

Figure 7.3 The BlueLab Tool Chain

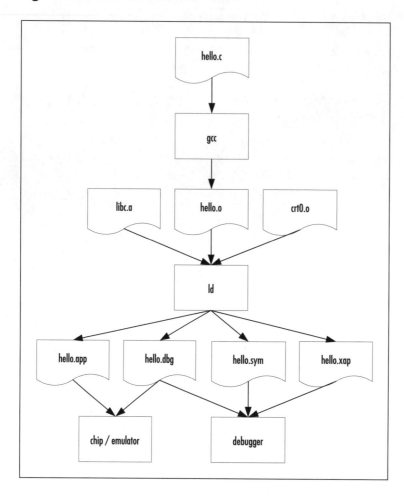

All you have done so far is build a "Hello World" program—this is *not* a BlueCore image, and you can't download it to the Casira yet. But you can use it to play with the debugging tools.

Running an Application under the Debugger

The debugger allows you to set breakpoints as well as single-step your code, and has many of the functions you find in a typical modern debugging environment. Code executes on the PC, but if you need to use functions from the BlueCore chip, such as the Radio or PIO, these are handled by the attached Casira.

Start off appdebug.jar by double-clicking the **appdebug.jar icon** in the **C:\BlueLab20\bin directory**. You should see the debugger window as shown in Figure 7.4.

Figure 7.4 Debugger Main Window

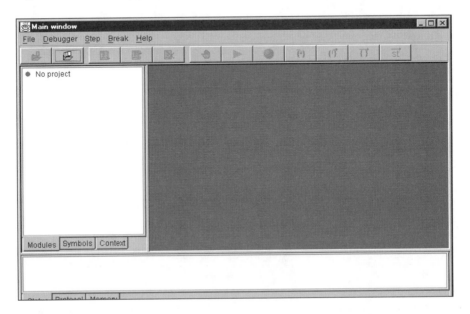

Select **File | Open project**, and load **hello.sym**. Once the project has loaded, you can browse the application downloaded using the Modules and Symbols tabs. Click a module name to see that module. Right-click a symbol to see the different places it appears.

Without communications, the debugger will report a problem and will fail to start. You can modify the comm port settings on the chip using PSTool, and editing the UART: baud rate. The Host Interface must be BCSP. To adjust the PC baud rate to match the Casira, select **File | Preferences** and click the **Comms** tab.

To run the program under the debugger, click the **Start Debugger** button. This opens communications to the Virtual Machine, lets you set break points, and allows you to run the code. Now, run the code by pressing the **Run** button. You should see "Hello World" in the debug output window (see Figure 7.5).

The Hello World program will run, output "Hello world," and then exit. It's not exactly a killer application, but it does verify that you have successfully installed all the tools, and configured the Casira correctly.

Figure 7.5 Active Debugger Window

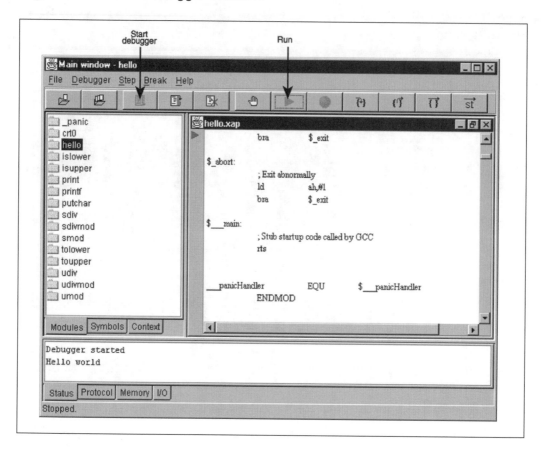

Using Plug-ins

The debugger can simulate code running on a BlueCore chip, and by communicating with the Casira can also use the radio and PIO ports on the BlueCore itself. Embedded applications are likely to run on custom hardware, so it may also be necessary to simulate extra hardware. For example, if you are creating a headset, a plug-in to simulate the buttons and lights on your headset will make it much easier to debug your headset application.

Simulating custom hardware is done by adding plug-ins to the debugger. The debugger is written in Java, so to create a plug-in, you just derive a new class to extend the existing Java class JComponent. Custom hardware will be controlled by the BlueCore chip's PIO pins, so plug-ins which simulate custom hardware must implement the PIOPlugin interface.

BlueLab includes an abstract *PIOPanel* class, which extends Jcomponent, and implements the PIOPlugin interface. It also provides useful functions for constructing and registering controls.

The following example is based on PIOPanel. The class implements two functions: tabName, which returns a string giving the name of the panel as it appears within the debugger, and the constructor function, which creates items that are displayed within the panel, positions them in the correct place, and informs the underlying PIOPanel about them. The items added to the panel must all implement the "Updater" interface:

```
public interface Updater
{
   void setEnabled(boolean show);
   void update(int on, int isout);
   void setDriver(PIODriver lis);
}
```

The updater interface specifies three functions that the control should support:

- **setEnabled** is called for each item in the panel whenever the panel becomes activated or deactivated. It is commonly used for graying out the controls.

- **update** is most useful for output items (lights). This interface function is called for each item in the panel whenever the PIO bits change state.

- **PIODriver** is used to drive PIO bits. This is needed to accept input from the user (e.g., a button press). An instance of "PIODriver" is passed to the item's "setDriver" function when the item is added to the PIOPanel.

If the hardware being simulated is just simple buttons or lights, then these can be added much more easily. The *PIOPanel* class provides utility functions that produce labels, buttons, and lights that are integrated into the panel in the correct way. These functions are:

```
// produces a simple text label, that is enabled in the correct manner.
public JLabel makeLabel(String label);


// produces a simple light, that is connected to one bit of the PIO port
```

```
public OnOffLight makeLight(int bit);


// produces a simple push-button, connected to one bit of the PIO port
public JToggleButton makeToggleButton(String label, int bit);
public JButton makeButton(String label, int bit);
```

Using these simple primitives, it is now possible to create the Headset plug-in panel. We begin by adding variables for each element of the panel, and creating them with calls to the **make*** functions. Then we use the initialization function to position the elements on the panel in a pleasing arrangement. This is achieved through the use of the standard Java Swing functions. A simplified version of the headset code is shown in the following:

```
// The new class 'Headset' is derived from the class 'PIOPanel'
public class Headset extends PIOPanel
{
  // The labels
  private JLabel volumeLabel = makeLabel("Volume");
  private JLabel powerLabel = makeLabel("Power");
  private JLabel[] labels = { volumeLabel, powerLabel };


  // The Light
  private OnOffLight powerLight = makeLight(9);


  // The Buttons
  private JToggleButton talkButton = makeToggleButton("!!Talk!!", 2);
  private JToggleButton upButton = makeToggleButton("Up", 4);
  private JToggleButton downButton = makeToggleButton("Down", 5);


  // A function to return the name of the panel
  public String tabName()
  { return "Headset"; }


  // The constructor - contains initialization code
  public Headset()
  {
```

```
    // bracket the initialization function a try/catch block
    try
    { jbInit(); }
    catch(Exception e)
    { e.printStackTrace(); }
}

private void jbInit() throws Exception
{
    // We want everything laid out on a grid
    setLayout(new GridLayout());

    // Set the alignment of the labels
    for(int i = 0; i < labels.length; ++i)
    { labels[i].setHorizontalAlignment(SwingConstants.RIGHT); }

    // Add the items to the panel
    add(talkButton, new GridConstraints(0, 1, 1, 1, 0.0, 0.0,
        GridConstraints.CENTER, GridConstraints.HORIZONTAL,
        new Insets(4, 8, 4, 8), 0, 0));
    add(volumeLabel, new GridConstraints(1, 1, 1, 2, 0.0, 0.0,
        GridConstraints.WEST, GridConstraints.NONE,
        new Insets(4, 8, 4, 4), 0, 0));
    add(upButton, new GridConstraints(2, 1, 1, 1, 0.0,0.0,
        GridConstraints.CENTER, GridConstraints.HORIZONTAL,
        new Insets(4, 4, 4, 8), 0, 0));
    add(downButton, new GridConstraints(2, 2, 1, 1, 0.0, 0.0,
        GridConstraints.CENTER, GridConstraints.HORIZONTAL,
        new Insets(4, 4, 4, 8), 0, 0));
    add(powerLabel, new GridConstraints(3, 1, 1, 1, 0.0, 0.0,
        GridConstraints.WEST, GridConstraints.BOTH,
        new Insets(4, 8, 4, 4), 0, 0));
    add(powerLight, new GridConstraints(4, 1, 1, 1, 0.0, 0.0,
        GridConstraints.CENTER, GridConstraints.NONE,
```

```
        new Insets(4, 4, 4, 8), 0, 0));
    // Everything should start off disabled
    setEnabled(false);
    }
}
```

BlueLab includes example plug-ins for a Headset, Telephone button grid, a 16-bit port expander using the I2C bus, a seven segment display and an output trace which reflects the state of the PIO pins. Rather than try to write plug-ins from scratch, you should pick the example closest to your application's needs and modify it as necessary.

Debugging under BlueLab

The Memory tab at the bottom of the main debugger window will show all active memory regions including their start and extent. If any address has a blank value, it means that address isn't acceptable. To follow a pointer from the variable window, just right-click it. This moves the memory window to that location.

If the application crashes, the debugger will stop just after the offending instruction. The call stack will show in the middle of the Context panel at the left of the main window. As you double-click the call stack, the source and variable displays are updated to that stack context.

Running an Application on BlueCore

To run a final application on the Casira, you must merge the application with a full Bluetooth stack. The Casira development kit arrives preloaded with a firmware image which allows the Casira to run the lower layers of the Bluetooth stack.

Figure 7.6 shows how an application image differs from the default Casira image. The application image has extra protocol stack layers: Logical Link Control and Adaptation Protocol (L2CAP), RFCOMM and Service Discovery Protocol (SDP). These are the protocol stack layers required to support the serial port profile, and are also used to support simple profiles based on the serial port profile such as the Headset profile. These stack layers are written by Mezoe and are collectively called *BlueStack*. BlueLab provides a royalty-free version of these stack layers for use on BlueCore chips. Above the BlueStack layers, a Connection Manager handles management of RFCOMM connections. The Connection Manager library is provided with BlueLab to make it easier to manage connections, but it is not compulsory to use it: if it does not meet the needs of your application, you can write your own Connection Manager.

Figure 7.6 Default Image and Image with Application

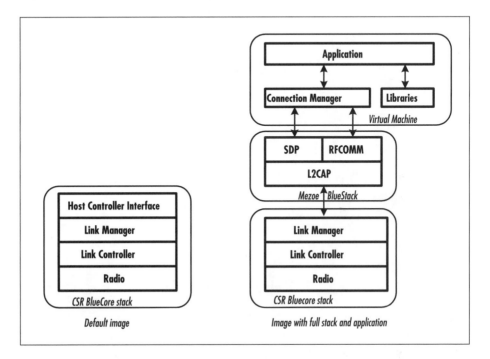

At the top of the application stack is the VM. The Virtual Machine allows the Connection Manager, Libraries, and application software to run in a protected memory space. Application software is compiled into Virtual Machine opcodes. As this is run, the Virtual Machine checks each instruction for invalid memory access. In this way, the Virtual Machine guarantees that your application software cannot interfere with correct running of the Bluetooth protocol stack.

When you are running applications under the debugger, you must have RFCOMM present on the Casira to drive the radio. However your application will actually be running under the debugger on a PC, so you do not want an image with your application built into it. The answer is to load the Casira with a "null" image—this is a firmware image that contains the Virtual Machine, but has no valid application. Note that if you have version 2.1 or later, you can have an image with an on-chip application installed; the on-chip application will automatically be disabled when the debugger is connected.

Developing & Deploying…

Virtual Machine Scheduling

The on-chip scheduler only allows a limited number of Virtual Machine instructions before giving another process some time. This means that you can not rely on an application running on the Virtual Machine to react quickly. This can be demonstrated by using VM code to toggle a PIO line. Consider the following code fragment:

```
while (1)

{

  val ^=4;

  PIOset (OUTPUT_BITS, val);

}
```

You should not write real code like this, as a continuous *while* loop is very bad for power consumption and can stop the chip from going into sleep states, but it is a useful routine to illustrate the scheduling of the Virtual Machine.

The *while* loop should execute, endlessly toggling the PIO line. If the line was connected to an LED, we would expect to see it shining brightly, as it flickers faster than the human eye can follow. In fact, if you follow the PIO line on an oscilloscope, you will see that what happens is the *while* loop toggles the PIO line at 3 KHz for 3 ms then remains in the last state for a while before another 3 ms of switching. (The exact time between bursts of switching varies depending on the other processes running.)

When writing applications for the Virtual Machine, you must bear in mind that your code will run fairly slowly since it's being interpreted. The preceding toggling speed equates to an equivalent clock speed of, at best, 40 KHz. Of course, the chip's real clock runs much faster, but your application effectively sees a slower clock because it is running through the delays caused by guarding the Bluetooth protocol stack.

You must also allow for the delays caused by other tasks being scheduled, as shown by the gaps in toggling the PIO line in the previous example.

Despite all these delays, it is still possible to write many useful applications, and even implement complete profiles under the Virtual Machine.

To program the Casira with a null image, simply go to the null project in the apps directory and run **make bc01**.

```
$ make bc01
```

This calls the command line version of the BlueFlash utility and downloads the image to the Casira for you. (You can also download images to the Casira across the SPI interface using a GUI version of the BlueFlash utility. Just run up BC01flash.exe and follow the instructions.)

Debugging Using VM Spy

Debug output from the application can be viewed using the VM Spy utility. To begin using VM Spy, complete the following steps:

1. Make sure the debugger isn't running, and nothing else is using the PC serial port.

2. Ensure the Casira serial cable is connected to the PC.

3. Make sure the Casira is configured to use BCSP.

4. Run **VMSpy.exe**.

5. Select **Connect**.

6. Select the COM port and baud rate that match the Casira configuration.

Figure 7.7 shows the VM Spy window (this figure also shows the VM data window which is explained in the next section, "Using VM Packets"). If the VM Spy window doesn't open, check to make sure the serial cable is connected correctly, that the Casira is configured for BCSP, and that no other applications are using the COM port.

VM Spy connects to the Casira, and debugging output (from BCSP Channel 11) is displayed in the main window. The window has several buttons which can be used to control the debugging session:

- **Disconnect**   This button disconnects from the Casira, but leaves the debugger window open.

- **Log**   This button allows a session to be logged to a file.

- **VM Data**   This button activates a window showing traffic on the VM data channel (BCSP Channel 13). Of course, this only works if the application makes use of the VM data channel. The bottom of the VM

Data window includes an edit box which can be used to send commands to the Casira.

- **Quit**   This button shuts down the debugger.

Figure 7.7 The VM Spy Window

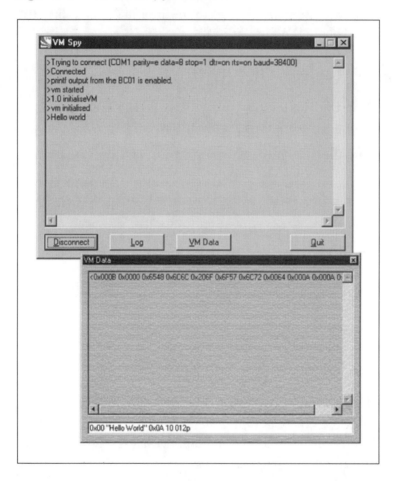

Using VM Packets

Applications running under the Virtual Machine can use BSCP Channel 13 to communicate with a host. The user application can send and receive packets of 16-bit data. For the final product, you will need to write software on the host to form the other end of the connection, but while developing embedded applications, Channel 13 can be a useful debugging tool. Applications which do not use

BCSP Virtual Machine packets can still communicate with Virtual Machine packets. (See Figure 7.8.) On USB and H4 they are sent over the Host Controller Interface (HCI) using the manufacturer's extension command.

Figure 7.8 Sending and Receiving Packets across Channel 13

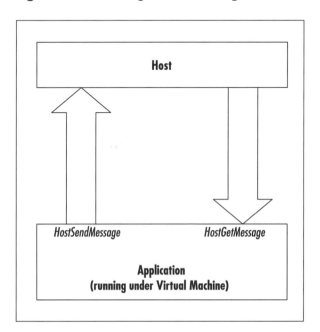

Incoming packets from the host cause a VM_EVENT_HOST event. The packets can then be retrieved using the *HostGetMessage* function. If there is no packet waiting, *HostGetMessage* returns NULL, otherwise a pointer to a new block of dynamic memory containing the packet is returned. This memory *must* be freed by the application once the application has finished with the packet.

The *HostSendMessage* function is used to send a message to the host. The application uses *malloc* to allocate a block of memory for the packet, and fills it in with the packet. Then *HostSendMessage* is passed a pointer to the memory block. The application can not access the memory block after the call, and should remove all references to it.

The Virtual Machine packet format is very simple (see Figure 7.9). The packet begins with a 16-bit word length field, which gives the total length of the packet, including the header. Note that the length is in 16-bit words, not in bytes.

Figure 7.9 Format of a Virtual Machine Packet

Length 16 bits	Sub-type 16 bits	Data (Length - 2) x 16 bits

The second field is a 16-bit sub-type word. The sub-type must be set to a value between $0 - 127$ (0x00 – 0x7f). The sub-type is useful to indicate the type of the packet to the code at either end.

The rest of the packet can contain any 16-bit data.

The code fragments that follow show how the *HostSendMessage* and *HostGetMessage* can be used.

```c
#include <host.h>    /* HostSendMessage and HostGetMessage */
#include <stdlib.h> /* malloc */

...


/* Send a small packet to the host */
uint16* data = (uint16 *) malloc(3 * sizeof(uint16));
if(data != NULL)
{
   data[0] = 3;       /* length */
   data[1] = 0x7e;    /* sub-type */
   data[2] = 0x1234; /* data */
   HostSendMessage(data);
   data = NULL;       /* removing reference to memory block */
}


/* receive a packet from the host */
if((data = HostGetMessage()) != NULL)
{
   /* do something with the data here */
   free(data);
}
```

The VM Data window of VM Spy can be used to send VM packets to test an application. The edit box at the bottom of the VM Data window is used to send commands to the Casira on BCSP Channel 13. The line can be used to input hexadecimal, decimal, or octal numbers. The line can also take character strings delimited with a quotation mark (").

The first entry on the edit window line is the sub-type number. This is followed by the contents of the packet. VM Spy will automatically calculate the packet length and fill in that field for you, so you don't need to worry about the length field.

Packing Format in Messages

The XAP2 processor on BlueCore works with 16-bit words. This means that single byte parameters are packed into 16-bit words. There are a few other rules to bear in mind when interpreting data structures from BlueCore:

- 8-bit values are sent as a 16-bit word, padded by setting the most significant byte to 0x00.

- 16-bit words are sent the least significant byte first.

- 24-bit words are sent as a 32-bit long word, padded by setting the most significant byte to 0x00. The most significant word is sent first.

- 32-bit long words are sent as two 16-bit words with the most significant word first.

- Pointers are sent as two bytes with their values set to [0x00 0x00].

- Data referenced by a pointer is appended to the primitive. If a primitive contains more than one pointer, the dereferenced data is appended in the same order that the pointers appear in the primitive.

- Where a primitive contains a pointer to *uint8* data, the dereferenced data is appended to the primitive and is sent as consecutive bytes (i.e., no padding bytes are inserted).

- Arrays are sent as a series of elements with the lowest indexed element first.

For example, consider the message CM_CONNECT_AS_MASTER_REQ:

```
CM_CONNECT_AS_MASTER_REQ:
  uint16 length = 0x10
  uint16 type = 0x6
```

```
/* Security */
uint16 use.authentication = 1
uint16 use.encryption = 1


/* BD address */
uint24 bd_addr.lap = 0xAABBCC
uint8  bd_addr.uap = 0x5B
uint16 bd_addr.nap = 0x0002


/* Target UUID */
uint16 target = 0x1108 /* Headset */


/* Master timeout */
uint16 timeout = 0xDDEE


/* Park parameters */
uint16 park.max_intval = 0x800,
uint16 park.min_intval = 0x800


/* Sniff parameters */
uint16 sniff.max_intval = 0x800
uint16 sniff.min_intval = 0x800
uint16 sniff.attempt = 0x08
uint16 sniff.timeout = 0x08
```

This message would be packed as shown in Figure 7.10.

Using the BlueLab Libraries

BlueLab offers a variety of libraries which provide functions to support basic C functions, BlueCore hardware, and Bluetooth applications (see Figure 7.11 for a graphical overview).

When linking, all object files are used, and then missing symbols are imported from the libraries. Each symbol is taken from the first library (in command-line order) which provides that routine. This means that the application's makefile must list libraries which override a routine before the libraries with default versions.

Figure 7.10 Message Packing Format for CM_CONNECT_AS_MASTER_REQ

Byte 1	Byte 2	Byte 3	Byte 4	Byte 5	Byte 6	Byte 7	Byte 8
0x10	0x00	0x06	0x00	0x01	0x00	0x01	0x00
uint16 length 0x10		uint16 type = 0x6		uint16 authentication = 1		uint16 encryption=1	
0xAA	0x00	0xCC	0xBB	0x5B	0x00	0x02	0x00
uint32 bd_addr.lap = 0xAABBCC				u int8 bd_addr.uap = 0x5B		uint16 bd_addr.nap = 0x0002	
0x08	0x11	0xEE	0xDD	0x00	0x08	0x00	0x08
uint16 target = 0x1108		uint16 timeout = 0xDDEE		uint16 park.max_intval=0x0800		uint 16 park.min_intval = 0x800	
0x00	0x08	0x00	0x08	0x08	0x00	0x08	0x00
uin t16 sniff.max_intval = 0x800		uint16 sniff.max_intval = 0x800		uint16 sniff.attempt = 0x08		uint16 sniff.timeout = 0x08	

Figure 7.11 Library Overview

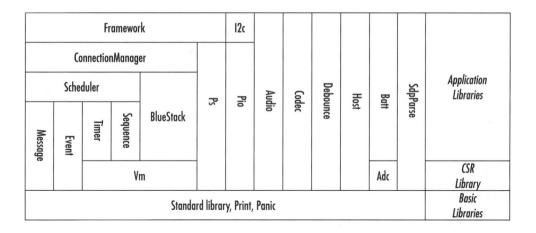

This makes it important that libraries are linked in the correct order. Each library should be listed before any others which appear after it in the list that follows.

The scheduler relies on the message and timer libraries. Some applications require the scheduler, but may not need both of those libraries. In that case, the libraries can be replaced with their stub versions which take less code and data space. Obviously, if messages and timers are stubbed out, then messages or timers can't be used.

Developing & Deploying…

Support for ANSI C

The XAP2 processor on BC01 is a 16-bit architecture with no direct support for 8-bit values.

As a result, the "char" type is a 16-bit quantity. While this is permitted by the C standard, care must be taken with code which assumes 8-bit characters.

Both "short" and "int" are 16-bit, while "long" is 32-bit. 32-bit quantities incur a significant performance overhead and should be avoided wherever possible. 64-bit quantities are not supported ("long long" is mapped to a 32-bit integer).

As is the case with most embedded systems, floating point values and floating point arithmetic are not available.

The amount of RAM on BC01 is limited, and memory must be shared between the Bluetooth stack and the application. RAM is divided into "pools" using fixed block sizes which limits the maximum size of a block that can be allocated. Finally, the memory management mechanism limits the application to holding at most 12 dynamically allocated blocks of memory. The size constraints also apply to the amount of stack space available to the application.

Basic Libraries

The basic libraries provide facilities required to run and debug C code:

- **Standard library**   Provides a selection of functions defined by the ANSI/ISO standard: *assert, limits, stdarg, stdio, stdlib, string, memory, printf, sprintf, vprintf, vsprintf, putchar, malloc, free, calloc, realloc, atoi, strcat, strcpy, strncpy, strcmp, strncmp, strchr, strrchr, memchr, strlen, memset, memcpy, memmove, bcopy, bzero, memcmp8, strdup.* These are provided in libc.a which is always linked in.

- **Panic**   Provides small utility routines which panic the application if conditions aren't met. Provided in libpanic.a with header file <panic.h>.

- **Print**   A simple header file which enables printing of debug messages when DEBUG_PRINT_ENABLED is defined.

CSR Library

The CSR library provides facilities specific to the BlueCore chip and the Virtual Machine. All of these routines are provided in libcsr.a. You can either include the corresponding header files (<event.h>, ...) selectively or use <csr.h> which will include all of them.

- **Event**   Enable and poll for application events.

- **Vm**   Reads the millisecond timer; *VmWait* suspends the VM execution until an event occurs; this library also supports sending and receiving BlueStack primitives.

- **Ps**   Accesses the on-chip persistent store: *PSstore* sets a key and *PSretrieve* reads a key.

- **Pio**   Provides access to the PIO pins on the BlueCore chip. *PIOset* sets a line; *PIOget* reads it. *PIOsetDir* and *PIOgetDir* can be used to change the line's direction.

- **Audio**   Allows an application to play audio sequences.

- **Codec**   Adjusts attenuation for the pulse-code modulation (PCM) compression/decompression (codec).

- **Debounce**   Provides debounced reading of PIO inputs; useful for connecting to push buttons or keys.

- **Host**   Supports communications with the host over BCSP Channel 13 using *HostGetMessage* and *HostSendMessage*.

- **Adc**   Allows an application to read values from the analog-digital converter (ADC). This is used by the battery library.

The Application Framework, Connection Manager, Scheduler, Timer, BlueStack, I2C, Message, and SDPparse libraries are interpreted, as are parts of the Standard Library. The rest of the libraries run in native mode and do not have to go through the Virtual Machine's interpreter.

Application Libraries

The application libraries (listed in the following) provide support for applications running on BlueCore. The source for these libraries is in src/lib. They can be rebuilt and installed by typing **make install** in that directory. This allows source level debugging in library code as well as application code.

Debugging...

PIO Pins

PSKEY_PIO_PROTECT_MASK stops you from setting values for PIO pins which are masked out, allowing pins used by the Casira to be protected. You should not tamper with this PS key.

- 0 – Used to control external hardware on Class 1 modules
- 1 – Used to control external hardware on Class 1 modules
- 2 – External RAM bank switch (optional); USB control
- 3 – Controls the LED on Microsiras
- 4 – USB control/reset
- 5 – USB on some modules (check your data sheet)
- 6 – Some packaging schemes use this for power (check your data sheet)
- 7 – Some packaging schemes use this for power (check your data sheet)

Lines 4 and 5 are connected to hardware interrupts, so if you need interrupts you must use these lines.

Lines 6 and 7 are best for connecting to custom hardware—as long as they aren't connected to a power line in the packaging of the BlueCore chip you plan to use!

Line 5 can be used if you want an interrupt line.

If you're not using USB line 2 is available; on most modules, line 3 is also available.

On some Casiras (revision F), line 4 is connected to a reset line and can cause resets when held low for longer than the value specified by PSKEY_HOSTIO_UART_RESET_TIMEOUT. As a result, this line is best avoided.

- **Timer** Manages queues of functions to call after specified delays, checks for any that are due to be run, and calculates the shortest period which can be passed to *VmWait* before the next check is required. Most significant applications use the scheduler to manage this. Use *timerAdd* to add a new timer.

- **Sequence**   Built on top of the timer library, it provides routines which can orchestrate a timed sequence of calls to designated user functions.

- **Message**   Manages queues of messages (in dynamically allocated buffers).

- **Scheduler**   Orchestrates the tasks which form the timer, message, and event libraries. Calls timer routines and *VmWait*; dispatches to appropriate handlers when events are triggered.

- **BlueStack**   Header files which define Bluetooth primitives.

- **Connection Manager**   An example connection manager using RFCOMM.

- **SdpParse**   Utility functions for unpacking an SDP record.

- **Framework**   Library to support the example applications supplied with BlueLab. For example, the headset framework adapts the framework library for use with the example headset supplied with BlueLab.

- **I2c**   A sample library which uses the PIO routines to support devices on the I2C bus.

- **Battery**   Provides periodic battery readings from a test pin.

A series of example applications are supplied with BlueLab. These include adaptations of the application framework which provide complete implementations of the Headset profile and Audio Gateway profile.

There are also examples of using Libraries, including the I2C Library, host communications, the Sequence Library, the Timer Library, General Purpose Input Output (GPIO), and a program to flash LEDs.

Rather than write your own applications from scratch, you should adapt the examples supplied, which will greatly speed up development time.

Using Tasks and Messages

The message library provides a mechanism for asynchronously posting messages between tasks. The scheduler library will automatically run tasks which have messages pending (the scheduler also runs tasks which have events pending). Messages have a type property and may also contain a user-defined payload.

Tasks and Message Queues

Messages are posted to *MessageQueues* which are owned by *Tasks*. A Task which owns a non-empty MessageQueue will be run by the scheduler. In the current implementation, the binding between Tasks and MessageQueues is static; a *MessageQueue n* is owned by *Task n*.

The DECLARE_TASK macro declares a task, and takes a Task identifier as an argument, which identifies the task's MessageQueue. For example:

```
DECLARE_TASK(4)
{
        void * msg = MessageGet(4,0) ;

        ...
```

Note that the task is declared with the same identifier, 4, that is used in the call to MessageGet. The argument to DECLARE_TASK must be an integer; it cannot be another macro. There are no restrictions upon which MessageQueues a task can post to.

Task and MessageQueue identifiers range from 0 to 15 although 0 and 1 are reserved (see Table 7.1).

Table 7.1 Reserved Task/Message Identifiers

Task/Message Identifier	Task Name
0	Connection Manager
1	Application Framework (e.g., Headset Framework)

Creating and Destroying Messages

Messages are dynamically allocated. All messages have a type property and some may also contain a payload. Both of these properties are specified when using the *MessageCreate* function. The code that follows shows how a message can be used to transfer a block of *uint16*s to a task:

```
#define TRANSFER_MSG 100

...

void sendMsg(uint16 * data,uint16 length)
{
  uint16 * msg = (uint16*) MessageCreate(TRANSFER_MSG,length)
  memcpy(msg,data,length) ;
  MessagePut(6,msg) ;
}

...
```

```
DECLARE_TASK(6)

{

  MessageType type ;

  void * msg = MessageGet(6,&type) ;

  if (msg)

  {

    switch (type)

    {

      case TRANSFER_MSG :

        uint16 * data = (uint16 *) data ;

        break ;

      ...

    }

    MessageDestroy(msg) ;

  }

}
```

Any task can use the *sendMsg* function to send data to the application framework (Task 1). Note that the type of the message does not appear in the message payload. Instead, it is set after creation using *msgSetType* and read after retrieval using *MessageGetType*.

It is important to delete messages using the *MessageDestroy* function rather than free. Messages are dynamically allocated which means that they come out of the very limited dynamic-block budget. This means it is important to ensure that messages are consumed as soon as possible after being produced. Put another way, messages are intended to be a signaling mechanism, not a data-buffering mechanism.

Using the MAKE_MSG Macro

Functions that use the message library declare a message with type X and structure X_T where X identifies the library. For example, messages for the *Connection Manager open* are defined as follows:

```
#define CM_OPEN 13    /* declare a message type for CM_OPEN */

typedef struct

{

  uint16 blah ;     /* declare the structure for messages to CM_OPEN */

  ...
```

```
} CM_OPEN_T ;
```

This leads to code that looks like:

```
void doOpen(void)
{
  CM_OPEN_T*msg = (CM_OPEN_T
*)MessageCreate(CM_OPEN,sizeof(CM_OPEN_T));
  msg->blah = ...
}
```

The MAKE_MSG macro can be used to reduce typing and minimize opportunities for mistakes. This macro creates a variable named *msg* of the requested type. So the preceding code can be replaced with the following call:

```
void doOpen(void)
{
  MAKE_MSG(CM_OPEN);
  msg->blah = ...
}
```

Connection Manager

The Connection Manager handles all the layers of the Bluetooth protocol stack from RFCOMM downwards. Without a Connection Manager, you would need to establish ACL links, configure the links for RFCOMM, set up and configure L2CAP links, and finally set up an RFCOMM link. With a Connection Manager, you can have all the layers you need set up and configured with a single call.

Most applications which send data will want to use RFCOMM connections, but for those who need to get in at a lower level, the BlueLab Connection Manager allows your application to send L2CAP packets as well as RFCOMM packets. (L2CAP is the lowest level of the Bluetooth Protocol stack that an application will send data to, since all user data on Bluetooth links has to be sent as L2CAP packets.)

Packets are sent on a connection, and every connection has to lead to some peer device, so, naturally enough, before any packets can be sent, the Connection Manager must be paired with a peer device.

The section on tasks and message queues mentioned that Task/Message Identifier 0 is reserved for the Connection Manager, and Task/Message Identifier 1 is reserved for the Application Framework. The practical effect of this is that whenever your application sends a message to the Connection Manager, it will send it to

MessageQueue 0, and whenever you get a message back from the Connection Manager, it will come back on MessageQueue 1. This rule on message queue numbers applies whether the message is control information, or data packets.

The Connection Manager's messages are all declared in cm_rfcomm.h.. The Connection Manager itself is implemented in the CM_RFCOMM library: libcm_rfcomm.a.

Developing & Deploying…

Receiving Messages from Multiple Sources

Some tasks will have to receive messages from several sources. One example is the application framework, which sits between an application and the Connection Manager and has to communicate with both.

Message types are just integers, so when the framework gets a message of type 5, it could have trouble deciding whether the message is a "data_indication" from the Connection Manager or a "close_request" from the application! There are two approaches to solving this problem:

1. Choose message type numbers so there is never any overlap between message type numbers going to the same task.

2. Ensure that the payloads of messages sent to the framework always contain a "source" field which is filled in before the message is sent.

Many embedded messaging systems provide a mandatory "source" field on all messages. This solves the problem of messages from multiple sources, but wastes valuable memory from the scare dynamic-block budget, so BlueLab leaves it up to the application programmer to decide when these identifiers are appropriate. In many cases, it will be possible to solve the problem using unique message type numbers, thus minimizing message size and saving memory.

Initializing and Opening the Connection Manager

The libraries which make up BlueStack and implement the Bluetooth protocol stack are compulsory to have in the system. This is because the basic protocol stack is essential to implement any Bluetooth product. To make sure that the protocol stack runs properly it is started up for you automatically.

The Connection Manager is not part of the Bluetooth protocol stack. It's a separate library which you can choose to use or not. Because the Connection Manager is not a compulsory part of the system, it isn't started up automatically. If you want to use the Connection Manager, then you must initialize and open it by making a few calls.

First, your application initializes the Connection Manager by sending it a CM_INIT_REQ message (see Figure 7.12). The Connection Manager will respond with a CM_INIT_CFM message once it has successfully registered with BlueStack. These messages just start the Connection Manager running, so neither message has any parameters.

Figure 7.12 Message Sequence Chart for Initializing and Opening the Connection Manager

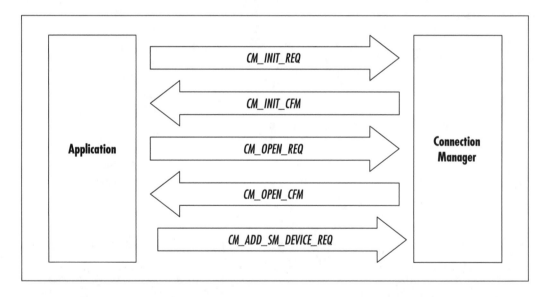

You could create and send the initialization message like this:

```
MAKE_MSG(CM_INIT_REQ);
PutMsg(msg);
```

But to make it even easier, the file rfc_init.c is supplied with BlueLab. This gives you a function *CmInit*, which makes and sends the message. So, if you link rfc_init into your build, all you need to do is use this call:

```
CmInit();
```

Now that the Connection Manager is running, the next stage is to tell the Connection Manager some information about your application.

BlueCore chips usually arrive with the Class of Device (CoD) set to Miscellaneous (all zeroes). This is probably not going to be appropriate for your application. For instance, if you are writing a headset application, you want the class of device to be set to **Audio** for the Major Device Class, and **conforms to the Headset profile** for the Minor Device Class. It is important to get this set correctly because the Class of Device is sent out in inquiry responses, and is then used by other applications to find devices they can connect with. It is possible to filter out inquiry responses based on the Class of Device information they contain. So, if your Class of Device doesn't accurately reflect your application's capabilities, then other applications may not even report your device's presence to the user.

You also need to let the Connection Manager know what Service Record you want used to describe the services provided by your application. Once you have done this, the Connection Manager can take care of handling service discovery queries without needing any more intervention from your application.

Your application passes the Class of Device and Service Record information to the Connection Manager in a CM_OPEN_REQ call, whereupon the Connection Manager responds with CM_OPEN_CFM. The CM_OPEN_REQ is sent as follows:

```
CM_OPEN_REQ( uint8 * serviceRecord,
             uint16 sizeServiceRecord,
             uint32 classOfDevice);
```

The *serviceRecord* parameter is a pointer to an area of dynamically allocated memory containing the service record which describes your application's services. The service record must contain a blank entry for the RFCOMM channel to be used for your application's service—in other words, a universal unique identifier (UUID) of 3 followed by an unsigned integer (UINT). The channel will be filled in by the Connection Manager. The *SizeServiceRecord* parameter is the size of the complete service record, and the *classOfDevice* parameter specifies the class of device to be used when responding to inquiries.

Having opened up the Connection Manager and told it about your application, you could just stop there, but you have the option of going on and using the Security Manager features, too. You can tell the Security Manager there are some devices you trust, and the Security Manager will store information about those devices in its Trusted Devices database. Once a device is registered as

Trusted in the Security Manager database, the Security Manager can automatically carry out all authentication procedures and allow a device to connect without further authorization from your application.

To use the Security Manager, your application sends a CM_ADD_SM_DEVICE_REQ with details of the device you want to add to the Security Manager's trusted devices database.

```
CM_ADD_SM_DEVICE_REQ (BD_ADDR_T addr,

                      uint8 link_key[SIZE_LINK_KEY],

                      Bool_t trust )
```

The *addr* parameter gives the Bluetooth Device Address of the device being added to the Security Manager database. The *link_key* parameter, meanwhile, gives the link key for that device, and the *trust* parameter is a Boolean value: TRUE if the device is trusted, FALSE if it is not. If you don't have a link key at this stage, you will have to skip this step for now. Later on you can go through pairing to get a link key, then call the Security Manager.

In addition to the preceding messages, you will need to start the timer subsystem and the scheduler. These calls go on either side of the call to initialize the Connection Manager as follows:

```
/* Initialize timer subsystem so the application can use timers */
TimerInit();


/* Initialise the connection manager */
CmInit();


/* start Virtual Machine scheduler to call application's tasks */
Sched();
```

You should not send the CM_OPEN_REQ until the CM_INIT_CFM is received, so you will need to wait until the message comes in. You need a message handler to check the message queue and process the event when it arrives. The following code fragment illustrates how this can be done.

```
void * msg;          /* incoming messages require a void msg pointer */
MessageType type;    /* we need to know what type of message was sent.
                        This type may be different in each application,
                        but the messages will not be very different from
```

```
                              those already defined by the Connection Manager.
                    */

/* Get the message, if any, from our queue so that we can process it.
   Notice that only one message is processed at a time.
*/
msg = MessageGet(1, &type);

if(msg)
{
  switch (type)
  {
    /* Connection manager library is ready, so send CM_OPEN_REQ */
    case CM_INIT_CFM :
      MAKE_MSG(CM_OPEN_REQ);
      msg->serviceRecord = opCreateServiceRecord(
                              &msg->sizeServiceRecord);
      msg->classOfDevice = HEADSET_COD; /* set your device's class */
      MessagePut(0,msg);
      break ;

    /* you will need a case statement for each event you can receive */
    case CM_XXXXXXX:
      some message handling code goes here
      break;

    /* Always a good idea to track unhandled primitives */
    default :
      PRINT(("rfc Unrecognised msg type %x\n",type));
      break;
  }
MessageDestroy(msg);
```

Now you know how your application can start up the Connection Manager, tell it about its services, and register devices you trust. This is all very necessary, but so far all you've done is configure the Connection Manager: not a single packet has been sent on the radio. The next sections will explain how to use the Connection Manager to communicate with other devices.

Inquiry

Before you initiate a connection, you might want to look around to find what other Bluetooth devices are in the neighborhood. At the user interface level this procedure is called *Device Discovery*, but in the Core Bluetooth Specification, you'll find it referred to as *inquiry*. Since your application is dealing with a Bluetooth protocol stack, you use the technical term not the user interface term, so you call the process *inquiry*.

An inquiry can be requested with CM_INQUIRY_REQ (see Figure 7.13). Your application will need to specify the overall length of the inquiry (the timeout) and the maximum number of unique responses required. The Connection Manager may perform more than one inquiry for you in the specified timeout. If the maximum number of responses is reached, the inquiry is terminated and your application is sent an inquiry complete returned with the appropriate status flag.

Figure 7.13 Message Sequence Chart for Conducting an Inquiry

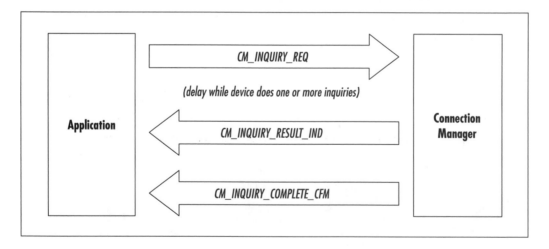

An inquiry gets you back information like a Bluetooth Device Address and the Class of Device, but if you are displaying information on devices to a user, you might want to know a bit more about them. You have the option of asking the Connection Manager to automatically go and get the user-friendly name of each device that responds to your device. This will take some time, as it involves setting up a connection to each device you haven't seen before. Setting up connections will also take up power and shorten your battery life, so you should only ask the Connection Manager to do this if your application will use the information.

To get BlueCore to perform an inquiry, use the following call:

```
CM_INQUIRY_REQ ( uint8 max_responses, uint16 inq_timeout, uint32 class_

    of_device, uint16 remote_name_request_enabled);
```

The *max_responses* parameter gives the maximum number of unique inquiry responses that can be received. The *inq_timeout* parameter is the timeout (in seconds) for the inquiry process, so this gives the maximum length of the inquiry. The *class_of_device* parameter acts as a filter: only inquiry responses with this Class of Device will be passed up from the Connection Manager to the application. The *remote_name_request_enabled* parameter is a flag indicating whether to perform a remote name request for each inquiry result not seen before.

The application can wait pending the arrival of a CM_INQUIRY_RESULT_IND or CM_INQUIRY_COMPLETE_CFM. By waiting on an event, the application allows the scheduler to allocate all its time to other tasks until the inquiry indication events occur. The CM_INQUIRY_RESULT_IND carries the results from the inquiry as follows

```
CM_INQUIRY_RESULT_IND (HCI_INQ_RESULT_T inq_result,

                    uint8 *handles[HCI_LOCAL_NAME_BYTE_PACKET_PTRS]);
```

The *handles* parameter is an array of handles corresponding to pointers to the name of the remote device as discovered by the remote name request. The *inq_result* parameter is the Inquiry result which is structured as follows:

```
typedef struct
{
    BD_ADDR_T                       bd_addr;
    page_scan_rep_mode_t            page_scan_rep_mode;
    uint8_t                         page_scan_period_mode;
    page_scan_mode_t                page_scan_mode;
    uint24_t                        dev_class;
    bt_clock_offset_t               clock_offset;
} HCI_INQ_RESULT_T;
```

These parameters are straight out of the Bluetooth Core Specification for HCI Inquiry Result Event (see part H:1 of the Specification for more details).

When all of the inquiry results are in, your application will get the CM_INQUIRY_COMPLETE_CFM as follows:

```
CM_INQUIRY_COMPLETE_CFM (inquiry_status_t status)
```

The *status* parameter lets you know why the inquiry completed. It is set to *CmInquiryComplete* if the user specified timeout for the inquiry has expired, *CmInquiryCancelled* if the inquiry was terminated before it finished, or *CmInquiryMaxResponsesReached* if the inquiry finished because it had reached the number of responses you specified.

At this point you may be thinking, "Why would I want an inquiry to finish before it had collected as many responses as possible?". There are two reasons, both to do with the limited resources you have. Firstly, you want to set a timeout because if you leave the device permanently inquiring, it will use up power and shorten battery life. Secondly, you may have to limit the number of responses because you need to store and process responses. Since you don't have an infinite amount of memory available there's a limit to how many responses you can process at one time.

Pairing

After the inquiry process, your application will have found some devices it could connect with, but there's one more step you should go through before creating a connection: *pairing*.

The pairing process creates a link key which can be used to encrypt communications on the Bluetooth link. The link key can also be used to authorize a device—that is, to check that the device is really the one you want to connect with, not just somebody trying to fool you into sending them all your private data. Figure 7.14 shows the process of creating a link key.

First you need to ask the Connection Manager to pair with a device using the CM_PAIR_REQ which is structured as follows:

```
CM_PAIR_REQ ( role_t role,
              Delay timeout,
              bool_t authentication,
              BD_ADDR_T bd_addr );
```

The *role* parameter is set to CM_MASTER or CM_SLAVE, and identifies which role the device is taking. The *timeout* parameter gives the delay before the attempt to pair is abandoned. The *authentication* parameter is a Boolean flag which is TRUE if authentication should be used and FALSE otherwise. The *addr* parameter is the Bluetooth Device Address of the remote device to pair with (this only applies when initiating pairing by attempting to create a connection).

Figure 7.14 Message Sequence Chart for Pairing

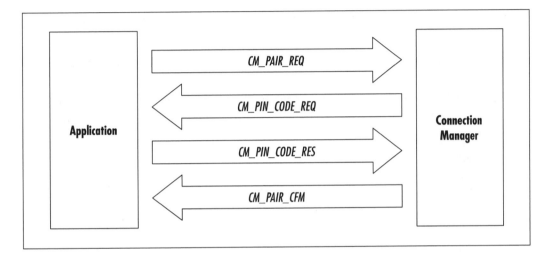

The shared link key is created using a PIN code which must be input separately at either end of the link. For devices without a user interface, the PIN code can be preprogrammed. These are called *fixed PINs*. Devices with fixed PINs have to be sold with a note to the user of the PIN code so that they can enter the same PIN in whichever device they want to pair with.

The Connection Manager needs to get the PIN code from your application. To do this, it will send you a PIN request CM_PIN_CODE_REQ as follows:

```
CM_PIN_CODE_REQ (BD_ADDR_T bd_addr );
```

The PIN code request carries a Bluetooth Device Address which you can use to look up the PIN code if you have PIN codes for various devices stored. If you don't have the PIN code stored, you may need to ask the user for a PIN code. You can use the Bluetooth Device Address to let the user know which device is asking for a PIN code. (If you stored the user-friendly name of the device along with it's Bluetooth Device Address, you could display the user-friendly name to the user instead of the Bluetooth Device Address.)

However you get hold of the PIN code, your application should send it to the Connection Manager in a CM_PIN_CODE_RE response as follows:

```
CM_PIN_CODE_RES (BD_ADDR_T addr,

                 uint8 pin_length,

                 uint16 pin[8]);
```

The parameters are fairly obvious: *addr* is the address of the device we are trying to pair with, *pin_length* is the length of the PIN key in bytes, and *pin* is an array containing the PIN code. One thing which is not immediately obvious is that you can reject the PIN code response just by setting the *pin_length* to zero. This works because the Bluetooth Specification does not allow you to use a zero length PIN, so this illegal value is taken as an indication that you don't want to supply a PIN for this device.

If the pairing is successful, the Connection Manager will store the address and link key associated with the paired peer device, and issue a confirmation giving the status of the pairing operation (see Figure 7.14).

```
CM_PAIR_CFM( pair_status_t status, BD_ADDR_T addr, uint8 link_key

    [SIZE_LINK_KEY]);
```

The *status* parameter is set to *CmPairingComplete* if successful or *CmPairingTimeout* if unsuccessful. The *addr* parameter is the Bluetooth Device Address of the device we have paired with. The *link_key* parameter is the link key to use with that device.

The link key will be needed later for authentication and encryption. You could store the link key in your application, but it is more efficient to use the CM_ADD_SM_DEVICE_REQ to pass the link key and device details to the Security Manager.

Now that you've learnt all about pairing, it's time to break the news that it isn't actually compulsory! You could skip past pairing and go straight to making a connection. However, if you don't create a link key then you wont be able to use encryption and authentication, so your connection will be unsecure. Because Bluetooth links can be intercepted, it is highly recommended you use encryption.

Connecting

Finally, your application is at the stage where it can request a data connection. The messages used to do this are shown in Figure 7.15.

If your application is initiating a connection as a master, then you need to send a CM_CONNECT_AS_MASTER_REQ message to the Connection Manager as follows:

```
CM_CONNECT_AS_MASTER_REQ (cm_auth_config_t use, BD_ADDR_T addr, uint16
    target, Delay timeout, cm_park_config_t park, cm_sniff_config_t sniff)
```

Figure 7.15 Message Sequence Chart for Connecting as Master

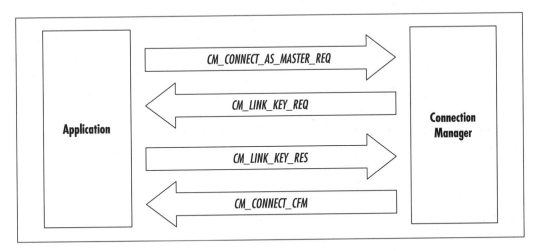

The *use* parameter configures authentication and encryption. The *addr* parameter gives the Bluetooth Device address of the device you want to connect to. The *target* parameter provides the UUID of the service your application wants to use; this information will be used for an SDP search. The *timeout* parameter gives a delay to wait before abandoning the connection attempt. The *park* parameter configures the *park* parameters to use on the connection. The *sniff* parameter configures the *sniff* parameters to use on the connection.

CM_CONNECT_AS_SLAVE_REQ is used to configure the BlueCore chip to accept connections as a slave. This will start page scanning, using parameters supplied as follows:

```
CM_CONNECT_AS_SLAVE_REQ ( cm_auth_config_t use,
                          BD_ADDR_T bd_addr,
                          uint16 ps_interval,
                          uint16 ps_window,
                          Delay timeout,
                          cm_park_config_t park,
                          cm_sniff_config_t sniff);
```

The *use* parameter configures authentication and encryption. The *addr* parameter is the Bluetooth Device Address to connect to. The *ps_interval* parameter specifies the Page Scan interval. The *ps_window* parameter specifies the Page Scan window. The *timeout* parameter gives a delay to wait before abandoning connection attempt. The *park* parameter gives the parameters for configuring park mode. The *sniff* parameter gives the parameters for configuring sniff mode.

Both CM_CONNECT_AS_MASTER_REQ and CM_CONNECT_AS_SLAVE_REQ take as parameters structures for configuring authentication, park, and sniff. These structures are as follows:

```
typedef struct
{
  uint16 authentication; /* 1 if connection is authenticated 0 if not
*/
  uint16 encryption; /*1 to enable encryption, 0 to disable encryp-
tion*/
}cm_auth_config_t;


typedef struct
{
  /* parameters for park mode negotiation */
  uint16 max_intval; /* maximum beacon interval in slots */
  uint16 min_intval; /* minimum beacon interval in slots */
}cm_park_config_t park;


typedef struct
{
  /* parameters for sniff mode negotiation */
  uint16 max_intval; /* maximum sniff interval, in slots */
  uint16 min_intval; /* minimum sniff interval, in slots */
  uint16 attempt;    /* sniff attempt length in slots */
  uint16 timeout;    /* sniff timeout length in slots */
}
cm_sniff_config_t sniff;
```

The following function illustrates how these parameters are filled in. It sends a message to the Connection Manager requesting a connection as master, but similar code would be used to fill in the parameters when connecting as a slave.

```
static void connect_as_master(uint16 timeout)
```

```
{
  MAKE_MSG(CM_CONNECT_AS_MASTER_REQ) ;

  /* Security */
  msg->use.authentication = 0 ;
  msg->use.encryption = 0;

  /* BD address */
  msg->bd_addr.lap = SLAVE_LAP;
  msg->bd_addr.uap = SLAVE_UAP;
  msg->bd_addr.nap = SLAVE_NAP;

  /* Target UUID */
  msg->target = 0x1108;   /* Headset */

  /* Master timeout */
  msg->timeout = timeout ;

  /* Park parameters */
  msg->park.max_intval = 0x800;
  msg->park.min_intval = 0x800;

  /* Sniff parameters */
  msg->sniff.max_intval = 0x800;
  msg->sniff.min_intval = 0x800;
  msg->sniff.attempt = 0x08;
  msg->sniff.timeout = 0x08;

  MessagePut(0,msg);
}
```

If the *use* parameter requested that the connection should use authentication or encryption, then a link key is needed. If your application has called CM_ADD_SM_DEVICE_REQ to register the device on the other end of the link, then the Security Manager already has link keys, and it can handle authentication and encryption without further intervention from your application.

Figure 7.15 shows the case where a link key is needed, but the application has not called CM_ADD_SM_DEVICE_REQ to pass the link key and device details to the Security Manager. In this case, the Connection Manager has to come to your application and ask it for a link key using the CM_LINK_KEY_REQ message as follows:

```
CM_LINK_KEY_REQ (BD_ADDR_T addr);
```

The *addr* parameter is the Bluetooth Device Address of the device we're trying to authenticate with. Your application has a link key for this device, so you should send it to the Connection Manager in a CM_LINK_KEY_RES message.

```
CM_LINK_KEY_RES(bool_t accept, BD_ADDR_T addr, uint8
key_val[SIZE_LINK_KEY]);
```

The *accept* parameter is a Boolean flag which signals whether to accept or reject the link key request. The *addr* parameter is the Bluetooth Device Address of the device we're trying to authenticate with, and the *key_val* parameter is the link key for that device.

If you don't have a link key, you have two options: you can either start pairing so you generate a link key, or you can set the accept flag to FALSE and reject the connection attempt.

The CM_CONNECT_CFM message is used to inform the application of the status of a connection attempt when it has succeeded or failed. It's structure is as follows:

```
CM_CONNECT_CFM (connect_status_t status, BD_ADDR_T addr)
```

The *status* parameter gives the result of the connection attempt. Possible values include:

CmConnectComplete Success

CmConnectTimeout Timed out

CmConnectCancelled Error during RFCOMM (or SDP) negotiation

CmConnectDisconnect Disconnect after *connectComplete*

The *addr* parameter is the Bluetooth Device Address of the device which is the target of the connection attempt.

Once you have set up a basic ACL link, your application could add a SCO link by using a CM_SCO_CONNECT_REQ. There must be an ACL link present and not in park mode for this call to succeed.

```
CM_SCO_CONNECT_REQ (BD_ADDR_T addr, uint16 pkt_type)
```

The *addr* parameter gives the Bluetooth Device Address of the device which the SCO connection will be opened to. The *pkt_type* parameter gives the type of SCO packet to use on the connection. The Connection Manager is intended for simple applications, so it only supports a single SCO link. The BlueCore chip itself supports up to three SCO links, so there is no hardware limitation on establishing SCO links. However, the Connection Manager was written this way because it was thought unlikely that an embedded on-chip application would need to use more than one bi-directional voice link.

The CM_DISCONNECT_REQ message is used to destroy a link. If a SCO link is destroyed, the underlying ACL link will still exist.

```
CM_DISCONNECT_REQ ( link_type_t link_type, BD_ADDR_T addr)
```

The *link_type* parameter is the type of link being destroyed, RFCOMM or SCO. The *addr* parameter gives the Bluetooth Device Address of the device at the other end of the connection being destroyed.

Sending Data

Once a connection has been established, data may be sent to or received from the peer. CM_DATA_REQ is used to transmit data; CM_DATA_IND is used to indicate incoming data. CM_DATA_CFM is used to indicate to the library client how many more packets can be sent before flow control is asserted.

The *data* parameter is a pointer to a dynamically allocated data block. The *length* parameter, meanwhile, gives the length of the data:

```
CM_DATA_REQ ( uint8 * data, uint16 length);
```

The *addr* parameter gives the Bluetooth Device Address of the device which data is to be transmitted to. The *length* parameter gives the length of the data block, and data points to the dynamically allocated data block. This must be freed by the client:

```
CM_DATA_IND( BD_ADDR_T addr, uint16 length, uint8 *data);
```

The *tx_credits_left* parameter gives the number of transmit credits that the application has left under the RFCOMM credit-based flow control scheme:

```
CM_DATA_CFM ( uint16 tx_credits_left )
```

Using Other Messages and Events

The Connection Manager supports three indication messages which are used to asynchronously indicate when a connection status changes, or when an error occurs.

The Connection Manager uses the CM_CONNECT_STATUS_IND message to inform the client of changes in the status of an RFCOMM connection. This is structured as follows:

```
CM_CONNECT_STATUS_IND ( connect_status_t status, BD_ADDR_T addr)
```

The *status* parameter is set to *CmConnectComplete* or *CmConnectDisconnect*. The *addr* parameter is the Bluetooth Device Address of the device whose link status is being reported.

The Connection Manager uses a similar indication to let your application know about changes in the status of a SCO link.

```
CM_SCO_STATUS_IND (connect_status_t status );
```

The *status* parameter is just the same as for the CM_CONNECT_STATUS_IND: it is set to *CmConnectComplete* or *CmConnectDisconnect*. The Connection Manager uses the CM_SCO_STATUS_IND message to inform the client of the establishment or loss of a SCO link. There is no need for the *addr* parameter, as you can only establish one SCO link at a time.

```
CM_ERROR_IND ( cm_error_t error,  BD_ADDR_T addr);
```

The *error* parameter identifies the error which occurred while performing an operation related to the remote device with Bluetooth Device Address *addr*. An error indication may be generated if the client application attempts to:

- Issue a connection request while the Connection Manager is not idle.

- Issue a pairing request while the Connection Manager is not idle.

- Send data before a connection is established.

- Issue a cancel request while the Connection Manager is idle.

The Connection Manager also provides a cancel request. This is used to cancel any pairing or connection activity in progress, so it takes no parameters. There is no confirmation for this message. However, a pairing or connection confirm with a status of CM_cancelled may be generated as a result of a cancellation.

```
CM_CANCEL_REQ();
```

Deploying Applications

The most direct route to deploying an application is to generate a complete image, including the firmware, and to program it in to your device over SPI. This is the approach used during development. Alternatively Device Firmware Upgrade (DFU) tools are available from CSR (see www.csr.com) which allow you to produce an image of the application and, optionally, any application persistent store data. This image can be loaded using the DFU protocol over USB, H4, or BCSP.

Why would you want to go to the extra trouble of producing an image suitable for loading using the device firmware upgrade tools? There are several reasons:

- End users can use the DFU tools to upgrade their devices.

- The DFU protocol works over USB, H4, or BCSP, so your end-user products do not need the extra circuitry to support the SPI interface.

- The DFU process permits signing and verification of application images. This means you can stop end users from downloading images other than the ones you provide. This allows you to control which applications run on your products, stopping anyone with a copy of BlueLab from hacking your devices.

Device Firmware Upgrade is not possible with RFCOMM firmware. The reason for this is that there is not enough code space on a BlueCore chip to support both RFCOMM and the bootloader used by DFU.

Debugging...

Using Event-Driven Code to Save Power

Applications running under the Virtual Machine should be event-driven. You should avoid using polling loops. If you must poll for a value then use a timer event to wake up your application periodically. This is more efficient than constantly running loops, as it will allow the chip to place itself in low-power mode whenever possible.

Summary

This chapter has shown how to create, debug, and download embedded applications for the BlueCore single chip Bluetooth device.

The BlueCore Bluetooth stack takes care of managing RFCOMM links. You just have to write applications to run on top of RFCOMM. Your applications will run under an interpreter called the Virtual Machine (VM) which will safeguard the Bluetooth protocol stack, allowing it to keep its prequalified status.

You can run your BlueLab applications on a PC under a debugger. This allows you to develop and debug your applications in an environment with all the usual debugging facilities. When your application runs on the chip, VM Spy can be used to communicate on BCSP Channel 13—this is the only way of debugging on the chip.

By using the libraries and sample applications supplied with BlueLab, you can speed up application development.

Device Firmware Upgrade (DFU) tools are available which allow field upgrade for applications which do not use RFCOMM. The bootloader required for DFU will not yet fit on builds with RFCOMM, so applications using RFCOMM cannot be upgraded with the DFU tools.

Solutions Fast Track

Understanding Embedded Systems

☑ Embedded systems commonly have many tasks running simultaneously. Since the processor can only run one line of code at a time, a scheduler swaps between tasks running a few instructions from each in turn.

☑ On BlueCore, your application task is called through an interpreter referred to as the Virtual Machine, which interprets a few of your instructions each time it is called. This interpreter means that even if you write code in an endless loop, the other tasks in the system will still get to run. The Virtual Machine's interpreter also stops you from accessing areas of memory which are needed for other tasks.

☑ Tasks communicate by sending messages to one another, using areas of memory which are set up as queues. The first message in the queue is the first out, so these are sometimes called FIFOs (First In First Out).

☑ Application software can interact with hardware using interrupts. There are two pins on BlueCore which will generate an interrupt when they change state. An application can register to be notified when these interrupts happen.

☑ When you close a switch, the contacts usually bounce off one another. This bouncing causes the switch to oscillate, making and breaking a connection. This means that if a switch (such as a pushbutton, or keypad) is connected to an interrupt line, you will get many interrupts as the switch closes. BlueLab provides debounce routines.

Getting Started

☑ To create embedded applications to run on CSR's BlueCore chip, you need BlueLab and a Casira. The Casira must be configured to run BCSP.

Running an Application under the Debugger

☑ The PC is connected to the Casira with a serial cable and an SPI cable.

☑ The Casira must be loaded with a null image containing an empty version of the Virtual Machine.

☑ Applications running under the debugger on the PC can then use facilities on the Casira, so they can access PIO pins and the BlueCore chip's radio while still having full PC debugging facilities.

Running an Application on BlueCore

☑ You must make a special firmware build linking your application with a Virtual Machine build to run your application on the Casira.

☑ Your application should be fully debugged before you build it for BlueCore, since on-chip debugging facilities are very limited.

☑ You can communicate with the Virtual Machine on BCSP Channel 13 using VM Spy.

Using the BlueLab Libraries

☑ A selection of libraries provide ANSII C support as well as access to the Bluetooth protocol stack, PIO pins, and various operating system facilities such as scheduling, timers, messaging, and so on.

Deploying Applications

☑ If you do not have RFCOMM in your build, you can upgrade devices in the field using the Device Firmware Upgrade (DFU) tools. Otherwise, you must program the flash using an interface similar to the SPI interface.

Frequently Asked Questions

The following Frequently Asked Questions, answered by the authors of this book, are designed to both measure your understanding of the concepts presented in this chapter and to assist you with real-life implementation of these concepts. To have your questions about this chapter answered by the author, browse to **www.syngress.com/solutions** and click on the **"Ask the Author"** form.

Q: Why does the Casira use BCSP instead of the H4 UART interface from the Bluetooth 1.1 Core Specification?

A: The H4 UART interface was designed for chips separated by about 3 mm of copper on a circuit board. When the ends of the serial interface are separated by a few feet of serial cable, errors can occur. BCSP protects against those errors. It also provides separate flow control for voice and data, which is not possible when using the 1.1 H4 UART Specification. Finally, BCSP provides a debug channel which is essential for developing and debugging embedded applications on BlueCore chips.

Q: Where does the output from *printf* or *putchar* go when the application is running on the chip?

A: STDIO is routed over BCSP and appears on the Channel 13 debug datastream. You can view it with the VM Spy utility. If you are running H4, the BCSP Channel 13 appears as a manufacturer extension.

Q: If the Virtual Machine slows my application down, why do I have to run applications under the Virtual Machine?

A: Your application could alter the way the Bluetooth protocol stack runs by taking too many system resources, such as processor time and memory. The VM checks all memory accesses and jumps, thus safeguarding the memory that the Bluetooth protocol stack needs. Without the Virtual Machine, the Bluetooth protocol stack could have its performance compromised, which would affect its qualified status.

Chapter 8

Using the Palm OS for Bluetooth Applications

Solutions in this chapter:

- **What You Need to Get Started**

- **Understanding Palm OS Profiles**

- **Updating Palm OS Applications Using the Bluetooth Virtual Serial Driver**

- **Using Bluetooth Technology with Exchange Manager**

- **Creating Bluetooth-Aware Palm OS Applications**

- **Writing Persistent Bluetooth Services for Palm OS**

- **The Future of Palm OS Bluetooth Support**

- ☑ **Summary**

- ☑ **Solutions Fast Track**

- ☑ **Frequently Asked Questions**

Introduction

Of all the PDAs on the market, it is probably Palm, Inc.'s devices that have made the most use of short-range communications. Previously, this has been limited to line of sight beaming via the infrared (IR) interface, but with version 4.0 Palm OS support was in place for Bluetooth wireless technology and line of sight limitations became a thing of the past. Palm, Inc. has said that it will begin to ship Bluetooth accessories in the near future (some are already available to developers), and it plans to integrate Bluetooth technology into its handheld devices before too long. A number of Palm OS licensees have also expressed interest in shipping a Bluetooth solution.

However convenient handhelds may be, it's undeniably awkward trying to juggle more than one device while you're on the move. Adding Bluetooth wireless technology to a Palm device frees users from the necessity of trying to physically line up two devices while they're mobile. It also allows up to eight devices to communicate at once. The Bluetooth system is omni-directional and its radio waves can pass straight through solid objects.

Bluetooth technology includes traditional Palm OS applications like Internet usage and "beaming" easier in mobile environments, but it also creates interesting opportunities for new applications. *Object push* opens up the possibility of spontaneous communication: you only need to walk into range of a server to see its information pop up on your Palm device's display. Of course with new communication channels come new security and user experience concerns. Security and ease of use are prime concerns of the new Bluetooth support.

This chapter will give you an insight into Palm OS Bluetooth support, enabling you to port your existing Palm OS applications to use Bluetooth technology, or explore a whole new vista of applications which were not practical with previous communication technologies. Examples make it clear exactly how things are done, so you can start using Palm OS for Bluetooth applications right away.

What You Need to Get Started

Before you start work on your first Palm OS Bluetooth application, there are a few tools you will need. Fortunately, if you are currently a Palm OS developer, you probably have many of these tools already, and those you don't have are easily available from the Palm, Inc. Web site at www.palmos.com.

Bluetooth support in the Palm OS is an extension to Palm OS 4.0, and is made up of several Palm Application files (.prc files) that may be included in a device's

ROM image, or may be installed with the HotSync install tool and run from RAM. In order to begin using Bluetooth technology, you will need to have a Palm OS device with at least 4 MB of memory that is running Palm OS version 4.0 or greater. Alternatively, if you wish to develop using the Palm OS Emulator, often the easiest and fastest way to create new application, you can obtain a 4.0 ROM image, and the 4.0 Software Development Kit (SDK), from the Palm Resource Pavilion at www.palmos.com/alliance/join. The Palm OS Emulator is available for download from the Development Support area of the Palm, Inc. Web site at www.palmos.com/dev/tech/emulator. You may also find it useful to download the Palm Reporter application, which allows you to see real-time traces from your application.

In addition to a Palm 4.0 device, you will need to have the Bluetooth Support Package installed. The Bluetooth Support Package consists of several .prc files that work together. For the moment, don't worry about understanding what each individual piece does, simply make sure that you have them all installed. The easiest way to know if your Palm device has Bluetooth support installed is to go into the "Preferences" application and check to see if "Bluetooth" appears in the list of preference screens in the upper-right corner. This indicates that at least part of the Bluetooth Support Package has been installed. If you find that you have trouble using Bluetooth technology later on, you may wish to double-check that all the files in the package are installed by going to the Info screen in the launcher (from the menu, choose "App" then "Info") or by simply reinstalling all of the .prc files in the package. Unless the device you are using has Bluetooth technology built-in, it is unlikely that the installed ROM image will include Bluetooth support. The latest version of the Bluetooth support .prc files, along with the Bluetooth header files and several pieces of example code, can be found in the Bluetooth area of the Palm Resource Pavilion at www.palmos.com/dev/tech/bluetooth. Developers can also find information on how to obtain early releases of Palm OS Bluetooth development hardware at this site.

In addition to the tools listed here, you will also want to have a copy of the Palm OS 4.0 SDK documentation, also available on the Palm, Inc. Web site. You may find that it is useful to have the 4.0 documentation on hand as you read through this chapter, since there may be references to Palm OS functions calls and data structures with which you are not yet familiar.

Finally, before you get started, you should know that the function definitions and data structures used in the code examples in this chapter are not final. As this text is being written, the Palm OS Bluetooth solution is still in the alpha phase, and while the overall model and methods are not expected to change, some characteristics and arguments of individual API calls, along with some file names, may vary from what is presented here. The code examples presented here should be seen as a basis from which to work, but may require slight modification in order to compile. Refer to the Palm OS documentation and header files for the final word on the API.

Understanding Palm OS Profiles

This section will present an overview of the different profiles supported in the Palm OS Bluetooth Support Package. If you are not familiar with the general concept of profiles, you way wish to go back and review Chapter 2 before continuing.

The Palm OS currently supports five Bluetooth profiles defined in the Bluetooth 1.1 Specification. As shown in Figure 8.1, these profiles are:

- Generic Access Profile
- Serial Port Profile
- Dial-up Networking Profile
- LAN Access Profile
- Object Push Profile

Figure 8.1 Bluetooth Profiles Supported by the Palm OS

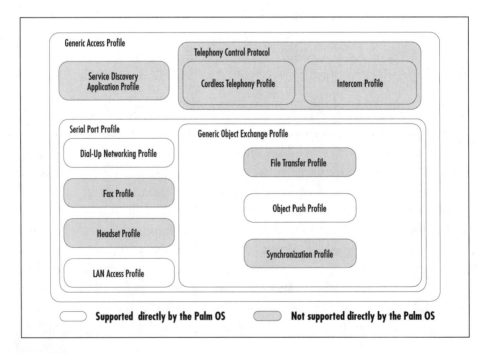

All the profiles help to ensure interoperability by providing common rules that all Bluetooth devices follow. It is vital to follow these rules as they form part of the Bluetooth qualification process. (Products must pass qualification to obtain the free license to use Bluetooth technology.)

Of these profiles, the *Generic Access Profile* (GAP) is unique. Unlike the other profiles, which describe a method for accomplishing a specific user goal, the GAP is a general look at the overall process of carrying out a Bluetooth transaction without regard to the nature of that transaction, and is background for all the other profiles. As such, there is no one place in the Bluetooth Support Package that the GAP is exposed, rather the values and language specified by the GAP are built into the Bluetooth Library and other Bluetooth components. GAP's main goal is to create a friendly and consistent user experience, a goal that is also considered critical in the Palm OS. We will see how the Bluetooth Support Package tries to help application developers maintain easy and consistent experience across applications.

The Bluetooth Support Package includes a new virtual serial driver (a VDRV for short), similar to the IrComm virtual serial driver you may already be familiar with, which provides support for the *Serial Port Profile*. Both Device A and Device B roles of the profile are supported. Existing OS components that make use of serial services such as Point-to-Point Protocol (PPP), HotSync, and the Telephony Manager are ready to take advantage of the Bluetooth VDRV, and other serial-based applications can easily be updated to make use of the Bluetooth VDRV. We will explore the use of the Bluetooth VDRV in great depth later in this chapter.

The Network Library (NetLib) supports the Data Terminal role of both the *Dial-up Networking* and *LAN Access Profiles*. After installing the Bluetooth Support Package, you'll notice that the Connection panel in the preferences application will allow users to choose Bluetooth technology as a transport when configuring a connection to a local network, phone, modem, or PC. The OS uses these settings to determine which profile to use when NetLib is opened. Since applications that use NetLib are unconcerned with how it creates its underlying transport, the use of the Dial-up Networking and LAN Access Profile is transparent to NetLib-based applications. An e-mail application, for example, that was developed using NetLib running over a normal modem can be used with Bluetooth technology when the user configures the Network panel to use a Bluetooth device. Since the application is unaware of the use of the Dial-up Networking and LAN Access Profiles, we will not spend too much time talking about them.

> ## Debugging…
>
> ### Using NetLib with Bluetooth Technology
>
> The Bluetooth protocol stack uses a good bit more heap space than a simple serial driver does. Because of this additional heap usage, you may run into problems if your application is already on the edge of causing a stack overflow, or running out of heap space. Running out of heap space will most likely cause your application to receive NULL back from a memory allocation operation. A well-written operation will always test for failure when allocating memory, and fail gracefully if the needed memory chunk can't be allocated. Testing with the Palm OS Emulator is a good way to watch for stack overflow conditions; the emulator will tell you when your application is running close to stack boundary conditions.

The Bluetooth Support Package also includes the Bluetooth Exchange Library. This new Exchange Library implements the *Object Push Profile*, much in the same way that the Exchange Manager supports IR-based Object Exchange Protocol (OBEX) push. You may have noticed that the Exchange Manager in OS 4.0 has been extended to handle multiple transports. Using these new features, it is easy to update legacy Exchange Manager-based code to take advantage of Bluetooth technology (in some cases by changing only a single line of code). New functions allow Bluetooth savvy applications to better handle multiple recipients, and create a better user experience. We will spend a bit of time going over some of these new functions and give some suggestions on how to update your application.

Choosing Services through the Service Discovery Protocol

You may have noticed that support for the *Service Discovery Application Profile*, a major part of many platforms' user experience, is absent from the Palm OS's list of supported profiles. It is important to note that supporting the Service Discovery Application Profile is very different from supporting the Service Discovery Protocol (SDP), which the Bluetooth specification mandates and for which Palm OS offers full support. The aim of the Service Discovery Application

Profile is to define how information gained through the Service Discovery Protocol might be presented to the user, and presents two basic usage models: *Service Browsing* and *Service Searching*.

In the Service Browsing model, the user would see a list of available devices (the result of a inquiry) and be able to open each device and look through the list of services that that device presents. After browsing, the user would presumably pick the device and service that they wish to utilize. Palm, Inc. does not endorse this model because they believe that the application, not the end user, should be responsible for knowing which service it needs to communicate with, and for being able to find that service. When I sit down at a PC, for example, and type an IP address into an application, I don't get a list of all of the possible services I can connect to on the remote server as well as a query about which one I wish to connect to. Rather, the application knows that it is a Web browser or a Ping application, and it knows how to find and connect to the appropriate service; if the host does not offer the service, I get an appropriate error message. The same should be true with Bluetooth technology; applications should be responsible for knowing which services they want to use and for knowing how to connect to them.

In the Service Searching model, the user (or application) selects which service they wish to use and then are presented with a list of available devices that present that service. From a user-experience point of view, this is clearly a better model. Unfortunately, this model still causes a problem. The most obvious time to do a service search is during the discovery process, an operation which most users find takes too long already. You could conceivably cache the service lists of remote devices, but this cache would need to be quite large to be useful and it would be difficult to know when your cache was out-of-date. On a large device that has lots of CPU time and battery power to waste making regular inquiries in the background, Service Searching might be a good model, but on a small device it seems like overkill. Rather, it seems to make more sense to use the Class of Device (CoD) information returned during inquiry to do the same kind of service-based filtering. While the information in the CoD is less specific than the information available through SDP, using CoD is probably sufficient in most cases and can actually shorten the total discovery time since devices can be eliminated before a name request is done. As we will see later on, the Palm OS offers a robust model for CoD-based filtering during discovery. Finally, if a developer decides that he or she really wants to use the Service Discovery Application Profile, all of the tools necessary to implement the desired parts of the profile are available to the application.

If none of the profiles cover what you are trying to do, don't despair. The Palm OS also provides a robust API that allows you direct access to the SDP,

RFCOMM, and Logical Link and Control Adaptation Protocol (L2CAP) layers of the Bluetooth stack, along with calls to allow you to manage the Bluetooth-specific concerns like discovery and piconet creation.

Updating Palm OS Applications Using the Bluetooth Virtual Serial Driver

Using the Bluetooth Virtual Serial Driver allows existing serial-based applications to quickly be updated to take advantage of Bluetooth technology, and is an easy way to create new Bluetooth-enabled applications. Virtual serial drivers in the Palm OS are individual .prc files of type *vdrv* and are used throughout the new Serial Manager interface, much the same way as traditional physical serial ports are used. The Bluetooth VDRV is included with the Palm OS Bluetooth Support Package. This section will focus on the unique aspects of using the Bluetooth VDRV; for information on the general use of the new Serial Manager, refer to the Palm OS documentation directly. Figure 8.2 shows a basic overview of how Bluetooth technology fits into the Palm OS communications architecture.

The Bluetooth VDRV, in accordance with the Serial Port Profile, runs on top of the RFCOMM protocol layer. It is worth noting that the VDRV does not implement RFCOMM itself. The RFCOMM protocol layer is implemented in the Bluetooth Library and can be accessed directly through the Bluetooth Library API (discussed in depth later in the chapter). The VDRV itself is "glue code" that allows Bluetooth functionality to be accessed though a more traditional API. Using the VDRV also gives you an advantage in writing multi-transport applications. Since there are only a few differences between using the IrComm VDRV and the Bluetooth VDRV, much of your code will not need to be altered in order to use both transports.

Gluing new technology underneath an old interface always presents some challenges and there are a few limitations to using the Bluetooth VDRV that you should be aware of. In order to achieve certain performance optimizations, the Bluetooth VDRV opens the Bluetooth Library with a slightly different configuration than is normally used when an application opens the Library. As such, the Bluetooth VDRV and the Bluetooth Library cannot be opened by the application at the same time. Since NetLib and the Telephony Manager can be configured to use the Bluetooth VDRV, the Bluetooth Library and the VDRV may not be available when these other components are in use. Applications are also limited to using a single instance of the Bluetooth VDRV at any given time.

Figure 8.2 How Bluetooth Technology Fits into the Palm OS Communications Architecture

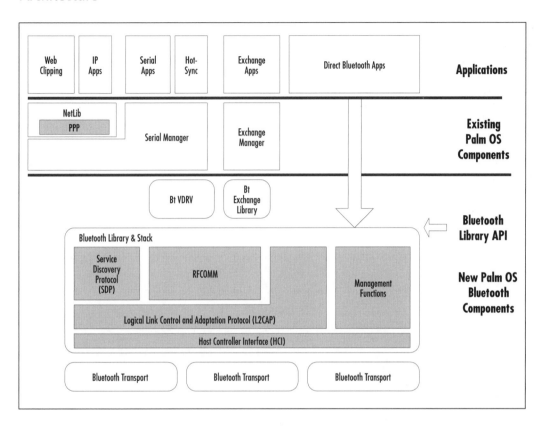

One of the challenges of mapping Bluetooth underneath a traditional serial API is that traditional serial ports are single-channel and non-addressed in nature, while the Bluetooth system is a multiplexing, address-based protocol stack. A traditional serial port driver can simply initialize its local hardware, start talking and hope that there is a cable in place and someone listening on the other side, while Bluetooth technology needs to know which device and which service on that device it is going to talk to; it must also actively create the underlying baseband connection. Since most Bluetooth radios are not capable of simultaneously listening for an inbound connection and trying to create an outbound connection, an instance of the Bluetooth VDRV also needs to know whether it is initiating or accepting the connection.

Since a traditional serial API does not present a mechanism for passing all of this extra information, Palm OS 4.0 has added a new call, *SrmExtOpen()* (found in SerialMgr.h), to the New Serial Manager API. The *SrmExtOpen()* call allows an application to pass down additional configuration data, along with a driver-specific configuration structure. *SrmExtOpen()* must be used to initialize the Bluetooth VRDV—passing the Bluetooth VDRV into the older *SrmOpen()* call will simply cause the call to fail.

The top level configuration structure that is passed into the *SrmExtOpen()* function for the Bluetooth VDRV is defined in the SerialMgr.h file as the following:

```
typedef struct SrmOpenConfigType {
    UInt32 baud;           // Baud rate that the connection is to
                           // be opened at.  The Bluetooth VDRV
                           //ignores this value.
    UInt32 function;       // Designates the function of the
                           // connection.
                           // Non-OS components should set this value
                           // to zero.
    MemPtr drvrDataP;      // For the Bluetooth VDRV, a pointer to an
                           // instance of RfVdOpenParams.
    UInt16 drvrDataSize;   // For the Bluetooth VDRV,
                           // sizeof(RfVdOpenParams).
    UInt32 sysReserved1;   // System Reserved.
    UInt32 sysReserved2;   // System Reserved.
} SrmOpenConfigType;
```

When using the Bluetooth VDRV, the *drvrDataP* element should be filled in with a pointer to an instance of the *RfVdOpenParams* structure. This is a Bluetooth VDRV-specific structure, and applications should be sure that they are dealing with the Bluetooth VDRV before passing the pointer. The *RfVdOpenParams* structure, along with several supporting structures, is defined in RfCommVdrv.h. Later, we'll see examples of how these structures are used. First, let's take a look at the structures themselves.

```
typedef struct {
    RfVdRole     role; // client or server?
    Boolean      authenticate; // force link authentication
    Boolean      encrypt; // force link encryption
```

```
    union {
        RfVdOpenParamsClient      client; // client parameters
        RfVdOpenParamsServer      server; // server parameters
    } u;
} RfVdOpenParams;
typedef enum {
    rfVdClient,       // RFCOMM client
    rfVdServer       // RFCOMM server
} RfVdRole;
```

As mentioned earlier, most Bluetooth radios are not capable of receiving inbound connections while trying to create outbound connections. For this reason, it is necessary for an application to indicate whether it wishes to initiate or accept the Asynchronous Connectionless Link (ACL) and RFCOMM connections. Palm OS refers to these roles as the *client role* and the *server role*, respectively. The application indicates its preference by setting the corresponding value for the role element in the *RfVdOpenParams* structure and filling the appropriate role-specific parameter structure inside the union. The *authenticate* and *encrypt* values are used to specify the security requirements for the link; if these requirements cannot be met, the link will be dropped.

```
typedef struct {
    BtLibSdpUUIDType    uuid; // UUID of the service to be advertised
    Char*               name; // optional readable name of the service
} RfVdOpenParamsServer;
```

When the VDRV is opened in the server configuration, it will register for an RFCOMM channel and advertise that channel via SDP. This creates a simple service record utilizing the Unique Universal Identifier (UUID) and name string defined in the *RfVdOpenParamsServer*. If the application wants to create a more robust service record, it should use RFCOMM and SDP directly through the Bluetooth Library (BtLib) API.

UUIDs are used to uniquely identify an application, or more specifically, the protocol the application expects to communicate with. If the application is willing to handle the possibility that it may get a connection to an incompatible application, and the application will only be used between two Palm OS devices, the *uuid* can be set to 0. This will cause the VDRV to use a predefined UUID unique to the Palm OS. If the server chooses to set the *uuid* to 0, the client should do so as well.

Since all actions involved in a server open are local, the open call should only fail if there is a resource conflict.

```
typedef struct {
    BtLibDeviceAddressType      remoteDevAddr; // the device to connect to
    RfVdClientMethod            method;        // how to determine remote
                                               // RFCOMM channel

    union {
        BtLibRfCommServerIdType channelId; // method ==
                                           // rfVdUseChannelId
        RfVdUuidList            uuidList;  // mettod ==
                                           // rfVdUseUuidList
    } u;
} RfVdOpenParamsClient;
typedef enum {
    rfVdUseChannelId,      // use an RFCOMM channel id
    rfVdUseUuidList        // use SDP to find a channel based upon a
                           // service UUID.
} RfVdClientMethod;
typedef struct {
    UInt8       len;           // length of table == number of UUIDs
    BtLibSdpUUIDType*  tab;   // table of UUIDs
} RfVdUuidList;
```

To open the VDRV in the client configuration, a more complex structure must be passed in to *SrmExtOpen()*. The *remoteDevAddr* parameter indicates the 48-bit Bluetooth device address of the remote device the VDRV should connect to. The application might determine what address to use by making a call to *BtLibDiscoverSingle()* in the BtLib API (discussed later), or by taking an address from a Connection Manager Profile that uses Bluetooth technology. If *remoteDevAddr* is set to 0, the VDRV will perform a device discovery and ask the user to specify a remote device during the open. After creating an ACL connection to the remote device, the VDRV attempts to establish an RFCOMM connection. The application must indicate which RFCOMM channel the VDRV should use. The channel is determined by using SDP to look up the Channel ID of the remote service. While the application is welcome to use the SDP function calls in the BtLib API to obtain the Channel ID (and the *rfVdUseChannelId* method to pass in), the VDRV presents an easier method. By using the *rfVdUseUuidList* method, the application can simply pass in the UUID of the service it wishes to utilize. Passing in more than one UUID will cause the VDRV to run through the list until it finds a service it can use. The VDRV will look for a service record with the given service UUID, and if a record is found, it will then search for the RFCOMM Channel in the record's protocol descriptor list (if multiple protocol descriptor lists are contained in the record, the VDRV will use the first RFCOMM channel it comes across). Setting the method to *rfVdUseUuidList* and setting *len* to 0 will cause the VDRV to look for the predefined Palm OS UUID (discussed earlier).

Since a client-open may block for several seconds while the ACL connection is brought up, the VDRV may display some UI to allow the user to see the connection progress.

Creating a VDRV Client-Only Application

Let's move on to looking at a real VDRV client-only application. Such an application might be useful when you know that the Palm device will always be playing a client-based role, and therefore never need to accept a connection. Let's imagine that we are creating an application for controlling home appliances, using the (entirely imaginary) Bluetooth Based Blender Remote Control Profile (B3RCP for short). Since, as we all know, B3RCP is based on the serial port profile, it is appropriate to use the VDRV. Furthermore, since we know that the Palm device will always initiate the connection to the blender (after all, appliances don't generally initiate contact with the remote control), the Blender-control application is a good example of a client-only application. For the purpose of this example, we will assume that the B3RCP is a well-known protocol, and that a UUID of 07004F16-3776-11D5-83CE-0030657C543C has been established as a service ID for B3RCP services. For your own applications, you will need to use established UUIDs for the profile you are using, or create a new UUID yourself using one of the many UUID (sometimes called GUID) generation tools that are commonly available on the Web.

Let's look at the code fragment that performs the VDRV open call.

```
#include <PalmOS.h>
#include <BtLib.h>
#include <BtLibTypes.h>
#include < RfCommVdrv.h>
```

The structure *BtLibSdpUUIDType* consists of a size indicator and an array of bytes that form the UUID itself. The size of all UUIDs not declared directly in the Bluetooth specification is btLibUuidSize128.

```
#define uuuidB3RCP  \
        {btLibUuidSize128,{0x07,0x00,0x4f,0x16,0x37,0x76,0x11,0xd5,  \
                        0x83,0xce,0x00,0x30,0x65,0x7c,0x54,0x3c}}
UInt16    gPortId;
Err OpenPortAsClient( void )
{
    Err                 err;
    SrmOpenConfigType   config;
    RfVdOpenParams      rfparams;
    BtLibSdpUUIDType    remoteServiceID = uuuidB3RCP;
    // To be on the safe side, set all of the parameter structures to 0
    // before starting:
    MemSet( &config, sizeof(config), 0);
    MemSet( &rfparams, sizeof(rfparams), 0);
    config.function = 0;    // non-OS components must use zero
    config.drvrDataP = (MemPtr)&rfparams; // driver specific params
    config.drvrDataSize = sizeof(RfVdOpenParams);
    // All other elements of the SrmOpenConfigType structure are ignored
    // by the Bluetooth VDRV, so skip to filling in VDRV specific info:
    rfparams.role = rfVdClient;  // we are the client side
    // We don't care about security but the appliance may insist on it:
    Rfparams.encrypt = false;
    Rfparams.autheniticate = false;

    // Use the discovery function in the Bluetooth Library to get the
    // remote device address:
    err = GetAddressFromUser( &rfparams.u.client.remoteDevAddr );
    if (err) return err;
    // Connect to the B3RCP server on the remote for this device.
    // Instruct the VDRV to find this device by looking for its Service
    // UUID:
```

```
        rfparams.u.client.method = rfVdUseUuidList;

        rfparams.u.client.u.uuidList.tab = &remoteServiceID;

        rfparams.u.client.u.uuidList.tab = 1; // no fallback services

        err = SrmExtOpen(

                sysFileCVirtRfComm, // specify the use of the Bluetooth VDRV

                &config,              // port configuration params

                sizeof(config),      // size of port config params

                &gPortId              // put the port id in a global

                );

        return err;

}
Err GetAddressFromUser( BtLibDeviceAddressType* addrP)

{

        Err error;

        UInt16 btLibRefNum = 0;

        BtLibClassOfDeviceType filter;

        // Find the Bt Library:

        if( SysLibFind( btLibName, &btLibRefNum) )

          {

          // Load the Library if it can't be found:

          error = SysLibLoad( sysFileTLibrary , sysFileCBtLib, &btLibRefNum);

          if( error  ) return error;

          }

        // Open the Library:

        error = BtLibOpen(btLibRefNum);

        if( error  ) return error;

        // Class of Device (CoD) is a value that devices return during the

        // discovery process.  A CoD value can be passed to the discovery

        // functions as filter, to keep devices in the wrong category from

        // showing up.  By setting the filter type to the values used by the

        // iBlend, the user will be restricted to a more appropriate subset

        // of discoverable devices.

        filter = btLibCOD_ServiceAny | btLibCOD_Major_Unclassified ;
```

```
// BtLibDiscoverSingleDevice() is defined in BtLib.h, and will be
// discussed in detail later in the chapter. Basically the call
// performs a discovery and asks the  user to select a device from
// the resulting list:
error = BtLibDiscoverSingleDevice( btLibRefNum, NULL, &filter, 1,
 addrP,

                                  false, false);


// You must always close the Library before returning, or the VDRV
// will not be able to open
BtLibClose( btLibRefNum );
return error;
}
```

> **WARNING**
>
> Applications and the VDRV use the Bluetooth Library in different modes. Because of this difference, the VDRV will not be able to open while the application is holding the Bluetooth stack open.

The main application block can now be coded to make a call to *OpenPortAsClient()*. If the call returns without error, the port is open and can be used as any normal serial port might be used. Closing the port will cause the RFCOMM and ACL connections to be dropped. In general, protocols that run over standard serial ports are responsible for defining their own stay-alive and timeout conditions. In general, this is true for Bluetooth VDRV ports as well, though if the ACL link is lost before *SrmClose()* is called, the *SrmSend()* call will return *serErrLineErr*.

Now, let's look at the problem from the other side.

Creating a VDRV Server-Only Application

As an employee of Frappé.com, you have been made the lead software engineer on the iBlend, the world's first Palm-device powered blender. Since the iBlend is a state-of-the-art home appliance, its feature set will clearly need to include support for B3RCP, allowing the user to make a margarita without the inconvenience of having to walk across the room. The iBlend will need to open the virtual serial port in the server role, which will require a slightly different open call.

```
Err OpenPortAsServer( void )
{
    Err                 err;
    SrmOpenConfigType   config;
    RfVdOpenParams      rfparams;
    BtLibSdpUUIDType    localServiceID = uuuidB3RCP;
    // Define a name for the service. This is optional but may be useful
    // for devices that support service browsing.
    Char*               serviceName = "Blender Control";
    // To be on the safe side, set all of the parameter structures to 0
    // before starting.
    MemSet( &config, sizeof(config), 0);
    MemSet( &rfparams, sizeof(rfparams), 0);
    config.function = 0;    // non-OS components must use zero
    config.drvrDataP = (MemPtr)&rfparams; // driver specific params
    config.drvrDataSize = sizeof(RfVdOpenParams);
    // All other elements of the SrmOpenConfigType structure are ignored
    // by the Bluetooth VDRV, so skip to filling in VDRV specific info.
    rfparams.role = rfVdServer;    // we are the server side

    // Insist on authentication, so that the mean neighbor next door can
    // not control your blender:
    Rfparams.encrypt = false;
    Rfparams.autheniticate = true;

    // Specify that the port should advertise itself in SDP with the
    // B3RCP UUID. Also provide a user friendly name for the service:
    rfparams.u.server.uuid = &remoteServiceID;
    rfparams.u.server.name = serviceName;
    err = SrmExtOpen(
```

```
            sysFileCVirtRfComm, // specify the use of the Bluetooth VDRV
            &config,            // port configuration params
            sizeof(config),     // size of port config params
            &gPortId            // put the port id in a global
            );
    return err;
}
```

The *OpenPortAsServer()* call will take care of setting up the server serial port for the main application on your iBlend. Note that setting up the port as a server does not cause the driver to go out and create an ACL or RFCOMM connection, it merely sets the port up as a listener. Like a normal serial port, the VDRV will not alert the application when an incoming connection is established, the application will simply begin to receive data from the port. Like any protocol that runs over a serial port, B3RCP must handle session establishment and termination. The port will also accept the first inbound connection it receives, as long as that connection meets the security requirements set in the *RfVdOpenParams* structure. If the protocol or application above the serial port requires additional security, it's up to that layer to implement it.

Now we have seen an example of both a client-only and a server-only use of the VDRV. At this point, you may be saying to yourself, "That's all great and everything, but I'm writing a Palm-to-Palm application. I need to be able to be both client and server!" Fortunately, this is easy. The simplest way to handle this case is to open the serial port as a server when your application is opened. When the user does something that requires a connection (i.e., pushes a start button, starts to generate input, and so on), close the serial port and reopen it as a client. You will have to somehow convey to your users that only one person should start the connection, but this is a commonplace enough idea that most users should get it without too much hassle.

Once the port has been opened, it behaves like any other Palm OS serial port. This means that you can use the same code and Serial Manager calls that you use with your existing serial application. By adding a few simple routines to open the port, you can make your legacy application Bluetooth-aware.

You should now know everything you need to know to create your first Palm OS Bluetooth application. Alternatively, you may have found that the VDRV doesn't suit your Bluetooth technology needs—it is, after all, only an emulation layer. The rest of the chapter will cover the use of the Exchange Manager and the Bluetooth API.

Using Bluetooth Technology with Exchange Manager

If you're interested in using Bluetooth technology to transfer records, or if having a constant data flow is not important to your application (as in a turn-based game), the Bluetooth Exchange Library might be the perfect tool for you to use. The Exchange Library allows applications to send data blocks without having to worry too much about the underlying transport. Unlike sockets and virtual serial drivers, the Exchange Manager is a concept unique to Palm OS. Gavin Peacock, the engineer at Palm, Inc. who came up with the Exchange Manager, explains that the need for the Exchange Manager comes from the lack of a file system in the OS (OS 4 does support a file system for use with expansion cards, but the user is unaware of it). In other OSs, if the user wants to send a file over a given transport, they save the file somewhere and then go to the application responsible for that transport (i.e., the e-mail application, the IR exchange application, and so forth) and specify the file they want to send. In Palm OS, the Exchange Manager creates a singular API that brings all of the available transports to each application, avoiding the need to deal with file systems and transport-specific applications. The Palm OS SDK documents go into the use of the Exchange Manager in great detail; we'll concentrate here on new issues that are of particular relevance to using the Bluetooth Exchange Library.

The Bluetooth Exchange Library is so easy to use, your application might already be set up to use it. The Exchange Manager in Palm OS 4.0 introduced a new URL send scheme, known as the *exgSendScheme*. Rather than referring to a specific transport, the send scheme instructs the Exchange Manager to allow the user to pick which of the installed transports they wish to utilize. The Bluetooth Exchange Library registers itself for the *exgSendScheme*, so if you've already updated your application to take advantage of the *exgSendScheme*, it should work with Bluetooth technology as soon as you have installed the Bluetooth .prc files. If you haven't yet updated your application to use send, the Address Book code in the SDK contains a good example of how *exgSendScheme* is used. If you know that your application only wants to use Bluetooth technology, you can indicate this by using the *btExgScheme* ("_btObex") instead of the *exgSendScheme*. The result will be the same as using the *exgSendScheme*, except that the user won't be offered a choice of transports.

Once the Bluetooth system has been chosen as the transport, the Exchange Library will automatically perform a discovery in order to determine the address of the remote device it should connect to. If you already know the Bluetooth device address you wish to connect to, you can indicate this in the URL by sticking the address in the URL you pass in with the exchange socket as follows:

```
Char *urlBase = "_btObex://12.34.56.78.9A.BC/filename.ext"
```

In reality, you would probably build this URL string dynamically, instead of hard coding it. The first section of the URL defines the scheme, as discussed earlier. The second section of the URL is a character representation of the Bluetooth device address of the target device. If you have the device address stored in a *BtLibDeviceAddressType* structure, you can easily convert the address to a string by calling *BtLibAddrBtdToA()* (this function can be called without first opening the Bluetooth Library). This kind of usage might be useful in situations where the application keeps some kind of "buddy list" of devices, making a discovery unnecessary. In addition to a single device address, the second section of the URL can also use the meta-addresses "_single" and "_multi", which indicate that the Exchange Library should perform a discovery and prompt the user to select one or multiple devices, respectively. For multiple recipients, the URL addressing convention is to separate the recipient's Bluetooth device addresses with a comma, as follows:

```
Char *urlBase = "_btObex://11.22.33.44.55.66,77.88.99.AA
    .BB.CC/filename.ext"
```

The last section of the URL is the name and extension you wish the file to have when it is sent to the remote device.

In some applications, such as a chess game, you may wish to have a discovery occur on the first move, but then always use the same device address for each move afterwards. This can be accomplished using a new ExgMgr call control call named *exgLibCtlGetURL*. The Bluetooth Exchange Lib is the first to implement this control, but it is expected that other Exchange Libraries that use addresses (such as SMS) will be updated to use it soon. The purpose of the *exgLibCtlGetURL* control is to allow the application to retrieve an exchange sockets URL *after* the Exchange Library has filled it out. The call can be made any time after a successful *ExgPut()*, *ExgConnect()* or *ExgAccept()* call, and before *ExgDisconnect()* is called.

When invoking the exgLibCtlGetURL control, the *valueP* parameter passed to *ExgControl()* should be a pointer to a *ExgCtlGerURLType* structure, which is defined as:

```
typedef struct _ ExgCtlGerURLType {

    ExgSocketType *socketP;

    Char *URLP;

    UInt16 URLSize;

} ExgCtlGerURLType;
```

Obviously, *socketP* is a pointer to the *ExgSocket* you are trying to get the URL for, *URLP* is a pointer to the character buffer where the URL will be stored, and *URLsize* is the size of the buffer. If the call is successful, the URL buffer will be filled in, and so will the length of the URL (including the mandatory NULL terminator). If the application wants to dynamically allocate the URL buffer space, it can first make the call with the URLP set to NULL and the *URLSize* parameter set to 0. In this case, the call will simply return the URL size so that the application can allocate an appropriately-sized buffer to retrieve the URL with. Once the application has retrieved the URL, it can utilize the same URL with future *ExgSockets* to indicate that it wants to use the same exchange scheme and remote device. The Tic-Tac-Toe application in the Palm OS Bluetooth developer kit provides an excellent example of an application that makes use of the Bluetooth Exchange Libraries' URL scheme to create a two-player game.

On the receiving side, the application is generally unaware of which exchange transport is being used.

In certain scenarios, such as the chess game just described, the receiving app may wish to grab the sender's address from the URL for use in subsequent moves.

Creating Bluetooth-Aware Palm OS Applications

The VDRV and Exchange Manager simplify using Bluetooth technology by encapsulating it inside familiar and easy-to-use interfaces, but the simplification also hides functionality and increases overhead. If the Exchange Manager or the VDRV suit your needs, then you should certainly use them, but if your application requires direct access to Bluetooth protocol layers or management functions, then you will need to make use of the Bluetooth Library (BtLib) API. This section will cover the use of the Library and provide some examples of good coding practices.

The Bluetooth Library API is fairly large, consisting of over sixty calls, and can generally be divided into six sections:

1. **Common Library calls**   Calls common to all libraries: Open, Close, Sleep, and Wake.

2. **Management calls**   Used for Discovery, managing ACL links, and global Bluetooth settings.

3. **Socket calls**   Used to manage RFCOMM, L2CAP, and SDP communications.

4. **SDP calls**   Used to create and advertise service records to remote devices and to discover services available on remote devices.

5. **Service calls**   Allows application developers to create persistent Bluetooth services (daemons).

6. **Security calls**   Used for managing the Trusted (Bonded) Device database.

This section focuses on the Management and Socket sections of the API, with a brief discussion of how to advertise your application using SDP. You should find that the Bluetooth API offers extensive access to Bluetooth functionality while managing to keep things relatively simple. Using the Bluetooth Library directly requires a better understanding of Bluetooth technology than using the VDRV or the Exchange Library, but the Library handles most of the minutiae of the Bluetooth protocols.

Like many communications interfaces, the Bluetooth API is made up of both *synchronous* and *asynchronous* calls. The synchronous calls block while they do their work and return a result when they are complete. These calls are used when the operations involved are purely local, not involving the radio or remote Bluetooth devices, and thus can be assured to complete in a reasonable time frame. Asynchronous calls are used whenever the operation involves talking to an external entity such as the radio or a remote Bluetooth device. This is done because most Palm OS developers work in a single thread, and thus should not be blocked for a long period of time while waiting for a call to return. Asynchronous calls return almost immediately and then report their results through a callback that the application must register to receive. The header files identify the asynchronous calls by noting that they return a "Pending" result and by listing the events that you can expect the call to generate. You may notice that a few of the asynchronous calls, such as *BtLibStartInquiry()*, generate multiple events.

There are two types of events: *management events*, which contain the results of management API activities, and *socket events*, which contain information about activity on a particular L2CAP, RFCOMM, or SDP socket. Management events are sent to a management callback, which the application should register after opening the Library. Socket events are passed to the callback that is passed in when the socket is created. While these data structures are termed "ManagementEvents" and "SocketEvents," they should not be confused with the general Palm OS events type used in the application event loop; the Bluetooth Library events are separate and unrelated to Palm OS events or notifications.

> **NOTE**
>
> It is worth noting that there are a few things that a Palm OS application cannot do even when using the Bluetooth Library directly. The Library does not currently allow applications to put the Palm device or the remote device into park, hold, or sniff modes. While an application can request that a given link be authenticated or encrypted, for security reasons the application is not allowed to specify the authentication passkey or insist that a device be added to a list of trusted (or bonded) devices.

Using Basic ACL Links

Before you can use the Bluetooth Library, you must find the Library and open it. Opening the Library will cause the OS to initialize the Bluetooth stack and radio. Stack initialization is an asynchronous function, so immediately after opening the stack, you should register a management callback. When the initialization is complete (this requires about 50ms for most radios), the callback will receive a *btLibManagementEventRadioState* event, whose status field will indicate whether the initialization was successful. Most of the calls to the Bluetooth Library require that the radio be initialized, and making these calls before the *btLibManagementEventRadioState* event is received will result in an error. The Bluetooth stack supports re-entry from the callback, so any additional configuration you wish to do can be done from the callback when the radio state event is received. Here is a quick example of how to open and close the Library:

```
static UInt16 gBtLibRefNum = 0;
// AppStart should be called during application initialization:
static Err AppStart(void)
{
    Err error = 0;
```

```
// Find the Library, and save its reference number in a global:
error = SysLibFind( btLibName, &gBtLibRefNum);
if( error )
    {
    // Normally, if a Library can't be found, then the application
    // should simply load it. The Bluetooth Library, however, is pre-
    // loaded by the Bluetooth Extension at boot time. Failing to find
    // the Library indicates there is a problem, and the application
    // should warn the user. Here we will display an alert dialog
    // that has been defined in the application's resource file:
    FrmAlert(BtLibNotFoundAlert);
    return error;
    }
// Open the Library:
error = BtLibOpen(btLibRefNum);
// If the open returned an error, warn the user:
if( error )
    {
    FrmAlert(BtLibOpenFailedAlert);
    return error;
    }
else // ... otherwise register a management callback
    {
    BtLibRegisterManagementNotification(gBtLibRefNum,
        MyBtLibManagementCallbackProc, 0);
    }
return errNone;
}
```

AppStop should be called just before the application exists:

```
static Err AppStop( void )
{
    // Always unregister the management notifications before closing.
    // This prevents your callback functions from accidentally being
```

```
    // called after your app quits if the library is somehow kept open

    // (perhaps by another application) after your application exists:

    BtLibUnRegisterManagementNotification(gBtLibRefNum,

MyBtLibManagementCallbackProc);

    // Close the Bluetooth Library:

    BtLibClose(gBtLibRefNum);

     return errNone;

}

void MyBtLibManagementCallbackProc(BtLibManagementEventType *mEventP,

UInt32 refCon)

{

    switch(mEventP->event)

        {

        case btLibManagementEventRadioState:

            if (mEventP->status == btLibErrRadioInitialized)

                {

                // Do any additional initialization here.

                }

            else

                {

                // Warn the user that the initialization failed:

                FrmAlert(BtLibRadioInitFailedAlert);

                }

            break;

        // Handle other events here.

        }

}
```

If your application is going to receive *inbound* connections, you should check to make sure that the radio's accessibility mode has been set to allow connection and (if desired) discovery. The current accessibility mode can be obtained by calling *BtLibGeneralPreferenceGet()* and passing the *btLibPref_UnconnectedAccessible* value for the preference type. The accessible state of the device is determined by the user's settings in the Bluetooth Preferences Panel, and the application *should never* override this state without first asking the user's permission. If the application does get the user's permission to change the state, it can do so by making a call to *BtLibSetGeneralPreference()*. Calling *BtLibSetGeneralPreference()* does not change the user-defined preferences, but rather only temporarily overrides them; nonetheless, the application should record the original radio settings, and restore them before exiting. If the user has set Bluetooth technology to be OFF in the Preferences panel, the Library itself will prompt the user before allowing an application to change settings that affect the radio. The application should never attempt to override the OFF setting.

If you plan to have your application create *outbound* Bluetooth connections, you will probably want to perform a device discovery in order to allow the user to select the remote device(s) with which she wished to create a connection. The Bluetooth Library offers two similar calls that handle the entire discovery experience, including inquiry, name retrieval, and user selection. *BtLibDiscoverSingleDevice()* and *BtLibDiscoverMultipleDevices()* differ only in that the number of the devices the UI will allow the user to select, and the fact that *BtLibDiscoverSingleDevice()* returns the selected device directly while *BtLibDiscoverMultipleDevices()* returns the number of devices selected, which can then be retrieved by passing an appropriately sized array to *BtLibGetSelectedDevices()*.

The discovery calls are designed to create a standardized user experience while still offering enough flexibility to be useful to a wide range of applications. Some of these things are quite simple, like letting the application specify the instruction text on the user selection screen. A chess game might pass, for example, the string *Choose an opponent* while a printing application might want to ask the user to "Select a printer." One of the most useful features of the discovery calls is the ability to filter out any devices that do not belong to one of the classes specified by the application. Using this feature, a Palm-to-Palm game could prevent non-PDA devices from showing up in the list of discovered devices, thus limiting the users' choices to the appropriate class of device. If an application passes in multiple CoD descriptions, the application will show devices that fit any of the indicated classes. The following is an example of a discovery call that will display all smart phones and all classes of computers:

```
Err  DoDiscovery( BtLibDeviceAddressType* resultP )
{
    BtLibClassOfDeviceType  allowedDeviceClasses[2];
    // Each COD contains one or more service classes, along with a Major
```

```
    // and Minor Device Class:
    allowedDeviceClasses[0] = btLibCOD_ServiceAny | btLibCOD_Major_Phone
                             | btLibCOD_Minor_Phone_Smart;
    allowedDeviceClasses[1] = btLibCOD_ServiceAny |
                             btLibCOD_Major_Computer |
                             btLibCOD_Minor_Comp_Any;
    // Do the discovery. Use the default instruction, and stick the
    // result in the location that was passed in:
    return BtLibDiscoverSingleDevice(
                 gBtLibRefNum, // the Library reference number
                 NULL,      // use the default instruction text
                 allowedDeviceClasses,       // the filter list
                 2,                     // the filter list length
                 resultP,          // store the selection here
                 false,// don't use addresses instead of names
                 false);            // don't skip the inquiry
}
```

You may have noticed that the discovery call contains two arguments that haven't yet been mentioned, the last two arguments: *addressAsName* and *showLastList*. The *addressAsName* argument instructs the Library to skip name retrieval and instead display the numeric Bluetooth device addresses of each of the devices. This is mainly useful as a debug tool, since in general we try to shield the user from long dealing with long numeric addresses. The *showLastList* argument causes the Library to skip the inquiry phase and instead show the same list as the last discovery. These two discovery calls should be flexible enough to handle most applications' needs; if for some reason, however, an application requires something outside of the discovery calls supported activities, the application can implement it's own discovery procedure using the *BtLibStartInquiry()* and *BtLibGetRemoteDeviceName()* calls detailed in the BtLib.h file. Once the application has set the appropriate accessibility mode and gained the address of a remote device (or devices) it wishes to connect to, it can begin the process of establishing ACL connections.

Bluetooth piconets have a star formation; one master connected to up to seven active slaves. The Bluetooth specification talks about overlapping networks of two or more piconets called *scatternets* (see Figure 8.3). These, however, are not well-defined and none of the Bluetooth radios currently available are capable of creating or managing scatternet formations.

Figure 8.3 Piconets and Scatternets

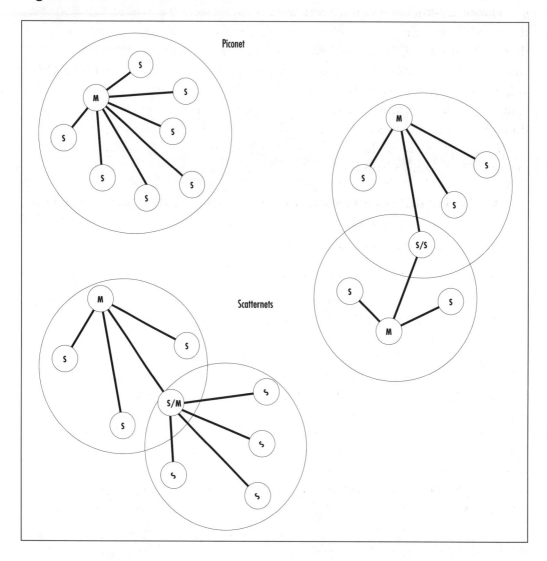

In single connection applications, where applications participate only in one-to-one connections or as a slave in one-to-many connections, ACL establishment is very simple. To receive an inbound ACL connection, the application should simply wait for the Management Callback to receive a *btLibManagementEventACLConnectInbound* event. This event will contain the address of the remote device, if the application wishes to reject the connection, it can call *BtLibLinkDisconnect()* in the callback. To create an outbound link, the application should call *BtLibLinkConnect()* with the address of the device it wishes to connect to, and wait for a *btLibManagementEventACLConnectComplete* to indicate whether the connection attempt was successful. By default, the initiator of a link is the master and the recipient of the link is the slave. When there is only one ACL connection, the role of the local device is unimportant, and the Palm OS will allow the master/slave switch to be performed. The OS will also change the accessibility mode to disable page scanning and inquiry scanning while a connection is in place, which will prevent unwanted connection attempts and increase the bandwidth available to the application.

If the application wishes to allow multiple connections, it should use the piconet calls found in BtLib.h:

- Err **BtLibPiconetCreate(**UInt16 *btLibRefNum,* Boolean *unlockInbound,* Boolean *discoverable*)

- Err **BtLibPiconetDestroy(**UInt16 *btLibRefNum*)

- Err **BtLibPiconetUnlockInbound(**UInt16 *btLibRefNum,* Boolean *discoverable*)

- Err **BtLibPiconetLockInbound(**UInt16 *btLibRefNum*)

The applications must first call *BtLibPiconetCreate()*. This call indicates to the Library that you want to create a multiple device piconet, and changes some of the policies that the OS uses. In order to have multiple ACL connections, a device must be the master of its piconet. Calling *BtLibPiconetCreate()* changes the OS policies to disable the master/slave switch on outbound connections (so that it remains master) and forces the master/slave switch on inbound connections (so that it becomes the master). If the device is already a slave in an ACL connection when *BtLibPiconetCreate()* is called, the call will return a pending response, and attempt to become the master of the link. The Bluetooth Lib will then generate a *btLibManagementEventPiconetComplete* event to inform the application whether or not the piconet creation was successful. If the device is a master in an ACL connection, or there are no ACL connections in place to begin with, *BtLibPiconetCreate()* will return a success response and no event will be generated.

Once a successful *BtLibPiconetCreate()* call has been made, up to seven simultaneous ACL connections can be established. Depending upon the usage model for your application, you may wish to have the piconet master actively create outbound connections, wait for inbound connections from remote devices, or both.

Outbound connections can be created at any time, simply by having the application call *BtLibLinkConnect()* with the address of each remote device with which it wishes to form a connection. Each call to *BtLibLinkConnect()* will generate a *btLibManagementEventACLConnectComplete* event with the result of the connection attempt in the status field. Similarly, calling *BtLibLinkDisconnect()* will cause the radio to disconnect an ACL link. Whenever a link is dropped, perhaps as the result of having called *BtLibLinkDisconnect()* or as the result of an action on the remote device or from range or interference problems, a *btLibManagementEventACLDisconnect* event will be generated. The status field of the event will give the reason for the disconnection.

In order to allow inbound connections, the piconet must be unlocked for inbound connections. Locking and unlocking the piconet affects the accessibility state of the radio. Unlocking the piconet causes the radio to periodically scan for inbound connections (a state called Page Scan mode in the Bluetooth core specification, or "connectable" in the Generic Access profile). When unlocking a piconet, the application can also specify that the radio should scan for and respond to discovery requests (called Inquiry Scan mode in the Bluetooth core specification or "discoverable" in the Generic Access Profile). Locking the piconets will make the device non-connectable and non-discoverable. If the piconet is full (i.e., if seven ACL connections are in place), the OS will also make the radio non-connectable and non-discoverable, even if the piconet is unlocked, until one of the ACL connections is dropped. After *BtLibPiconetCreate()* is called, the lock/unlock state of the piconet overrides the user's accessibility preferences or the accessibility mode set with *BtLibSetGeneralPreference()*. When the application calls *BtLibPiconetDestroy()*, the OS will return, sever all ACL connections and set the accessibility mode back to its original state. While the application is free to leave the piconet unlocked all of the time, you should be aware that since the radio will periodically have to spend time performing page and inquiry scans, the throughput on the ACL links of an unlocked piconet will be lower than the throughput of the links on a locked piconet. Bandwidth-conscious applications should leave the piconet locked most of the time.

Creating L2CAP and RFCOMM Connections

The L2CAP and RFCOMM protocol layers are exposed in the Bluetooth API through a sockets-based interface. The SDP interface uses the sockets-based API as well, but that will be discussed further in the following section. The application creates a socket by calling *BtLibSocketCreate()*, which allocates a socket structure and associates it with a protocol. *BtLibSocketCreate()* also takes a callback function pointer as an argument; this callback is associated with the socket and will receive all of the events for that socket. After a socket is created, it needs to be assigned a role. The application can make the socket into a listener for inbound connections by calling *BtLibSocketListen()*, or create an outbound connection by calling *BtLibSocketConnect()*. When an inbound connection occurs, a listener socket will spawn a new socket for that connection. It's worth noting that the ability to create and receive RFCOMM and L2CAP connections is entirely independent of the device's role in a piconet; a device that receives an inbound ACL connection may create an outbound L2CAP connection. It's really up to the profile or the application you are working with. In this section, we'll look at how to create and use sockets for L2CAP and RFCOMM communication.

Developing & Deploying…

RFCOMM versus L2CAP

Before we get too far into the Palm OS specific handling of L2CAP and RFCOMM, let's take moment to examine the two layers themselves. As you will have noticed by now, the RFCOMM is built on top of the L2CAP layer. In general, when we see protocols layered on top of each other, we assume that the upper layer protocol somehow extends the functionality of the protocol layer below it. For example, most of us are familiar with the fact that the IP layer of the TCP/IP stack is responsible for routing and delivering packets through a network, and that the TCP layer builds on top of IP to offer reliability and in-order delivery. This is not really the case for RFCOMM and L2CAP, however. RFCOMM and L2CAP are both what the OSI model describe as Data-Link layer protocols; which is to say that both are concerned with reliably delivering packets of data between two linked devices: in our case, a master and a slave. Neither L2CAP nor RFCOMM offer any kind of networking or routing functions. They are only capable of delivering data to devices with which there is a direct ACL link. Given these similarities, many people have wondered why both protocols exist in the Bluetooth stack. This is a very good question, without a very good answer. The short answer is that RFCOMM is a legacy of the original goal of Bluetooth technology: to create a wireless replacement for serial cables. If you look in the RFCOMM specification, you will see that the protocol deals heavily with physical line simulation, giving upper layers the ability to set and poll individual line states, just as they would with a physical serial port. In reality, however, very little use is made of

Continued

www.syngress.com

these serial port emulation features of the protocol, and in general, it is treated as a simple packet-based data-link layer. While most of the profiles in the 1.1 specification make use of the RFCOMM layer, over time I think we will see most new usage models run directly over L2CAP.

There are, however, a few differences between L2CAP and RFCOMM that may influence which one you decide to use. Since RFCOMM runs on top of L2CAP, RFCOMM has a slightly higher header overhead than L2CAP does (about 5 extra bytes), which decreases RFCOMM's total data throughput and MTU size. A more important difference is that RFCOMM provides flow control, while L2CAP does not. This means that an L2CAP channel is capable of pushing data at you as fast as the remote device can send it, and there is no way for the application to flow the L2CAP channel off. This is not really a problem; it simply means that applications or protocols that run on top of L2CAP must be able to handle the flow control themselves, while applications that run on top of RFCOMM can make use of its built-in flow control. Another important difference between RFCOMM and L2CAP is the way that inbound connections to listeners are handled. We will talk in more detail about the differences between L2CAP and RFCOMM listener sockets in a moment, but the main divergence to note is that an RFCOMM listener is only capable of supporting one connection at a time, while a L2CAP listener can receive an unlimited number of connections. For applications that only make use of single ACL links, the difference is probably not important, but for an application that wants to be a server in a seven-slave piconet, having to only register and advertise one socket can be a big convenience.

Of course, if your application involves functionality covered by a Bluetooth profile you will not have to make a choice of which layer to use as the profiles provide guidance on how to use the Bluetooth protocol stack.

To create a listener socket, first allocate a socket with your desired protocol by calling *BtLibSocketCreate()*, then register the socket as a listener by calling *BtLibSocketListen()*. Since listener sockets do not need to specify a remote device, they can be created any time after opening the Library, whether or not there are any ACL links in place. The *listenInfo* argument to *BtLibSocketListen()* is a pointer to a structure of type *BtLibSocketListenInfoType*, which contains protocol-specific listening information.

```
typedef struct BtLibSocketListenInfoType {
    union {
```

```
    struct {
        // The PSM (Protocol Service Multiplexor) identifies the
        // destination of an L2CAP channel. Predefined PSM values are
        // permitted; however, they must be odd, within the range of
        // 0x1001 to 0xFFFF, and have the 9th bit (0x0100) set to zero.
        // Passing in BT_L2CAP_RANDOM_PSM will automatically create a
        // usable PSM for the channel. In this case the actual PSM value
        //will be filled in by the call.
        BtLibL2CapPsmType        localPsm;
        UInt16 localMtu;
        UInt16 minRemoteMtu;
    } L2Cap;
    struct {
        // Service IDs are assigned by the RFCOMM protocol layer. The
        // serviceID assigned an RFCOMM listener socket is returned
        // in the serviceID field of the listen info:
        BtLibRfCommServerIdType serviceID;

        // BT_RF_MIN_FRAMESIZE <= maxFrameSize <= BT_RF_MAX_FRAMESIZE
        // Use BT_RF_DEFAULT_FRAMESIZE if you don't care
        UInt16    maxFrameSize;
        // Setting advance credit to a value other then 0 causes the
        // socket (upon a successful connection) to automatically
        // advance the remote device the set amount of credit.
        // Additional credit can be advanced once a connection is in
        // place with the BtLibSocketAdvanceCredit call.
        UInt8 advancedCredit;
    } RfComm;
    } data;
} BtLibSocketListenInfoType;
```

The *BtLibSocketListenInfoType* structure is interpreted based upon the protocol assigned to the socket that is becoming a listener. As you can see, slightly different information is used to register an RFCOMM listener than to register an L2CAP listener. L2CAP identifies available listeners by a Protocol Service Multiplexor (PSM), which can be thought of as being similar to an IP port. PSM values up to 0x1000 are reserved for use by the Bluetooth SIG. Values above 0x1000 can be used by applications, as long as the ninth bit (0x0100) is set to zero (the ninth bit is an escape bit to indicate a PSM longer than 16 bits, which the Palm OS does not currently support). While you are welcome to define your own PSM, the fact that there is no central registry for PSM values means that you cannot be assured you will be able to avoid conflicts with other applications on the device. A better idea is to pass in BT_L2CAP_RANDOM_PSM, which will cause the OS to assign an available PSM value to the listener. You can let remote applications know which PSM to connect to by advertising the PSM value with SDP, discussed in the next section.

The *localMtu* and *minRemoteMtu* values are used by L2CAP to negotiate the maximum packet size from the connection. Both *localMtu* and *minRemoteMtu* must be between BT_L2CAP_MAX_MTU and BT_L2CAP_MIN_MTU and *minRemoteMtu* must be less than or equal to *localMtu*.

The RFCOMM protocol uses a simple enumeration called a Server ID to distinguish its listeners. Unlike the L2CAP PSM value, an RFCOMM listener socket's Server ID cannot be chosen by an application. Rather, Server IDs are sequentially assigned by the OS. Like L2CAP listener socket's PSM values, after an application has created an RFCOMM listener socket, it should advertise the listener socket's Server ID using SDP. The RFCOMM listen parameters also include a *maxFrameSize* that defines the maximum frame size allowed for the channel, and should be between BT_RF_MIN_FRAMESIZE and BT_RF_MAX_FRAMESIZE. The RFCOMM listen parameters also contain an *advanceCredit* field that allows an application to specify a default amount of credit a remote device should be advanced upon connection (more on RFCOMM credit-based flow control in a moment).

Once a listener socket has been created, it will wait for connection attempts until the socket is closed with the *BtLibSocketClose()* call or until the Library is closed (as a precaution, applications should always close all sockets before they close the Library, since another application may hold the Library open even after you close it). When an L2CAP or RFCOMM connection attempt is made, the appropriate listener socket's callback will be sent a *btLibSocketEventConnectRequest* event. The socket must call *BtLibSocketRespondToConnection()* during the callback to accept or reject the inbound connection. After responding, the listener socket will receive a *btLibSocketEventConnectedInbound* event; the status field indicates whether or not the connection was successfully negotiated. If the connection was successful, the listener socket will spawn a new connection socket, which will be identified in the *btLibSocketEventConnectedInbound* event structure. The new connection socket will share a callback with its parent listener socket (you can identify which socket an event is for by looking at the socket field of the event structure).

To create an outbound connection, the application should first allocate a socket by calling *BtLibSocketCreate()*, and then create a connection with that socket by calling *BtLibSocketConnect()*. Like *BtLibSocketListen()*, *BtLibSocketConnect()* takes a pointer to a structure that indicates protocol-specific parameters.

```
typedef struct BtLibSocketConnectInfoType {
    BtLibDeviceAddressTypePtr remoteDeviceP;
    union {
        struct {
            BtLibL2CapPsmType remotePsm;
            UInt16 minRemoteMtu;
            UInt16 localMtu;
        } L2Cap;
        struct {
            BtLibRfCommServerIdType remoteService;
            UInt16      maxFrameSize;
            UInt8 advancedCredit;
        } RfComm;
    } data;


} BtLibSocketConnectInfoType;
```

As you can see, most of the information contained in the *BtLibSocketConnectInfoType* is analogous to information in the *BtLibSocketListenInfoType*, and like the *BtLibSocketListenInfoType* is interpreted based upon the protocol of the socket passed to the *BtLibSocketConnect()* call. The *minRemoteMtu*, *localMtu*, and *maxFrameSize* fields are used by the lower layers to negotiate the maximum packet size for the connection, and the *advancedCredit* is used by RFCOMM to automatically advance flow control credits upon connection. The *remotePsm* and *remoteService*, for L2CAP and RFCOMM sockets respectively, are used to determine which listener socket to connect to on the remote device. If the desired service on the remote device has a statically assigned L2CAP PSM value (not recommended, see earlier), the PSM value can be defined directly in the application. In most cases, you will want to use SDP to find the PSM or Server ID for the remote service. After *BtLibSocketConnect()* has been called, the socket callback will receive a *btLibSocketEventConnectedOutbound* event, with a status field that indicates whether or not the connection was successful.

Once a connection socket, inbound or outbound, has successfully been established, data can begin to flow. The application can send data by calling *BtLibSocketSend()*, and will receive data through *btLibSocketEventData* events sent to the sockets callback. *BtLibSocketSend()* will cause a *btLibSocketEventSendComplete* event to be generated when the data has been successfully transmitted. In order to minimize memory consumption and processing time, the Bluetooth Library does not buffer outbound *or* inbound data. This means that applications are responsible for handling their own buffering. When an application calls *BtLibSocketSend()*, it should consider the memory block indicated by the data pointer to be owned by the Bluetooth Library until the application receives a *btLibSocketEventSendComplete* event. Changing or freeing the memory block during this time can corrupt the data being sent, or even crash the device. Since the Library does not buffer data, only one call to *BtLibSocketSend()* can be pending at any given time; additional calls will result in a "busy" error. Since the Library does not buffer inbound data, the application must handle the data indicated in a *btLibSocketEventData* immediately, either by processing the data immediately or by copying and storing it for future processing. Once the *btLibSocketEventData* callback has returned, the event data pointer is no longer valid.

In the case of RFCOMM connection sockets, in order to receive data, the application must first advance credits by calling *BtLibSocketAdvanceCredit()*. Each RFCOMM flow control credit represents one packet on that channel. Advancing 10 credits indicates to the remote device that your application is ready to receive up to ten packets. Credit advances are cumulative, so making three calls to *BtLibSocketAdvanceCredit()* with a value of 5 credits would extend a total of 15 credits to the remote device. The credit count for a socket is decremented each time that the socket receives a packet. When the credit count reaches zero, the remote device is blocked from sending data on the channel. You should look at the total available buffer space your application has available and divide by the channel's maximum receivable packet size (that is, the Maximum Receivable Unit [MRU]) for the socket (found by calling *BtLibSocketGetInfo()*), and rounding down to find the number of credits your application should initially advance. When your application has processed data from its buffer, it can advance credits corresponding to the size of the processed data divided by the channel MRU. A maximum of 256 total credits can be advanced at any given time.

Handling your own buffering is not as much work as it might seem. In most cases, a few simple queue structures will suffice. The following is an example buffering code from a shared white board application. In this case, the application keeps only one queue for buffering outbound data; inbound data does not need to be buffered since it is handled immediately by drawing to the screen. Since space is limited, instead of giving the source code for an entire Palm application, this section will focus on a few important functions that can be used in a Bluetooth-aware application. For example, instead of putting in an entire OS event loop, the example only shows a pen event handler, which is called from the main event loop. For the purpose of this example, we will assume the existence of some standard queue functions that allow us to create and manage a normal first-in-first-out queue. We will also assume that the application has already managed to open the Library and create an L2CAP connection.

```
struct _DrawDataType {
UInt16 from_X;
UInt16 from_Y;
UInt16 to_X;
UInt16 to_Y;
} DrawDataType;
// Globals
UInt32 btLibRefNum;
#define TX_QUEUE_MAX_SIZE 50
QueueType   txQueue;
BtLibSocketRef connectionSocket;
#define INVALID_PEN_COORD 0xFFFF
UInt16 lastLocalPen_X = INVALID_PEN_COORD;
UInt16 lastLocalPen_Y = INVALID_PEN_COORD;
// TxQueueInit is called from AppStart
Err TxQueueInit( void )
{
    // Initialize the TX queue, using the defined queue size and the size
    // of our data elements:
    return QueueInit( txQueue, TX_QUEUE_MAX_SIZE, sizeof(DrawDataType));
```

```
}
// TxQueueInit is called from AppStop
Err TxQueueDeInit( void )
{
    return QueueDeInit( txQueue );
}
Boolean ConnectionUp(void)
{
    BtLibL2CapChannelIDType channel;
    if ( btLibErrNoError == BtLibSocketGetInfo(btLibRefNum,
            connectionSocket, btLibSocketInfo_L2CapChannel,
            &channel, sizeof(channel)))
        return true;
    else
        return false;
}
Boolean SendPending(void)
{
    Boolean sending = false;
    BtLibSocketGetInfo(btLibRefNum, connectionSocket,
                        btLibSocketInfo_SendPending, &sending,
                        sizeof(sending));
    return sending;
}


// HandlePenEvent is called by the form event handler for pen down, pen
// move, and pen up events:
Boolean HandlePenEvent(EventPtr eventP)
{
    Err error;
    switch (eventP->eType)
        {
        case penDownEvent:
            if (ConnectionUp())
```

```
                {
                lastLocalPen_X = eventP->screenX;
                lastLocalPen_Y = eventP->screenY;
                }
        break;
    case penUpEvent:
        lastLocalPen_X = INVALID_PEN_COORD;
        lastLocalPen_Y = INVALID_PEN_COORD;
        break;
    case penMoveEvent:
        {
        DrawDataType penData;
        // If the last pen value is valid, than a connection is in
        // place. Otherwise ignore the event:
        if(lastLocalPen == INVALID_PEN_COORD)
            break;

        penData.from_X = lastLocalPen_X;
        penData.from_Y = lastLocalPen_Y;
        penData.to_X = eventP->screenX;
        penData.to_Y = eventP->screenY;
        // Draw the local pen stroke on our screen:
        DrawData (&penData);
        // Enqueue the draw data in the TxBuffer:
        error = QueueEnqueue(txQueue, &penData);
       if(error)
          {
          // The Tx queue has overflowed. Handling this is application
          // dependant, so we'll just display an error and break:
          FrmAlert(TxQueueOverflowAlert);
          break;
          }

   // Attempt to send now. If there is already a send pending, the
```

```
                // call will return an error, but we don't care because the send
                // complete callback will see that there is pending data in the
                //queue:
                AttemptSend();
            break;
            }
        // Always return false when handling pen events so that the OS gets a
        // chance to handle them too:
        return false;
}
void DrawData(DrawDataType dataP)
{
 WinDrawLine( dataP->from_X, dataP->from_Y, dataP->to_X, dataP->to_Y);
}
Err AttemptSend( void )
{
    Err error;
    UInt32 numToSend = GetNumToSend();
    UInt8 *dataP;
    UInt32 dataSize;

    if(numToSend == 0) return errNone;

    dataP = (UInt8*) QueueHeadPtr(txQueue);
    dataSize = numToSend * QueueElementSize(txQueue);

    return BtLibSocketSend(btLibRefNum, connectionSocket, dataP,
                            dataSize);
}
UInt32 GetNumToSend( void )
{
    UInt32 numPossible, channelMaxTxsize;
    Err error;
    // find the maximum size packet the socket can send
```

```
     error = BtLibSocketGetInfo( btLibRefNum, connectionSocket,
                  btLibSocketInfo_MaxTxSize, & channelMaxTxsize,
                  sizeof(channelMaxTxsize));
     // Make sure we didn't get an error:
     if (error)
        {
        ErrAlert(error);
        return 0;
        }
     // Find the maximum number of data structures that can be sent in one
     // packet:
     numPossible = channelMaxTxsize / sizeof(DrawDataType);
     // If numPossible == 0, then the minRemoteMtu used in establishing
     // the connection was too small. You should check the value here and
     // deliver some kind of appropriate error message.
     // The number of queue items the application should try to send
     // assume QueueSize() returns the in use size, not the max size:
     return min( numPossible, QueueSize(txQueue) );
}
// This is the callback associated with the connection socket:
void ConnSocketCallback(BtLibSocketEventType *sEventP, UInt32 refCon)
{
     UInt32 numDataElements,i;
     DrawDataType *rxDrawData;
     switch(sEventP->event)
        {
        case btLibSocketEventSendComplete:
            // Check the status of the event
            if( sEventP->status != errNone)
                {
                ErrAlert(sEventP->status);
                return;
                }
            // We can dequeue the sent data:
```

```
                numDataElements = sEventP->eventData.data.dataLen /
                                    QueueElementSize(txQueue));
            QueueDequeue( txQueue, numDataElements);
            // Send enqueued data if there is any:
            AttemptSend();
            break;
        case btLibSocketEventData:
            // We received data to draw. Check the status of the event:
            if( sEventP->status != errNone)
                {
                ErrAlert(sEventP->status);
                return;
                }
            numDataElements = sEventP->eventData.data.dataLen /
                                    QueueElementSize(txQueue));


            // Draw the received data:
            rxDrawData = (DrawDataType*)sEventP->eventData.data.data;
            for( i=0; i<numDataElements; i++ )
                {
                DrawData( &rxDrawData[i] );
                }
            break;


        // Handle other socket events here.
        }
    return;
}
```

Pushing the buffering back to the application ensures that memory usage is as efficient as possible for any given application, without adding too much burden to the application developer.

The application should close the socket by calling *BtLibSocketClose()* when it is ready to close the socket connection, or when it receives a *btLibSocketEventDisconnected* event (at this point the connection is already lost, so the close call simply frees the socket). If the ACL connection is lost, a *btLibSocketEventDisconnected* event will automatically be generated. Applications should be sure to close all sockets before closing the Bluetooth Library.

Using the Service Discovery Protocol

In the previous section, we discussed the need to use Service Discovery Protocol to find an application's PSM or Service ID, rather than relying on hard-coded values. The Bluetooth Library offers an extensive set of APIs for working with SDP. In this section, we will concentrate on the calls needed to advertise a basic service record for an L2CAP or RFCOMM listener socket and to retrieve connection information about L2CAP and RFCOMM listeners on a remote device.

Developing & Deploying...

About UUIDs

UUIDs are 128-bit values that are guaranteed to be unique across time and space without the need for a central registry. In the Service Discovery Protocol, UUIDs are used to represent services, types, or attributes. Bluetooth Profiles specify UUID values to identify the services that they describe; new Bluetooth applications will need to specify their own service UUID. A number of UUID (also called GUID) generators are available on the Web. One point of confusion that sometimes occurs with UUIDs involves the Bluetooth specification's reference to 16-bit and 32-bit UUID values. All UUIDs are in fact 128-bit values, but the Bluetooth specification has reserved a range of UUID values. Since all values in this range have the same base address, the specification uses 16-bit and 32-bit values as a shorthand method to represent 128-bit values within the reserved range. The Bluetooth Library allows applications to specify 16- and 32-bit UUID values and handles the conversion and comparison between the different representations. It's worth noting that only the Bluetooth specification is allowed to specify 16- and 32-bit UUID values—all other applications must use 128-bit UUID values.

As we noted earlier in the chapter, the Palm OS does not support a service-browsing usage model; applications are expected to know what kind of service they want to communicate with. Services in Bluetooth technology are identified by one or more UUID values. When a service advertises more than one service class UUID, the various UUIDs are assumed to have a hierarchical nature. If you are writing an application that will only talk to other instances of itself (a game, for example), a single service UUID for the application is probably sufficient.

Advertising a Basic Service Record for an RFCOMM or L2CAP Listener Socket

Let's begin by looking at what we need to do in order to create and advertise a basic service record for an RFCOMM or L2CAP listener socket. We'll assume that the application has already opened the Bluetooth Library and created a listener socket (the socket's protocol is unimportant). A new application needs to declare a new service UUID, so we'll use a UUID generation tool to generate one, in this case 7FD82E36-47E8-11D5-83CE-0030657C543C.

```
#define myServiceUUID \
{btLibUuidSize128,{0x7F,0XD8,0x2E,0x36,0x47,0XE8,0x11,0XD5,0x83,0xCE, \
0x00,0x30,0x65,0x7C,0x54,0x3C}}
// Globals
UInt32 btLibRefNum;
BtLibSocketRef listenerSocket;
BtLibSdpUUIDType myServiceUUIDList = myServiceUUID;
UInt8 myServiceUUIDListLen = 1;
Char *myServiceName = "MyService";
BtLibSdpRecordHandle myServiceRecordH;
// AdvertiseSocket is called after the application has successfully
created a
// listener socket.
Err AdvertiseSocket( void )
{
    // Error checking for this code is fairly straight forward, and the
    // error codes are well documented in BtLib.h, so we'll leave the
    // error checking out for the sake of space.
    // First we need to allocate a service record to advertise:
    BtLibSdpServiceRecordCreate( btLibRefNum, myServiceRecordH);

    // Now we will use a very useful call to fill in the important
    // information for our service record:
    BtLibSdpServiceRecordSetAttributesForSocket(btLibRefNum,
```

```
                    listenerSocket, &myServiceUUIDList,
myServiceUUIDListLen,
                myServiceName, StrLen(myServiceName), myServiceRecordH);


    // If we wanted to add any additional info to our service record, we
    // would do so here by using BtLibSdpServiceRecordSetAttribute().
    // Now all that's left to do is to advertise the record:
    BtLibSdpServiceRecordStartAdvertising(btLibRefNum, myServiceRecordH);
    return errNone;
}
```

That's all there is to it! The *BtLibSdpServiceRecordSetAttributesForSocket()*
really does all the work for us, filling in the record with Service UUID, protocol listener information, and service name. There are, of course, a lot of
other attributes that can be contained in a service record (see the
BtLibSdpAttributeIDType), and we can add attributes to the record before
advertising it by calling *BtLibSdpServiceRecordSetAttribute()*. Before closing the
Bluetooth Library, the application should stop advertising the SDP record by
calling *BtLibSdpServiceRecordStopAdvertising()* and then free the service record
by calling *BtLibSdpServiceRecordDestroy()*.

Retrieving Connection Information about L2CAP and RFCOMM Listeners on a Remote Device

Now let's look at the other side of the picture: retrieving L2CAP or RFCOMM
connection information from a remote device. In order to get SDP information
from a remote device, we need to have established an ACL link to that device.
Retrieving SDP information is an asynchronous operation, so we need to provide the Bluetooth Library with a callback with which it can return the SDP
query results. The Bluetooth Library uses an SDP socket to specify the callback
that should be used to return information. Although the same API calls are used
to create SDP sockets as to create L2CAP and RFCOMM sockets, SDP sockets
have very little in common with L2CAP or RFCOMM sockets; they can not be
set up as listeners, or used to create connections. SDP sockets are really just a
convenient way for the Bluetooth Library to manage information about SDP
calls. A single SDP socket can handle one pending SDP request at a time, if your
application does not need to have multiple simultaneous SDP queries pending,

you can use a single SDP socket for all your SDP queries. The SDP socket should be closed before closing the Bluetooth Library.

Let's look at an example function that gets the L2CAP PSM for a service advertised with the same UUID that we used in the preceding example (in this case, we are assuming that the listener socket being advertised was an L2CAP socket).

```
#define myServiceUUID \
{btLibUuidSize128,{0x7F,0XD8,0x2E,0x36,0x47,0XE8,0x11,0XD5,0x83,0xCE, \
0x00,0x30,0x65,0x7C,0x54,0x3C}}
#define INVALID_SOCKET_REF 0xFFFF
// Globals
UInt32 btLibRefNum;
BtLibSocketRef sdpSocket = INVALID_SOCKET_REF;
BtLibDeviceAddressType remoteDevice;
BtLibSdpUUIDType myServiceUUIDList = myServiceUUID;
UInt8 myServiceUUIDListLen = 1;
BtLibL2CapPsmType remotePSM;
// GetRemotePSM is called after an ACL connection to a remote device has
// been established in order to retrieve PSM for the service we wish to
// connect to. That PSM can then be used in the connection
// structure passed in to BtLibSocketConnect().
Err GetRemotePSM(void)
{
    Err error;

    // First create an SDP socket if we haven't already created one.
    // This socket should be closed before closing the Bluetooth Library
    if (sdpSocket == INVALID_SOCKET_REF)
        {
        error = BtLibSocketCreate( btLibRefNum, &sdpSocket,
                        SdpSocketCallback, NULL, btLibSdpProtocol);
        if (error) return error;
        }
    // Now request the PSM info.  The call should return "pending".
    error = BtLibSdpGetPsmByUUID( btLibRefNum, sdpSocket, &remoteDevice,
                        &myServiceUUIDList,
myServiceUUIDListLen);
```

```
        return error;
}
void SdpSocketCallback(BtLibSocketEventType *sEvent, UInt32 refCon)
{
    switch( sEvent->event )
        {
          case btLibSocketEventSdpGetServerChannelByUUID:
            if( sEvent->status != btLibErrNoError )
                {
                // SDP was unable to find a service record for the UUID
                // list you specified.  This is most likely because your
                // application is not running on the remote device.  Warn
                // the user that they need to have the application running
                // on both devices.
                FrmAlert( RemoteAppNotFoundAlert );
                return;
                }
            // SDP found a service record with the UUID list you
            // specified.  Copy the PSM value into a global so it
            // can be used to set up the connection
            remotePSM = sEvent->eventData.sdpByUUID.param.psm;

            // You may wish to call the code that creates the L2CAP Socket
            // connection here, or wait for some user action.
            break;

        // Handle other socket events here if needed ...
        }
}
```

As you can see, retrieving remote RFCOMM and L2CAP listener information is pretty straightforward. If *BtLibSdpGetPsmByUUID()* or *BtLibSdpGetServerChannelByUUID()* are called with multiple items in the UUID list, the call will search for a service record that contains all of the service UUIDs in the list, although it will not insist that they appear in the same order in the record as they do in the list. If multiple records with the required UUIDs are found, the call will return the first one that it comes across.

These two cases should handle most applications' SDP requirements. However, if your application needs to make more extensive use of SDP, the Bluetooth Library contains calls that allow you to make more specific searches, retrieve and set any attribute value defined in the Bluetooth Specification, and, for the very gung-ho, deal with SDP records as raw data.

Using Bluetooth Security on Palm OS

Palm OS provides full support for Bluetooth authentication and encryption. What level of Bluetooth security is required for a link is up to each individual application, which corresponds to Bluetooth Security Level 2. Applications can cause link authentication or encryption to occur by calling *BtLibLinkSetState()* with *btLibLinkPref_Authenticated* or *btLibLinkPref_Encrypted*, which will generate a *btLibManagementEventAuthenticationComplete* or *btLibManagementEventEncryptionChange* event, respectively. It is up to an application to decide what to do if an authentication or encryption request fails. The OS will handle any pairing producers (such as asking the user for a passkey) that are necessary for authentication to occur.

It is worth noting that Bluetooth security is link level security and *does not* take the place of application level security (except perhaps on single application devices, which Palm OS devices are not). Bluetooth authentication simply ensures that the user is connected to the device they think they are connected to—it does not ensure that the remote device is authorized to use your service. Bluetooth encryption ensures that the data can not be sniffed over the air. It uses 128-bit encryption keys, but if this is not sufficient for your application, you are free to add an extra layer of security to your application, as some writers of financial software have indicated they are likely to do.

Writing Persistent Bluetooth Services for Palm OS

In general, a service, or *server daemon*, is a program that has a persistent presence on a device, performing its function as needed, often in the background. Unlike a client application, which normally begins operation directly in response to a user action, services generally initiate action in response to a non-user event: in our case, a communication event. In a resource-rich environment, such as a PC, services often run continuously in their own process. While this approach has advantages, particularly in terms of response performance, it means that the resources needed by these services are always in use. Having just a few services on a Palm device can quickly eat away at the device's limited resources.

Developing & Deploying…

Services and Bluetooth-Aware Applications

As with the Palm OS IR stack, the Bluetooth stack is not available for services while it is in use by an application. Although Bluetooth technology is a multiplexing protocol, our services will follow the same model as IR. Service notifications are simply not generated when the Bluetooth Library has been opened by an application. If an application opens the stack while a service is in use, the OS will generate an "all shutdown" message for the services. The decision not to allow services and applications to use the stack at the same time was made because, despite Bluetooth's multiplexing capability, there are complications that arise with remote device management when more then one application tries to use Bluetooth technology at the same time.

In order to avoid the problems associated with having truly persistent services, Palm, Inc. has had to rethink the services model in the Palm OS, allowing services to run on a more as-needed basis. Palm, Inc. took such an approach when implementing the OBEX service in the IR implementation. While the client side of OBEX starts up in response to a user action (the "beam" command), the service side of OBEX is brought up by the OS when an inbound IR connection is detected.

Using this mechanism, the IR implementation is able to avoid the overhead of the OBEX service and IR stack when they are not in use. This model has been highly successful, despite the tight timing requirements for responding to an IR connection request. The only hitch in IR service implementation is that, since the inbound connection triggers the OBEX service directly, third parties have been unable to develop new IR-based services. Since Palm, Inc. forged the way in the IR world, and thus set the usage direction, this has not been a major hindrance. However, given the diversity of usage expected for Bluetooth technology, support for multiple services has become an important part of providing a robust Bluetooth solution. The Bluetooth Services API attempts to take this logic a step farther and allow third parties to create Bluetooth applications with a persistent presence.

In order to efficiently support multiple persistent services at the same time, certain restrictions must be made. The principle restriction will be that only one service may be in-session at a time. In other words, although multiple services can be registered, once a given service begins a session, the other services become unavailable until it completes its session. When used in conjunction with some reasonable coding guidelines for the services, this restriction should allow the availability of multiple services without a significant impact on memory usage. The restriction has the added benefit of avoiding potential problems in which two services simultaneously attempt to display UI.

Creating a service is actually pretty straightforward. Essentially, services are simply pieces of code that register for and respond to *Bluetooth service notification*. Bluetooth service notifications are normal Service Manager notifications of type *BtLibServiceNotifyType* (btsv). The easiest way to create a service is by packaging the service in a normal application. When the application is launched in the "normal" manner (i.e., with *sysAppLaunchCmdNormalLaunch*), the application can display controls that allow the user to enable and disable the service, which can correspond to registering and unregistering for the Bluetooth service notification. It is best to register for the notification to be delivered be a launch command, rather than by a callback, since this avoids the need for locking the code resource (remember, the service notifications may be delivered while your application is not running).

The details pointer of a Bluetooth service notification is a pointer to a *BtLibServiceNofityDetailType* structure, which is defined as:

```
typedef enum {
    btLibNotifyServiceStartup,
    btLibNotifyServiceAllShutdown, // see err for reason
    btLibnotifyServiceNotInSessionShutdown
} BtLibServiceNotifyEventType;


typedef struct _BtLibServiceNofityDetailType {
    BtLibServiceNotifyEventType     event;
    Err err;
} BtLibServiceNofityDetailType;
```

The event element of the *BtLibServiceNofityDetailType* contains the event information that will allow your service to start up and shut down correctly. The state diagram in Figure 8.4 shows the basic flow for a service.

Figure 8.4 Service States

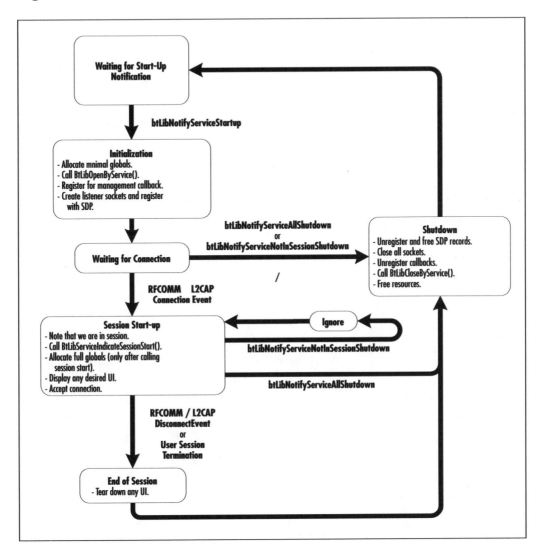

In general, a service sits in an uninitialized state, waiting for a
btLibNotifyServiceStartup notification. This notification is generated when the
OS detects an inbound ACL link, and the Bluetooth Library is currently not in
use by an application. The *btLibNotifyServiceStartup* notification is basically an
instruction to that service to initialize itself. Initialization should include allocating essential globals, opening the Bluetooth Library with the
BtLibServiceOpen() call, registering for an L2CAP or RFCOMM listener socket,

and advertising that socket via SDP. Since services are running in the context of another application, it is important to make sure that the owner ID of the services' globals is set to 0 (the system ID) using *MemPtrSetOwner()*. If the owner ID is not set, the memory will have the current application as its owner, and will be cleaned by the system if the current foreground application exits. It is also important that the service performs all its initialization during the notification callback or sub-launch, since the OS will allow the ACL connection to proceed once the notification is complete. During the startup phase, all of the registered services will be launched, which can place a strain on the system resources. In order to avoid overwhelming the system stack, services should initially allocate only the globals necessary to create and register a listener socket; additional memory can be allocated later when the service is actually in session. This helps avoid creating a big bump in memory usage during service initialization. During initialization, services should avoid displaying any UI, since multiple services may be running. Once the service is initialized and listening, several things can happen.

The service may receive a *btLibNotifyServiceAllShutdown* notification, which means that the service has timed out (the OS only allows the remote device to hang around for a limited amount of time without connecting to a service), the ACL link has been dropped (probably because the remote device didn't find the service it wanted in the SDP database), the device power has been cycled, or the foreground application has opened the Bluetooth Library (applications take precedence over services). The reason for the notification is not really important, but you can check the *err* parameter of the notification details if you really want to know. Whatever the reason, however, the service's response to the *btLibNotifyServiceAllShutdown* notification should be the same; the service should remove all of its advertised records, close its sockets, call *BtLibServiceClose()*, and free its allocated memory.

Alternatively, the service might receive a connection request on one of its listener's sockets. If this happens, the service is considered "in session" and should call *BtLibServiceInSession()*. Calling *BtLibServiceInSession()* causes the *btLibNotifyServiceNotInSessionShutdown* notification to be sent out. This notification instructs the services that did not call *BtLibServiceInSession()* to shut down, just as if they had received a *btLibNotifyServiceAllShutdown* notification. It's important to note that all services will receive the *btLibNotifyServiceNotInSessionShutdown* notification, so before calling *BtLibServiceInSession()* a service should set a value to remind itself that it is in session and should not respond to the *btLibnotifyServiceNotInSessionShutdown* notification. Once a service is in session, it can go ahead and allocate additional memory and display UI, if necessary. When a service's session is complete, it should clean up and call *BtLibServiceClose()*.

Developing & Deploying…

Creating New Services

While it is tempting to create a new service to solve a problem, in general you should avoid creating a new service unless it is absolutely necessary. When possible, it is always better to use an existing service. This approach decreases complexity and resource usage and probably makes your code a good bit simpler. For example, an instant-messaging type application is more easily created by registering with the Exchange Manager than by creating a new service. If you want to be able to invite people nearby to join your game, this is probably also more easily done with an Exchange Manager interaction than by creating a whole new service. New services should be restricted to applications that are not easily handled by existing services, like creating a Bluetooth keyboard driver or other applications where using OBEX is simply not possible.

The Future of Palm OS Bluetooth Support

Bluetooth is, of course, a very young technology, and will certainly see a fair amount of evolution over the next few years. Similarly, Palm OS's Bluetooth support will likely continue to evolve alongside the technology. In the near future, Bluetooth devices will address the issues of Layer 3 (Network level) support in the Bluetooth communication protocol stack. New specifications will define a network layer for communications between all the members of a piconet (not just master to slave), as well as inter-piconet communication issues. Roaming and scatternets will also be addressed. The eventual goal is the creation of true ad-hoc networks, self-configuring network groupings that grow and change as the user's environment changes. For Bluetooth technology to succeed in the long run, it will also need to address issues like discovery time (currently far too slow) and maximum throughput (to align with 3G technologies).

As much as possible, these changes will be integrated seamlessly into the Palm OS Bluetooth Library. New editions of the library will expand the Palm OS's Bluetooth capabilities, without compromising existing applications.

Summary

With version 4.0, Palm OS support has been put in place for Bluetooth wireless technology and line of sight limitations have become a thing of the past. Adding Bluetooth wireless technology to a Palm device frees users from the necessity of trying to physically line up two devices while they're on the road. Bluetooth technology makes traditional Palm OS applications like Internet usage and "beaming" easier in mobile environments and introduces opportunities for applications using object push communication.

The Palm OS Bluetooth Support Package currently supports five Bluetooth profiles that are defined in the Bluetooth 1.1 Specification: the Generic Access Profile (GAP), the Serial Port Profile, the Dial-up Networking Profile, the LAN Access Profile, and the Object Push Profile. The values and language specified by the GAP are built into the Bluetooth Library and other Bluetooth components. GAP's main goal is to create a friendly and consistent user experience, a goal that is also considered critical in the Palm OS. The other profiles describe a method for accomplishing a specific user goal.

The Bluetooth Support Package includes a new virtual serial driver (VDRV), which provides support for the Serial Port Profile. Using the Bluetooth VDRV allows existing serial-based applications to quickly be updated to take advantage of Bluetooth technology, and is an easy way to create new Bluetooth-enabled applications. The Bluetooth VDRV runs on top of the RFCOMM protocol layer (it does not implement RFCOMM itself—the RFCOMM protocol layer is implemented in the Bluetooth Library and can be accessed directly through the Bluetooth Library API).

One of the challenges of mapping Bluetooth technology underneath a traditional serial API is that traditional serial ports are single-channel and non-addressed in nature, while the Bluetooth system is a multiplexing, address-based protocol stack. Bluetooth technology needs to know which device and which service on that device it is going to talk to; it must also actively create the underlying baseband connection. Most Bluetooth radios are not capable of receiving inbound connections while trying to create outbound connections. For this reason, it is necessary for an application to indicate whether it wishes to initiate or accept the Asynchronous Connectionless Link (ACL) and RFCOMM connections. Palm OS refers to these roles as the *client role* and the *server role*, respectively. The application indicates its preference by setting the corresponding value for the role element in the *RfVdOpenParams* structure and filling the appropriate role-specific parameter structure inside the union.

When the VDRV is opened in the server configuration, it will register for an RFCOMM channel and advertise that channel via SDP. This creates a simple service record utilizing the Unique Universal Identifier (UUID) and name string defined in the *RfVdOpenParamsServer*. To open the VDRV in the client configuration, a more complex structure must be passed in to *SrmExtOpen()*. The *remoteDevAddr* parameter indicates the 48-bit Bluetooth device address of the remote device the VDRV should connect to. After creating an ACL connection to the remote device, the VDRV attempts to establish an RFCOMM connection. The application must indicate which RFCOMM channel the VDRV should use.

When a constant data flow is not important to your application (as in a turn-based game), the Bluetooth Exchange Library allows applications to send data blocks without concern for the underlying transport. Unlike sockets and virtual serial drivers, the Exchange Manager is a concept unique to Palm OS. Rather than referring to a specific transport, the new *exgSendScheme* send scheme of Exchange Manager in Palm OS 4.0 allows the user to pick which of the installed transports they wish to utilize. Once Bluetooth technology has been chosen as the transport, the Exchange Library will automatically perform a discovery in order to determine the address of the remote device it should connect to.

Palm OS provides full support for Bluetooth authentication and encryption. What level of Bluetooth security is required for a link is up to each individual application, which corresponds to Bluetooth Security Level 2. Bluetooth security is link level security and *does not* take the place of application level security. Bluetooth authentication simply ensures that the user is connected to the device they think they are connected to—it does not ensure the remote device is authorized to use your service.

Given the diversity of usage expected for Bluetooth technology, support for multiple services has become an important part of providing a robust Bluetooth solution. Having just a few services on a Palm device, however, can quickly eat away at the device's limited resources. Palm OS's new services model allows services to run on an as-needed basis, implementing the OBEX service in the IR implementation, the principle restriction being that only one service may be in-session at a time. Services are simply pieces of code that register for and respond to Bluetooth service notifications. New services should be restricted to applications that are not easily handled by existing services, or applications where using OBEX is simply not possible.

This chapter provides a comprehensive introduction to developing Bluetooth-aware software for Palm OS devices. From information on where to get the tools you need to get started, to advanced techniques for creating

Bluetooth services, this chapter walks developers through the new Bluetooth libraries in the Palm OS, and revisits existing communications APIs that have been enhanced with new Bluetooth-based capabilities. Developers learned tricks for using Bluetooth technology with the Serial and Exchange Manager APIs, as well as how to work directly with the Bluetooth Library.

Solutions Fast Track

What You Need to Get Started

☑ In order to begin using Bluetooth technology, you will need to have a Palm OS device with at least 4MB of memory that is running Palm OS version 4.0 or greater. Alternatively, you may wish to develop using the Palm OS Emulator, often the easiest and fastest way to create new application.

☑ In addition to a Palm 4.0 device, you will need to have the Bluetooth Support Package installed. The Bluetooth Support Package consists of several .prc files that work together. The latest version of the Bluetooth support .prc files, along with the Bluetooth header files and several pieces of example code, can be found in the Bluetooth area of the Palm Resource Pavilion at www.palmos.com/dev/tech/bluetooth.

☑ In addition, you will also want to have a copy of the Palm OS 4.0 SDK documentation, also available on the Palm, Inc. Web site.

Understanding Palm OS Profiles

☑ The Palm OS currently supports five Bluetooth profiles defined in the Bluetooth 1.1 Specification: the Generic Access Profile, the Serial Port Profile, the Dial-up Networking Profile, the LAN Access Profile, and the Object Push Profile.

☑ Generic Access Profile (GAP) is a general look at the overall process of carrying out a Bluetooth transaction without regard to the nature of that transaction, and is background for all the other profiles.

☑ The new virtual serial driver (VDRV) in the Bluetooth Support Package provides support for the Serial Port Profile.

☑ The Network Library (NetLib) supports the Data Terminal role of both the *Dial-up Networking* and *LAN Access Profiles*.

☑ The new Bluetooth Exchange Library implements the Object Push Profile, much in the same way that the Exchange Manager supports IR-based Object Exchange Protocol (OBEX) push.

☑ If none of the profiles cover what you are trying to do, don't despair—the Palm OS also provides a robust API that allows you direct access to the SDP, RFCOMM, and Logical Link and Control Adaptation Protocol (L2CAP) layers of the Bluetooth stack, along with calls to allow you to manage the Bluetooth-specific concerns like discovery and piconet creation.

Updating Palm OS Applications Using the Bluetooth Virtual Serial Driver

☑ Using the Bluetooth Virtual Serial Driver allows existing serial-based applications to quickly be updated to take advantage of Bluetooth technology. The VDRV itself is "glue code" that allows Bluetooth functionality to be accessed though a more traditional API. Using the VDRV also gives you an advantage in writing multi-transport applications.

☑ Virtual Serial Drivers in the Palm OS are individual .prc files of type *vdrv* and are used throughout the new Serial Manager interface, much the same way as traditional physical serial ports are used.

☑ Since most Bluetooth radios are not capable of simultaneously listening for an inbound connection and trying to create an outbound connection, an instance of the Bluetooth VDRV also needs to know whether it is initiating or accepting the connection. Since a traditional serial API does not present a mechanism for passing all of this extra information, Palm OS 4.0 has added a new call, *SrmExtOpen()* (found in SerialMgr.h), to the new Serial Manager API.

☑ A VDRV client-only application might be useful when you know that the Palm device will always be playing a client-based role, and therefore never need to accept a connection.

☑ Applications and the VDRV use the Bluetooth Library in different modes. Because of this difference, the VDRV will not be able to open while the application is holding the Bluetooth stack open.

☑ Setting up the serial port as a server does not cause the driver to go out and create an ACL or RFCOMM connection, it merely sets up the port as a listener. Like a normal serial port, the VDRV will not alert the application when an incoming connection is established, the application will simply begin to receive data from the port.

Using Bluetooth Technology with Exchange Manager

☑ You can make an Exchange Manager-based application Bluetooth-aware with just a few lines of code. The Bluetooth Exchange Library registers itself for the *exgSendScheme*, so if you've already updated your application to take advantage of the *exgSendScheme*, it should work with Bluetooth technology as soon as you have installed the Bluetooth .prc files.

☑ The Exchange Library allows applications to send data blocks without having to worry too much about the underlying transport.

☑ The VDRV and Exchange Manager simplify using Bluetooth technology by encapsulating it inside familiar and easy to use interfaces, but the simplification also hides functionality and increases overhead.

Creating Bluetooth-Aware Palm OS Applications

☑ If your application requires direct access to Bluetooth protocol layers or management functions, then you will need to make use of the Bluetooth Library (BtLib) API.

☑ Even when using the Bluetooth Library directly, a Palm OS application cannot put the Palm device or the remote device into park, hold, or sniff modes. Also, while an application can request that a given link be authenticated or encrypted, for security reasons the application is not allowed to specify the authentication passkey or insist that a device be added to a list of trusted (or bonded) devices.

☑ The Bluetooth Library API is fairly large, and can generally be divided into six sections: Common Library calls, management calls, socket calls, SDP calls, services calls, and security calls.

☑ If your application is going to receive *inbound* connections, you should check to make sure the radio's accessibility mode has been set to allow connection and (if desired) discovery. The accessible state of the device is determined by the user's settings in the Bluetooth Preferences Panel.

☑ If you plan to have your application create *outbound* Bluetooth connections, you will probably want to perform a device discovery in order to allow the user to select the remote device(s) with which she wished to create a connection. The Bluetooth Library offers two similar calls that handle the entire discovery experience, including inquiry, name retrieval, and user selection, *BtLibDiscoverSingleDevice()* and *BtLibDiscoverMultipleDevices()*.

☑ Bluetooth piconets have a star formation: one master connected to up to seven active slaves. Once a successful call *BtLibPiconetCreate()* call has been made, up to seven simultaneous ACL connections can be established. Depending upon the usage model for your application, you may wish to have the piconet master actively create outbound connections, wait for inbound connections from remote devices, or both.

☑ The L2CAP and RFCOMM protocol layers are exposed in the Bluetooth API through a sockets-based interface. The ability to create and receive RFCOMM and L2CAP connections is entirely independent of the device's role in a piconet.

☑ Applications or protocols that run on top of L2CAP must be able to handle the flow control themselves, while applications that run on top of RFCOMM can make use of its built-in flow control. Also, an RFCOMM listener is only capable of supporting one connection at a time, while a L2CAP listener can receive an unlimited number of connections. If your application involves functionality covered by a Bluetooth profile, you will not have to make a choice of which layer to use, as the profiles provide guidance on how to use the Bluetooth protocol stack.

☑ L2CAP identifies available listeners by a Protocol Service Multiplexor (PSM), which can be thought of as similar to an IP port. The RFCOMM protocol uses a simple enumeration called a Server ID to distinguish its listeners. You can let remote applications know which PSM and Server ID to connect to by advertising them with SDP.

☑ The Bluetooth Library offers an extensive set of APIs for working with SDP.

Writing Persistent Bluetooth Services for Palm OS

☑ The Palm OS allows services to run on an as-needed basis by implementing the OBEX service in the IR implementation. While the client side of OBEX starts up in response to a user action (the "beam" command), the service side of OBEX is brought up by the OS when an inbound IR connection is detected. Palm OS's IR service implementation is able to avoid the overhead of the OBEX service and IR stack when they are not in use.

☑ Although multiple services can be registered, once a given service begins a session, the other services become unavailable until it completes its session.

☑ Services are simply pieces of code that register for and respond to Bluetooth service notifications, normal Service Manager notifications of type *BtLibServiceNotifyType* (btsv). When the application is launched in the normal manner, it displays controls that allow the user to enable and disable the service, which can correspond to registering and unregistering for the Bluetooth service notification.

The Future of Palm OS Bluetooth Support

☑ In the near future, Bluetooth technology will address the issues of Layer 3 (Network level) support in the Bluetooth communication protocol stack. New specifications will define a network layer for communications between all the members of a piconet (not just master to slave), as well as inter-piconet communication issues.

☑ Roaming and scatternets will also be addressed.

☑ The eventual goal is the creation of true ad-hoc networks, self-configuring network groupings that grow and change as the user's environment changes.

☑ New editions of the Palm OS Bluetooth Library will expand the Palm OS's Bluetooth capabilities without compromising existing applications.

Frequently Asked Questions

The following Frequently Asked Questions, answered by the authors of this book, are designed to both measure your understanding of the concepts presented in this chapter and to assist you with real-life implementation of these concepts. To have your questions about this chapter answered by the author, browse to **www.syngress.com/solutions** and click on the **"Ask the Author"** form.

Q: How does RFCOMM credit-based flow control work with pre-Bluetooth Specification v.1.1 devices, since credit-based flow control was not mandatory before the 1.1 release?

A: When the Bluetooth Library cannot negotiate RFCOMM credit-based flow control, it will try to use the aggregate flow control defined in the earlier versions of the specification to emulate credit-based flow control behavior. In most cases, this technique is highly successful, but due to a design bug in the pre-1.1 specification, it is possible for an application communicating with a pre-1.1 device to receive more data than it has advanced credit for.

Q: Am I allowed to make calls back into the Bluetooth Library from within a library callback? In other words, does the Library allow re-entry?

A: Yes, but you will not get any more callbacks until the initial callback is released. In other words, don't block a callback waiting for another callback, because the second callback will not come until the first callback is allowed to return.

Q: I've noticed that the passkey request mechanism does not work properly sometimes when I am using the Telephony Manager over Bluetooth technology. What's going on?

A: When use you the Telephony Manager in synchronous mode, it completely blocks the UI thread, preventing the Bluetooth Library from requesting a passkey when necessary, and causing authentication to fail. This can be avoided by using the Telephony Manager in asynchronous mode, especially during the open, when an authentication is most likely to occur.

Q: Where can I get help with problems or report bugs that I find?

A: There is a Palm OS developer's mailing list set up for Bluetooth-specific concerns. You can find out more information on the Palm, Inc. Web site at www.palmos.com/dev/tech/support.

Chapter 9

Designing an Audio Application

Solutions in this chapter:

- Choosing a Codec
- Configuring Voice Links
- Choosing an Audio Interface
- Selecting an Audio Profile
- Writing Audio Applications
- Differentiating your Audio Application

- ☑ Summary
- ☑ Solutions Fast Track
- ☑ Frequently Asked Questions

Introduction

Bluetooth technology began in the labs of Ericsson, a major player in the mobile phone market, so it's not surprising that voice quality audio links play a large part in the capabilities of Bluetooth technology. According to *Semiconductor Business News'* market research report in its May 2001 edition, Cirrus Logic, which has a large share of the market for digital audio players and other portable consumer electronics, says it will begin building Bluetooth into its popular Maverick embedded processor. The Maverick processor features Internet appliances and Internet audio players.

Moreover, the Bluetooth specification will support the next generation of cellular radio systems for mobile telephony known as *third generation* (3G) that has been defined by the International Mobile Telecommunications 2000 (IMT-2000) program. The first group of audio/telephony profiles available for public with the current Bluetooth Specification v1.1 includes headset, intercom, and cordless phone.

Today, there are voice-command mobile phones and even voice-enabled Internet browsing, so audio applications and their capabilities can be a little too rich at times. Before writing an audio application, we need to understand the expectations of our target users. Do they want to transmit and receive near-CD quality audio? Do they want an acceptable range for home use with no extraneous sounds, clicks, or silences intruding? Do they want to listen to music, or hold a three-way phone conversation? We also need to know whether we are writing generic code to fit into bulky static devices such as stereos, or if we are producing a compact purpose-built system such as might slot into the strictly constrained resources of a tiny portable MP3 player. There are so many possible audio applications that we can't cover them all in detail, but this chapter will explain the basics and help you make intelligent decisions when designing your audio application.

First, we'll look at the choice of analog-digital-analog conversion schemes (Codecs). This section explains why Bluetooth technology supports several Codecs and explains how the different types perform in the presence of errors. We then go on to look at how Bluetooth links can support multiple voice channels along with simultaneous data capabilities. We explain the Synchronous Connection-Oriented (SCO) link and the three types of voice packet (High-rate Voice [HV]1, HV2, and HV3) it uses. This section explains how each packet type is transmitted at different rates and provides different amounts of error correction.

We examine the three audio profiles released with the first Bluetooth profile specification document, and briefly touch upon profiles that are soon to be released. Then we look in detail at how you might implement one particular profile: the Headset profile.

Finally, we present a few techniques you might use to differentiate your audio application and add value for the end user.

What you need to know before reading this chapter:

- Basic communications theory

- Bluetooth protocol stack component functions

- Generic Access Protocol procedures

- Host Controller Interface

Choosing a Codec

This section explains the different ways that Bluetooth systems encode voice for transmission on air. The product you are writing applications for may not allow you to choose a Codec, in which case you can safely skip this section. If you do need to choose a Codec type then it is worth taking time to understand what Codecs do, and why a choice of Codecs with different performance levels were incorporated in the Bluetooth specification.

There are several stages involved in getting from speech to the digital signals transmitted on a SCO link. The sounds we hear in human speech, music, and so on, are made up of pressure waves. A microphone converts those pressure waves into analog electrical signals. The analog signal from the microphone is fed into a Codec, which converts the analog signals of a voice signal into a digital signal to be transmitted over a communications medium. The digital signal is passed to the baseband for incorporating into a SCO packet; this packet is then sent to the radio for modulating onto a carrier for transmitting on air.

In the receive direction, the radio receives and demodulates the incoming digital signal, and passes it to the baseband. The baseband extracts the audio data and passes it to the Codecs. The Codecs take the digital signal and convert it to an analog signal for the speaker front end. Finally, the speakers, as we all know, take analog electrical signals and convert them into sound waves for us to hear.

In brief, microphone and speaker convert from sound waves to analog electrical signals. The Codec converts those analog signals into a digital format. The term *Codec* is an acronym that stands for "**co**der/**dec**oder."

Developing & Deploying...

Why Convert to Digital?

It is possible to directly modulate analog signals onto a radio without first converting them into digital format. This raises the question of why anybody would bother converting analog audio into digital formats to begin with.

There are several reasons, two of which include: digital signals tend to be more robust in the noisy environments; encoding into a digital format allows error detection and correction to be added to the signal. This means that digitally-encoded speech performs much better on noisy channels.

Of course, in the case of Bluetooth wireless technology, the baseband is designed to handle digital signals, so transmitting analog audio signals is just not an option, even if it was desirable.

If all that was required was converting between analog and digital, we could just use an Analog to Digital Converter (ADC) and Digital to Analog Converter (DAC). However, the Bluetooth specification enforces a low data rate for its voice channels: the SCO links carry just 64 Kbps. At this sort of low data rate, the Codecs are required to compress the audio signal as well as convert between analog and digital formats. The Bluetooth specification supports three different audio coding schemes on the air interface:

- Continuous Variable Slope Delta Modulation (CVSD)
- Log Pulse Code Modulation (PCM) coding using A-law compression
- Log PCM with μ-law compression

CVSD is a differential waveform quantization technique that employs a two-level adaptive quantizer (one bit). PCM uses a non-uniform quantization (a large number of progressively smaller quantization levels for low amplitude signals and fewer, coarser quantization levels for larger amplitude signals).

CVSD is more robust in the presence of bit errors than PCM. With an increase in the number of bit errors in a transmission, the perceptible voice quality of PCM drops rapidly—much more rapidly than the voice quality of

CVSD. On the other hand, PCM is simple, cheap, and more importantly, it is already used in a lot of devices. For error tolerance, we need CVSD, but for maximum compatibility with legacy systems, we need PCM. We'll look at both technologies in more detail later in this section.

The overall architecture of a Codec is illustrated in Figure 9.1. On the left, the front-end amplifiers adjust the levels between those required by the microphone and speaker and those required by the converters. ADC and DAC convert the audio signal from analog to digital format. Then some type of digital signal processing (DSP) performs the Codec function. This could be a generic DSP capable of performing many functions, or the Codecs could be implemented in dedicated circuitry.

Figure 9.1 General Block Diagram of Bluetooth Codec

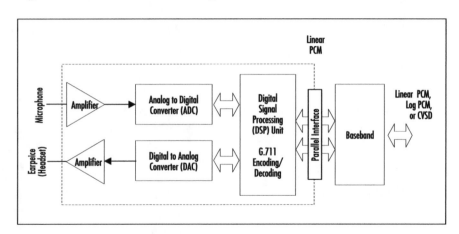

The output of the Codecs must be fed into the Bluetooth baseband. In Figure 9.1 this is shown as a direct input to the baseband (a technique commonly used in Bluetooth chips), but it is possible that the signal from the Codecs could be encapsulated in a Host Controller Interface (HCI) packet and fed across the Host Controller Interface. (This might be done, for instance, if a mobile phone with PCM Codecs were connected to a Bluetooth chip by the HCI.)

In the following sections, we shall look at the different Bluetooth Codecs in more detail.

Pulse Code Modulation

Pulse Code Modulation systems are commonly used in public and private telephone networks. In PCM systems, a waveform Codec takes samples of an analog-speech waveform and encodes them as modulated pulses, represented by logic 1 (high) and logic 0 (low). The sampling rate, or number of samples per second, is several times the maximum frequency of the analog waveform (human–voice) in cycles per second, usually at a rate of 8000 samples per second.

Configuring & Implementing...

Why Bluetooth Technology Uses Waveform Codecs

In addition to waveform Codecs, there are source Codecs that compress speech by sending only simplified parametric information about the voice transmission (as opposed to a compressed version of the voice transmission); these Codecs require less bandwidth. Examples of source Codecs include linear predicative coding (LPC), code-excited linear prediction (CELP), and multipulse, multilevel quantization (MP-MLQ).

So, if source Codecs require less bandwidth, why does the Bluetooth specification use waveform Codecs? There are two main reasons. First, the PCM Codecs specified in the Bluetooth specification follow existing standards. The International Telecommunication Union (ITU-T) coding techniques and Recommendation G.711 specify the waveform Codec providing tables to and from linear PCM and log PCM for both A-law and μ-law compression. Because these Codecs are used by existing standards, there is a large installed base of equipment (such as mobile phones) already using them. Second, these waveform Codecs provide better quality and imperceptible impairment according to Mean Opinion Score (MOS) testing.

Using PCM A-law or μ-law is optional. μ-law compression is used in North America and Japan, and A-law compression is used in Europe, the rest of the world, and international routes. The compression schemes are as described in the following (assuming $x(t)$ is the current quantized message, xp is the peak value of the message and $y(t)$ is the compressed signal output):

μ-Law Definition:

$$\text{Output} = y(t) = \frac{\text{sgn}\,(x\,(t))}{\ln\,(256)}\,\ln\left(1 + 255\left|\frac{x(t)}{xp}\right|\right);\quad \left|\frac{x(t)}{xp}\right| \leq 1$$

A-Law Definition:

$$\text{Output} = y(t)= \begin{cases} 16 \times \left(\dfrac{x(t)}{xp} \right) & \left| \dfrac{x(t)}{xp} \right| \le 0.0114 \\[4mm] \dfrac{x(t)}{5.473} \left(1 + \ln \left(87.6 \times \left| \dfrac{x(t)}{xp} \right| \right) \right) & 0.0114 \le \left| \dfrac{x(t)}{xp} \right| \le 1 \end{cases}$$

A general example of PCM coding is described in Figure 9.2. The input signal is quantized at 8KHz (meaning we take a sample every 0.125 milliseconds). For 255 code levels, we get 8 bits per sample. Therefore, we transmit 64 Kbps.

Figure 9.2 PCM Waveform Sampling

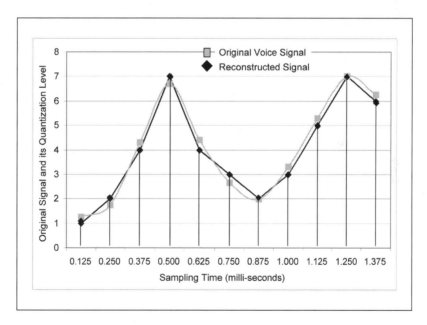

Continuous Variable Slope Delta Modulation

Continuous Variable Slope Delta Modulation was first proposed by Greefkes and Riemes in 1970. CVSD requires a 1-bit sample length compared to the 8 bits used in PCM, so more samples can be sent in the same bandwidth. As a result, CVSD is more tolerant of communications errors. Because of its error tolerance,

CVSD performs well in noisy channels, and for this reason, it has been widely used in military communications systems. The ability to tolerate errors is also what makes CVSD attractive for use in Bluetooth systems.

CVSD quantizes the difference in amplitude between two audio samples (that is, between the current input sample and the previous sample). The challenge is always to choose the appropriate step size $\delta(k)$. Small step sizes are better for tracking slowly changing low amplitude signals, but a larger step size is needed to accurately track a fast-changing high amplitude signal. This effect is shown in Figure 9.3.

Figure 9.3 The CVSD Operational Concept

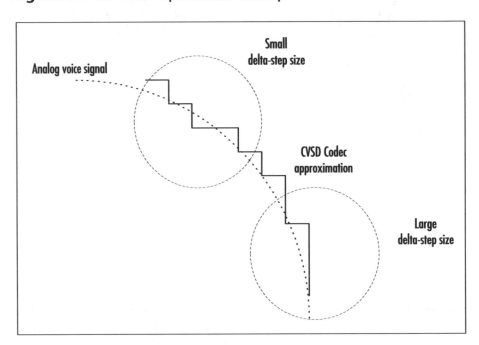

Let's consider a random input voice signal that we would like to convert from analog samples to digital format using CVSD. Figure 9.4 shows how this happens. As the input signal increases, bits set to 1 are transmitted. If the input signal decreases, bits set to zero are transmitted. In the first declining cosine slope of the signal, we can see how poorly the signal was quantized, but since it is an adaptive differential quantizer, it starts to adapt by changing the step size. Given this, if the signal characteristics remain the same, it will excel in following almost exactly the trace of the input signal.

Figure 9.4 The CVSD Waveform

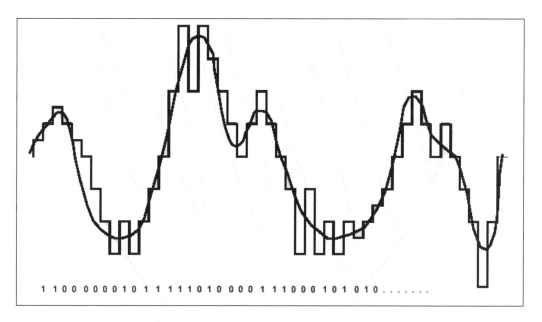

In the CVSD algorithm, the adaptive changes in step size, $\delta(t)$, are based on the past three or four sample outputs (for example, b*(k)*, b*(k-1)*, b*(k-2)*, b*(k-3)*) where it increases or decreases to catch up with the input signal as was shown in the example of Figure 9.4 earlier. The step size, $\delta(t)$, is controlled by the syllabic companding parameter, α, which determines when to increase $\delta(t)$ or allow it to decay. The step size decay time, β, is related to speech syllable length (sometimes called delay). The Bluetooth system specifies β to be 16 ms and the accumulator decay factor, *h*, to be 0.5 ms.

The accumulator decay factor decides the threshold of how quickly the output of the CVSD decoders decay to zero after an input; this determines how quickly the Codec will recover from errors in the received signal. Figure 9.5 shows flow diagrams of the algorithms for the encoder and decoder. The internal state of the accumulator depends upon the equations that follow.

$$\alpha = \begin{cases} 1, & \text{if the J bits in the last K output bits are equal} \\ 0, & \text{otherwise} \end{cases}$$

$$\delta(k) = \begin{cases} \min\left\{ \delta(k\text{-}1) + \delta_{min}, \delta_{min} \right\} & \alpha = 1 \\ \\ \max\left\{ \beta\, \delta(k\text{-}1), \delta_{min} \right\} & \alpha = 0 \end{cases}$$

Note step size delay (β) = 16 ms

Figure 9.5 The CVSD Encoder and Decoder Block Diagram

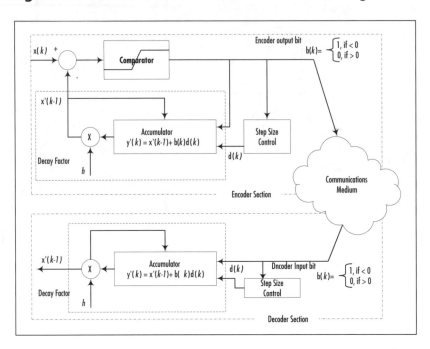

A standard called Mean Opinion Scale (MOS) testing is used to assess the subjective quality of voice links. A rating of 4 to 4.5 is considered *toll quality* (equivalent to commercial telephony). As MOS decreases, so quality decreases; a value of just less than 4 indicates communication quality with some barely perceptible distortion. Figure 9.6 compares MOS ratings for μ-law PCM and CVSD with various bit error rates on the channel.

> **NOTE**
>
> The term *toll quality* was first used about 22 years ago when T1 multiplexers first started transporting voice over private T1 lines. The original idea was that a private wide area network (WAN) could provide voice quality equal to that of the long-distance public switched telephone network (PSTN), which charged a toll for each minute of use using what is nowadays known as Voice Over IP (VoIP).

Notice CVSD performs as well as μ-law PCM in a clean communication medium. However, CVSD operates much better than μ-law PCM in the presence of bit errors. To be more specific, CVSD retains quite good MOS ratings at low bit error rates; however, it drops to a MOS rating of 3 (fair quality but tends to be annoying) at higher bit error rates. This robustness to bit errors (channel noise) makes CVSD an ideal solution for many wireless speech communication applications, including Bluetooth technology. But because PCM is cheap and already available in a lot of devices, we really need both.

Figure 9.6 MOS versus Bit Error Rate for CVSD and μ-Law PCM Codec

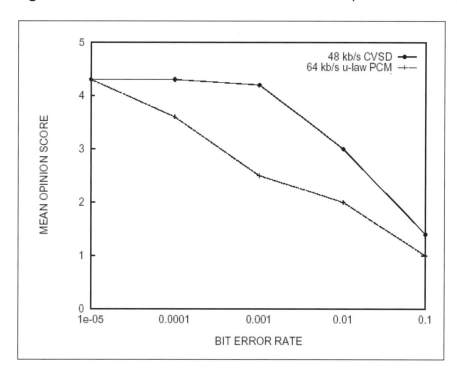

In this section, we have described CVSD and PCM Codecs, the circumstances that governed their design, and how robust their performance is in the presence of bit errors. You may be unable to choose Codecs because you are limited by what is available in your hardware systems. Your choice may be constrained by Bluetooth profiles as well, but you should now appreciate the performance impact of choosing a particular Codec. Now that we understand Codecs, we shall turn to the code you need to write and create your audio link.

Configuring Voice Links

The Bluetooth specification provides the means for devices to transfer data and voice simultaneously using Asynchronous ConnectionLess (ACL) channels for data and SCO channels for voice. The specification also allows up to three duplex voice (SCO) channels to be active simultaneously.

The specification provides these various capabilities by using a variety of packet types (High-rate Voice HV1, HV2, HV3, or Data-Voice [DV]). The application initiating the connection configures the voice link by choosing an HV packet type. The different packet types configure the link to occupy a different percentage of the channel bandwidth. This means that the choice of packet type determines whether space is left for other voice channels, and whether it is possible to transfer data while the voice channel is active.

As always, nothing is free—adding voice channels will severely impact your ability to transfer data. Furthermore, if you choose to use multiple voice channels, each channel will have less error protection, so performance will be worse on noisy channels. If you choose to send data at the same time as voice, you will also lose out on error protection on the voice links.

Because your application's configuration of the voice link will affect data rates and voice quality, it is important that you understand the implications of choosing different types. This section will take you through the capabilities of the different packet types, and explain their impacts on data rates and voice link quality in the presence of errors.

Choosing an HV Packet Type

Bluetooth technology uses a combination of circuit and packet switching technology to handle voice and data traffic. A circuit switched channel is a channel that provides regularly reserved bandwidth. Live audio needs circuit switched channels to guarantee regular delivery of voice information—the receive Codecs need a regular feed of information to provide a good quality output signal. The circuit switched channels are the Synchronous Connection-Oriented links—they occupy fixed slots assigned by the master when the link is first set up.

A packet switched channel is only active when data needs to be transmitted, and does not have reserved bandwidth. The packet switched channels in the Bluetooth system are the Asynchronous ConnectionLess links. If voice was sent on the ACL links, there would be no guarantee of regular bandwidth, and the quality of the received signal would suffer.

The various packets used on SCO links all provide the same symmetrical 64 Kbps between master and slave. Each packet type is sent in periodically reserved slots, but the different types require different spacings of reserved slots. Each SCO packet type, meanwhile, uses a different encoding for the payload data. The SCO packets (HV1, HV2, and HV3) are defined as follows:

- **HV1**   Carries 1.25 milliseconds (ms) of voice in 10 bytes. 1/3 Forward Error Correction (FEC) adds 2 bits of error correction for every bit of data, increasing the payload size to 30 bytes. HV1 packets are sent and received as single-slot packets in every pair of slots.

- **HV2**   Carries 2.5 ms of voice in 20 bytes. 2/3 FEC adds one bit of error correction for every 2 bits of data, increasing the payload size to 30 bytes. HV2 packets are sent and received as single-slot packets in two consecutive slots out of every four slots.

- **HV3**   Carries 3.75 ms of voice in 30 bytes. There is no error correction payload. HV3 packets are sent as single-slot packets in two consecutive slots in every six slots.

All of the SCO packets are single slot packets, and none of them carries a CRC, but we can easily see whether or not the packet types permit the flexibility to use FEC in the payload. In a noisy environment, there is no retransmission of SCO packets even if they contain errors, but the FEC scheme on the data payload protects the 80 percent of voice samples providing higher quality audio. However, the FEC encoding uses up space in the payload, so the packets that carry more error protection must be transmitted more often. In a reasonably error-free environment, FEC gives unnecessary overhead that reduces the throughput.

One more packet type can be used to carry audio data: the *data-voice* packet. This combines both ACL and SCO. The DV packet uses 2/3 FEC and a 16-bit CRC on the ACL data, but is without FEC on the SCO data. The DV packet carries 10 bytes of audio data, so it can be used to replace an HV1 packet—that is, it can be used on a SCO link where packets are sent every two slots.

Sending Data and Voice Simultaneously

One important question is how much voice links affect throughput of data. If we ignore the effect of errors and retransmissions, then it's quite a simple calculation (reference Table 9.1 for maximum throughput).

With no voice links present, it is possible to use the highest rate packets: DH5 packets. These use up to five 625 μs slots each and carry at most 339 bytes of the user's data. So, in 10 x 625μs we get a maximum of 339 bytes in each direction. This gives us 5424 bytes per second in each direction.

If we add an HV3 SCO link (the lowest load that a voice link can place on the system), then we will only have four slots in every six to transmit data. This means we cannot send five slot packets, and cannot send two consecutive three-slot packets. The most intelligent use of the available slots would be to send one three-slot DH3 packet (carrying, at most, 183 bytes of the user's data) and one single slot DH1 packet (carrying, at most, 27 bytes of the user's information). If the direction that sent the DH3 packet could be alternated, the bandwidth would be maximized, but both ends of the link would get the same share of the available bandwidth. Now in every 6 x 625µs, we get 183 bytes in one direction and 27 bytes in the other. Assuming the three-slot packets can be allocated so that the bandwidth averages out in each direction, our maximum data rate will average to 105 bytes transferred in each direction every 6 x 625µs. This gives us 2800 bytes per second in each direction, at 51 percent—this is almost half the maximum data rate without a SCO link present.

If we add an HV3 SCO link and just use single slot packets for data (which many basebands will do when an HV3 SCO link is active), then we get a lower throughput. In this case, we can send two DH1 packets (carrying at most 27 bytes of the user's information), giving 54 bytes in each direction every 6 x 625µs. This gives us 1440 bytes per second in each direction.

If we add two HV3 SCO links, then we only have two slots in every six available. At this point we could only send single slot packets. The best throughput we can get will be with DH1 packets carrying, at most, 27 bytes of the user's information. With just two slots out of every six available, we will be able to send one DH1 packet in each direction, giving 27 bytes transferred in each direction every 6 x 625µs.

If we add an HV2 link, then we only have two slots in every four available. At this point we could only send single slot packets. The best throughput we can get will be with DH1 packets, carrying, at most, 27 bytes of user's information. With just two slots out of every four available, we will be able to send one DH1 packet in each direction, giving 27 bytes transferred in each direction every 4 x 625µs. This gives us 1080 bytes per second.

If we add an HV1 link, then decide that we also want to transfer data, we could only transfer data by replacing the HV1 packets by DV packets. This payload carries a maximum of 9 bytes of the user's information (the 10 byte payload includes a byte of header information). The HV3 link uses every single slot, so we can send DV packets in every slot. This means we can transfer 9 bytes in each direction every 2 x 625µs. This gives us 720 bytes per second.

We have zero data throughput with three simultaneous voice channels because the DV packet type can only be used with a single voice link, and three HV3 links will use up every single slot.

While there is no user data throughput Link Manager Protocol (LMP), messages will take higher priority and will interrupt the voice links. This has to happen, otherwise there would be no way to send the LMP messages to tear down a voice link!

Table 9.1 Bluetooth Packet Type Maximum Throughput

SCO Packet Type	ACL Packet Type	Maximum Symmetric Throughput (bytes per second)	Percentage of Throughput without SCO
No SCO link	DH5	5424	100
HV3	DH3+DH1	2800	51.1
HV3	DH1	1440	26.5
HV2	DH1	1080	19.9
DV	DV	720	13.2
HV3 – two links	DH1	720	13.2
HV1	-	0	0

Using ACL Links for High-Quality Audio

So far, we have looked at voice links that use the HV packet types transmitted in reserved bandwidth provided by SCO links. The SCO links support the same sort of voice quality you would expect from a cellular phone. This is great for applications such as mobile phone headsets, but not acceptable for applications that require higher audio bit rates.

Obviously, with a maximum bit rate of 64 Kbps, a Bluetooth SCO link can't serve audio CD quality sound (1411.2 Kbps). For any high bit rate audio application (for example, a portable Bluetooth device playing MP3 music), the SCO channels will be inappropriate.

However, with suitable compression, it would be technically feasible to send high bit rate audio packets using asymmetric ACL channels. This allows us to get the maximum bandwidth from the Bluetooth link by using an asymmetric ACL link that can provide up to 723.2 Kbps, as shown in Table 9.2.

The SCO links provide guaranteed latency on the link, but do not retransmit lost or errored packets. By contrast, the ACL link provides guaranteed delivery of packets, but as this is done through retransmissions, there are no guarantees on latency (delay).

There are two levels of choice when configuring Bluetooth audio links. First, you must choose whether to use the Bluetooth audio Codecs and the SCO links, or send compressed audio across the ACL links. For real-time duplex voice communications, you should always choose the SCO links because of their guaran-

Case Study Example

Let's assume audio data streaming in a wireless point-to-point network, which includes a PC, a loudspeaker, and a subwoofer. The PC playing MP3-coded music is the piconet's master; the speaker and the subwoofer are both slaves.

Because we are listening to music, the SCO channel is too low quality, so we want to send packets across the ACL link. The ACL link is designed for bursty data, not for audio, so it will retransmit any packets which are subject to errors. This introduces delay into the link.

In order to cope with the delay, we need to buffer packets at the receiver—that way we can feed a steady stream of information to the MP3 decoder even if there are delays in the signal.

This has important implications for our application. We must ensure we use compression that allows all information to get through the channel even if there are errors. Though theoretically we have 732.2 Kbps to share between our slaves, in practice some of that capacity will be used up by errors and retransmissions, so our MP3 encoding must compress to less than the theoretical maximum channel capacity.

teed latency. For high bit-rate simplex audio such as that required for music, the SCO links will not provide the required quality and compressed audio must be sent across the ACL links.

Table 9.2 Bluetooth Communication Channel Support in Master-Slave Pairs

Channel			Maximum Data Rate	
Maximum Number	Type	Configuration	Transmitting	Receiving
7	Asynchronous (ACL)	Asymmetric data	732.2 Kbps	57.6 Kbps
			57.6 Kbps	732.2 Kbps
		Symmetric data	433.9 Kbps	
3	Synchronous (SCO)	Voice	3×64 Kbps	

Once you have chosen the link type suitable for your application, you must configure the link by choosing a packet type for it. For ACL links, you should always allow the baseband to choose the correct packet type for the current environment. To do this, you simply configure the link to use all data packet types, then the baseband automatically picks the best packet type for the current link quality. (This is done using Channel Quality Driven Data Rate [CQDDR]—for more details, see Chapter 1).

If you choose to use SCO links for your application, you should now have a good feel for how to select an audio packet type (HV1, HV2, or HV3). Basebands that support the DV packet type will automatically use it when an HV1 link is in use and there is user data to send.

Now that we understand how Bluetooth wireless technology transmits audio, let's examine the interfaces by which the audio signal gets into the Bluetooth subsystem.

Choosing an Audio Interface

Audio is not a layer of the Bluetooth protocol stack, it is a just a packet format that can be transmitted directly over the baseband layer. Figure 9.7 shows an example system such as might be used to implement an audio gateway in a cellular phone. Because the phone (the host) already has a processor, the upper layers of the Bluetooth protocol stack can be implemented on the host processor. The illustration shows the layers from the Bluetooth specification shaded in gray. There are two routes for audio: either a direct link between the baseband and the application layer, or through the HCI.

Figure 9.7 Audio Is Part of the Baseband Protocol Stack

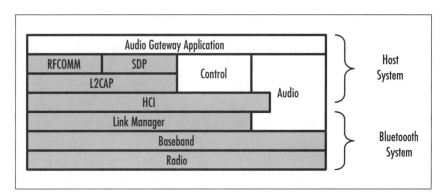

The only difference between the two routes through the system is that all packets passing through HCI experience some latency. The time taken for the

Bluetooth subsystem's microcontroller to transfer the audio data from the base-band into HCI packets introduces some delay, but this is imperceptible. However, there is a second factor that can cause severe delays and lead to loss of SCO packets: this is flow control of the HCI interface.

If the Universal Asynchronous Receiver Transmitter (UART) HCI transport is used, there is no way to separately flow control voice and data, so when data transport is flow controlled, the flow of voice packets across the HCI will also stop. Buffering in the baseband chip could be used to prevent loss of data, but in practice, since audio signals are time-sensitive, any late samples are simply discarded, leading to gaps in the audio signal. The problem does not arise if the USB transport is used for HCI, as this transport provides a separate channel for voice packets; however, USB requires complex drivers and is not appropriate for all products. To solve the problem of flow control affecting audio quality on serial links, the Bluetooth Special Interest Group (SIG)'s HCI working group is currently working on a new serial interface which will allow audio and data to be flow controlled separately.

Often, by the time the application developer gets involved, hardware choices have already been made—which means you really have no choice of audio interface, and must work within the limitations of what you have. However, if you are lucky enough to be involved in the choice early on, then in choosing a chip/chip set you should be aware of the potential impacts of choosing different interfaces to get audio into the Bluetooth subsystem. When you make a choice of silicon, be aware that not every chip/chip set supports audio, so obviously you need to work with a chip/chip set that does! Of those that do support audio, most provide direct access to the baseband. Some, however, do not support audio across HCI.

Selecting an Audio Profile

The Bluetooth specification is broken up into several parts. So far, we have looked at items covered by the Core Specification—this includes the radio base-band and the software layers which make up a Bluetooth protocol stack. The Core Specification has a second volume, which provides a series of profiles. The profiles give guidelines on how to use the Bluetooth protocol stack to implement different end-user applications.

The first version of the profiles document provides three different profiles covering audio applications: the Headset profile, the Cordless Telephony profile, and the Intercom profile. Within the Bluetooth SIG, there are working groups that are producing profiles to support further audio applications.

Many textbooks (such as *Bluetooth: Connect Without Cables*) will take you through the details of the profiles and protocol stack layers, and, of course, the Bluetooth specification itself provides the definitive guide to the subject. This section will just cover enough about the audio profiles to give you a taste of what's involved. Use this information to decide which profiles may be appropriate for your application.

The first thing to be aware of is that your choice need not be limited to one particular profile. If your product supports several services, it may be appropriate to implement more than one profile. Figure 9.8 illustrates this point: it shows a 3-in-1 Bluetooth phone, which implements the Headset, Cordless Telephony, and Intercom profiles. Let us examine each of these profiles in turn.

Figure 9.8 Audio Profiles and Link Establishment for Bluetooth-Enabled Devices

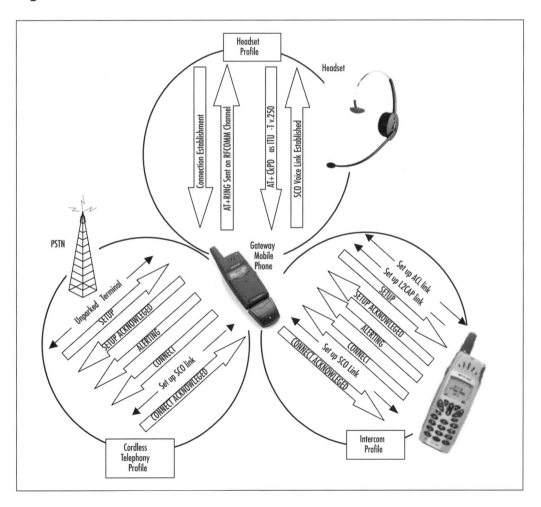

The Headset profile allows the audio signal from a telephone call to be transferred between an audio gateway (AG) and a headset. A mobile phone is a typical audio gateway, but any device that receives incoming audio calls could be used. Similarly, the headset side is usually a headset with microphone and speaker, but it would be possible for a laptop computer to implement the Headset profile and use its microphone and speaker to handle the audio part of a telephone call.

The Headset profile uses AT commands across an RFCOMM connection for control. First, an ACL link is established and a connection to RFCOMM is set up. Then an AT+RING command is sent on the RFCOMM connection to trigger a ring tone in the headset. The user pushes a button on the headset to pick up the incoming call. The button push is signaled to the phone using an AT_CKPD (keypad command). Once the button press information is received, a SCO link can be set up to carry the voice call between the headset and the audio gateway.

The Cordless Telephony profile allows incoming calls to be transferred from a base-station to a telephone handset. In many ways, the Cordless Telephony profile provides similar capabilities to a digital enhanced cordless telecommunications (DECT) telephone system, except that it is not possible to hand over an active call to a different base station. This means that the phone handset must stay within range of a single base station. The Cordless Telephony profile provides control of information in addition to the transfer of audio, so, for instance, a calling line identifier (CLI) can be sent to the phone handset so the user can see who is calling them before deciding to answer the call.

The Intercom profile allows telephone calls to be transferred across a Bluetooth link without involving a telephone network at all. Again, identifying information can be sent with the call so that the receiver can display the number of the device initiating the call. There have been some questions about whether the Intercom profile is really useful (the lowest power Bluetooth devices only operate within a 10 meter range, and at these distances, you may as well shout). However, devices with class 1 radio modules can achieve 100m ranges, and this means that the Intercom profile could provide telephony services within an office building where it is not always appropriate to shout!

The Cordless Telephony and Intercom profiles both use Telephony Control Protocol (TCS) commands for control. The first stage is to establish an ACL link. Figure 9.8 shows two ways in which this can be done—the cordless telephony example shows a connection being unparked, while the intercom example shows a fresh connection being established. In both cases, the first step is to send a SETUP message to indicate a new call is being established. The SETUP message is acknowledged and the device receiving the call begins generating a ring tone to tell the user that a call is coming in. So that the device originating the call knows the user is being alerted, an ALERTING message is sent back by the device receiving the call. When the user accepts the call a CONNECT message is sent to the device originating the call, this triggers the setup of a SCO link. Once the SCO link is in place, the CONNECT message can be acknowledged.

Debugging…

Trap-Link Supervision Timeout

While considering how different devices disconnect, it is worth thinking about one aspect of wireless connections which can trip up developers who are used to wired systems. Bluetooth is a wireless technology, and like other wireless technologies, it's always possible the link will fail because of interference or because mobile devices move out of range of each other.

When the link fails, it will cause a link supervision timeout at the Link Management layer. This means that the Link Manager has detected that it has not been able to send packets on the link for a preset timeout period.

The default link supervision timeout on a Bluetooth link is 30 seconds, so by then the user will probably have given up and terminated the connection themselves. You could set the link supervision timeout period so that the link will automatically disconnect *sooner* than the default 30 seconds. To do this, you use the HCI *Write_Link_Supervision_Timeout* command. When the link disconnects, the HCI will return a *Disconnection Complete* event, and this should cause the various protocol stack layers to disconnect.

When the link does disconnect, your application will be notified. At this point, you will need to tidy up any resources in use by the link—free memory, close down audio channels, and so forth—just as if the call had been terminated by the user. If your device has a visual user interface, it is a good idea to display a message to the user informing them that the link has failed. If your device has an audio interface, you must decide whether to generate some tone to indicate the link has failed, or just leave the user listening to silence.

By now, you should be realizing that the Intercom and Cordless Telephony profiles are very similar in the ways they establish the link, whereas the Headset profile uses a completely different mechanism. This is because the Intercom and Cordless Telephony profile are controlling the link with TCS commands, while the Headset profile controls the link with AT commands. The different control mechanisms mean that when the profiles disconnect, we again see similarities between Intercom and Cordless Telephony, but the Headset profile still behaves differently.

Figure 9.9 shows how the Intercom and Cordless Telephony profiles share the same disconnection procedure. First of all, the party that is to end the call, sends a disconnect signal to the client that replied with a release permission and waits for the SCO link release signal to tidy up the resources and avoid memory leakage. The Headset profile is slightly different because it sends an AT-based keypad control (AT+CKPD) command to the audio gateway first, and the audio gateway releases both the SCO links and the connection.

Figure 9.9 Audio Profiles and Link Release for Bluetooth-Enabled Devices

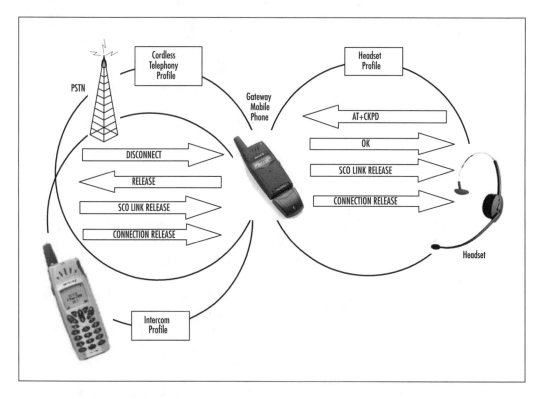

It is interesting to note that the Headset profile does not provide any commands for the headset to terminate the connection; however, if the headset just

drops the link, the audio gateway must be able to cope. So, if you wish to provide a disconnect facility for your users, then your code will be very simple: don't send any commands, just disconnect!

Figures 9.8 and 9.9 show example calls going in one direction. For the headset profile, the audio gateway side always initiates the call—the headset cannot initiate a voice call to the audio gateway, it can only accept an incoming call. With the Intercom and Cordless Telephony profiles, either device can initiate the call—so, for instance, with the cordless telephony profile, the base station can receive an incoming call from the PSTN and send that call to the phone handset, or the phone handset could initiate a call to the base station, and the base station would then pass that call out to the PSTN.

If you just want to transfer the audio part of a call without control information, then the Headset profile is small, simple, and definitely the one to use. If you need to initiate voice calls to other Bluetooth devices in the area, but are not passing them on to a network, then use the Intercom profile. If you are implementing a base station to pass voice calls to and from a telephone network, then you should use the Cordless Telephony profile.

Applications Not Covered by Profiles

You may have noticed that all of the three profiles previously described are oriented towards distributing telephony devices. All of these profiles use SCO links to carry the audio information. As we discussed earlier, that's fine for telephones, but not so good if you need high-quality audio for music.

If your application provides a service which is covered by the existing Bluetooth profiles, then you should implement the relevant profile. However, at the moment, there are many possible audio applications which are not covered by profiles. If your application fits in this class, then you will have to design a complete proprietary application yourself without guidance from a profile document.

A disadvantage of producing your own proprietary application is that it will only work with other products that use the same control systems. That's fine if you are implementing a closed system, but if you want to make some Bluetooth stereo headphones, then you'd probably prefer them to work with lots of different brands of stereos so that more people will buy them. The solution is to join in one of the working groups of the Bluetooth SIGs and get together with other manufacturers to come up with a profile that lots of devices can implement. To join a SIG working group, you must be an associate level member of the Bluetooth SIG (there is an annual fee for associate companies). Participating in a working group can also be quite time-consuming, often involving international travel to meetings, so this route will not suit everyone.

Another alternative is to look on the Bluetooth Web site and find out which working groups are producing new profiles. It may be that the profile you need is just around the corner. If that's the case, it may be worth your while to wait for the profile to be released rather than go to all the trouble of developing a proprietary system only to discover that it fails in the market because everybody else is using a standardized profile.

New Audio Profiles

The Bluetooth SIG has working groups who are developing new profiles. There is a car working group, which is due to release a hands-free profile soon and an audio/visual (AV) group, which is working on a series of profiles to provide distribution of low bit rate video and high-quality audio.

The hands-free profile being produced by the car working group is targeted at in-car, hands-free kits, but could also be used in other applications, such as call centers. The hands-free profile will allow the hands-free device to initiate calls to the audio gateway. This will be done by transferring dialing information using AT commands across an RFCOMM serial link. Because the hands-free profile uses AT commands to dial, it will be simpler to implement than the TCS-based profiles.

The AV working group is providing a variety of profiles which will allow Bluetooth systems to support standardized audio and video capabilities. These provide videoconferencing capabilities—note that a video capability suitable for videoconferencing is probably not satisfactory for distributing video for entertainment purposes. In short, you won't find this profile much good for watching movies! There is also an advanced audio distribution profile which supports higher quality audio than the basic SCO links. Distribution profiles provide standardized streaming channels to be set up and controlled to support audio or video distribution. There are also profiles defining how links should be controlled and how remote control should be provided.

In the future, more profiles will be released. Members of the Bluetooth SIG are notified by e-mail whenever new profiles become public.

Writing Audio Applications

In the previous section, we looked briefly at the various profiles available for audio applications. In this section, we'll look in more detail at how a particular profile could be implemented at application level. We shall use the headset profile as our example application, because it is the simplest of the audio applications. Even then, much of the application functionality will remain the same whichever profile you use. For example, all inquiry, paging, scanning, and service discovery are the same no matter which profile you implement. Similarly, the audio must be routed into the Bluetooth subsystem somehow, regardless of the audio profile chosen.

As we explained in the previous section, the headset profile is used to transfer the audio part of a call between an audio gateway and a headset. Figure 9.10 shows some examples of devices that implement the Headset profile: the Ericsson DBA-10 snap-on Bluetooth accessory provides Bluetooth system capability to the Ericsson T28 world phone. The combined phone and accessory act as an audio gateway. The Ericsson and GN Netcom headsets both implement the headset part of the Headset profile.

Figure 9.10 Bluetooth Devices that Use Audio Links (Ericsson Bluetooth Headset and Mobile Phone, GN Netcom GN9000 Headset)

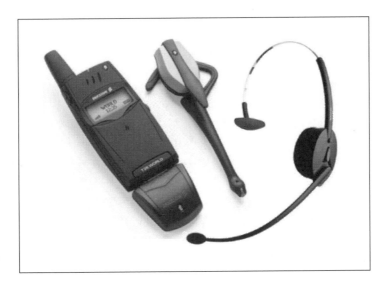

Discovering Devices

Whichever audio profile is being supported, the initial steps in establishing a link will be similar. The first step will be finding suitable devices in your neighborhood using the Bluetooth Device Discovery procedures.

Chapters 1 and 2 explained how Inquiry and Inquiry Scan modes are used to implement device discovery. For audio applications, it is also worth noting that the inquiring device can use an HCI command to filter inquiry responses by device class. The Frequency Hopping Synchronization (FHS) packets used to respond to inquiries, each contain a major and minor device class. For the Headset profile, we are only interested in devices with the Class of Device set as follows:

```
Major class of device = audio
Minor class of device = headset
```

The following pseudo-code shows how an application might implement device discovery:

```
// Device Discovery
Display "Discovering Devices" message to the user
Send HCI Set_Event_Filter command Filter_Type = inquiry result Filter
    Condition = devices with a major class of device = audio minor class
    of device = headset
Send HCI Inquiry command to initiate an inquiry
WHILE (HCI inquiry complete event not received)
{
        Receive and process inquiry response events
}
```

The exact code used will vary from system to system, but the procedure to set event filters, initiate an inquiry, and process the results until the inquiry completes, will remain the same. One possible variant would be to use *periodic inquiry* mode. This will set the lower layers to periodically perform an inquiry. Most audio applications will run on small battery-operated devices, and since periodic inquiries will drain the device's batteries, their use is not recommended for audio applications.

Of course, the inquiry won't get any results if there are no devices scanning, so to match the previous inquiry code, we need the inquiry scan pseudo-code that follows:

```
Send HCI Write_Inquiry_Scan_Activity
Send HCI Write_Scan_Enable scan mode = inquiry scan enabled
Start timer
Wait for timer to cause a timeout
Send HCI Write_Scan_Enable scan mode = inquiry scan disabled
```

As explained in Chapters 1 and 2, the inquiry scan activity should be set according to the requirements of the Generic Access Profile. Again, because of the power drain caused by scans, it is recommended that a device should not be left in Inquiry Scan mode for long. This is why the previous code runs a timer, and when the timer causes a timeout, it disables the inquiry scan.

The fact that Inquiry and Inquiry Scan only happen for short periods implies that you must be able to trigger them somehow from the user interface. Usually, the audio gateway performs inquiries and the headset scans for them. If the audio gateway is a phone, an inquiry can be triggered through the phone's menu system. A headset is more problematic since it will have a very limited user interface—buttons take up space and cost money, so you can't have many of them! The Ericsson headset has a single button that is pressed to switch the headset on and off. If you keep the button held down after switching it on, you go into Inquiry mode. Experience shows that some users find interfaces that have many functions attached to one button difficult to operate, but you must balance this against the size, weight, and cost penalties of adding more controls onto the headset.

Using Service Discovery

Once the audio gateway application has found a device that belongs to the audio/headset class of devices, it needs to find out how to connect to the headset service. To do this, it uses Service Discovery Protocol (SDP) and performs a service search for the headset service.

The pseudo-code that follows illustrates the steps an audio gateway would go through when using service discovery on a headset.

```
// Service Discovery
display "Discovering Services" message to the user
For (each device with audio as major class of device discovered during
     device discovery)
{
    send HCI_Create_Connection command to create an ACL link to device
    send HCI_Remote_Name_Request command to get user-friendly name for
      remote device
    create L2CAP link using PSM for SDP
    send SDP service search for headset service
    IF (headset service record returned)
    {
        store headset service record for device
```

```
        display device to user using user-friendly name
    }
    disconnect L2CAP and ACL links
}
```

An ACL link is created, and once the link is up, a remote name request is used to find the user-friendly name of the remote device. This isn't mandatory, but it will make your application a lot easier for users if you get this information for them.

A Logical Link Control and Adaptation Protocol (L2CAP) link is created across the ACL link. This must be created specifically for SDP, and uses a Protocol Service Multiplexor (PSM), which tells the remote device to connect the L2CAP link to its SDP server. Once the L2CAP link is established, it can be used to send SDP service search requests to retrieve the service record for the headset service. This record confirms that the remote device implements the headset profile, and gives version information, along with information required to connect to the headset service.

Once the service record is returned, it can be stored locally so that if the device is encountered in future service discovery, it does not have to be performed again. Any new information can also be displayed to the user, and as the link is now finished, it may be destroyed.

Leaving the link up wastes power, but establishing a link also takes up power, so there is a decision to be made about disconnecting links. In the preceding example, the L2CAP and ACL links were both disconnected, but there is a chance that the ACL link will be reused to connect to the headset service. This means that it might be advisable to wait a while before disconnecting the ACL link. Because of this, you might implement something like this:

```
Disconnect L2CAP link
Start timer
Wait for timeout
IF (connection to headset service has not been requested)
{
    disconnect ACL link
}
```

The L2CAP link is disconnected straight away because it was created with a PSM value for SDP. This means that the L2CAP link cannot be used for anything other than service discovery.

Connecting to a Service

Now we can finally get to the whole point of the application and connect to an audio service. The first step is to set up an ACL link—this could be a link leftover from the service discovery phase, or if that link was disconnected, it could be a new link set up by repeating the paging process. This connection is used to create an L2CAP link using the PSM value for RFCOMM. Next, an RFCOMM channel is set up to control the headset. The Channel ID for the headset was provided to the Audio Gateway in the headset's service record.

The RFCOMM connection is used to send the AT commands which control the headset service. The first command shown in the following is an AT+RING signal, which tells the headset to produce a ring tone. This ring tone alerts the user that a call is coming from the audio gateway.

The user should somehow accept the call—this could be done with a voice recognition system, but it will most likely be done by the user pressing a button on the headset. However, the user actually accepts the call with the keypad signal AT+CKPD, which is sent back to the audio gateway across the RFCOMM channel.

Now that the audio gateway knows the headset is willing to accept the call, it establishes an audio (SCO) link. This could optionally have been done earlier on, but audio links consume power, so it is better to wait until the last possible moment to set up the SCO link. The link must be configured, and our example shows an HCI *Write_Voice_Setting* command which sets the Codec format (A-law or μ-law PCMs and CVSD). The Codec does not have to be chosen at this point—this could have been set earlier on, or left at some default value. Once the Codec settings are configured as required, a SCO connection can be set up using the HCI *Add_SCO_Connection* command. The parameters for this command specify the connection handle of the ACL connection across which the SCO connection will be set up, as well as the packet type to be used on the SCO connection (HV1, HV2, and HV3).

Note that the audio gateway initiates the SCO connection, which means it chooses the Codec and HV packet type to be used on the link. Because the audio gateway chooses the Codec and packet type, the headset must be able to accept all Codecs and packet types. However, because the headset does not need to worry about deciding which type is appropriate, the headset application is much simpler to write.

Immediately after the *Add_SCO_Connection*, an *HCI_Command_Status* event is returned to acknowledge the command. When the SCO connection is established, an *HCI_Connection_Complete* event is received. If there were any problems with the connection, the status field will carry a reason for failure. The following pseudo-code illustrates this procedure.

```
//Connection
IF (headset was not found during service discovery)
{
    display message "no headsets found" message to the user
}
ELSE // at least one headset was found
{
    display message "please select a headset to connect with"
    IF(user selects a device)
    {
        display message "connecting with headset"
        send HCI_Create_Connection command to create an ACL link to
            the device
        create L2CAP link using PSM for RFCOMM
        create RFCOMM link to headset service using RFCOMM channel from
            headset service record
        send a ring signal using an AT+RING command
        IF (receive an AT+CKPD from headset)
        {
            send HCI Write_Voice_Setting
            send HCI Add_SCO_Connection to establish SCO link
            send any control commands required to route audio to user,
                set volume, etc.
            IF (HCI_Connection_Complete with status = success)
            {
                display message "Connected to headset"
            }
            ELSE
            {
                display message "could not connect to headset"
```

```
                    disconnect links and tidy up resources used
                }
            }
        }
    }
```

This example is simplified and does not cover security procedures. For an in-depth look at security, see Chapter 4. It is worth noting in passing, however, that a headset can be paired to the audio gateway, and it is possible to pair a headset with more than one device. If this was done, then the same headset could quickly and easily be used with a variety of audio gateways. For instance, while on the move, you could use your headset with a mobile phone, but in the office, the same headset might be used with a Bluetooth-enabled desk phone.

Using Power Saving with Audio Connections

Some of the Bluetooth-audio enabled devices might have very small batteries, because of both size and weight constraints, so optimizing power consumption is important. Sometimes an audio device will be idle for a long time—for example, after terminating the communications link or while waiting for an audio connection to be established. During these idle periods, it doesn't need to participate in the channel. We could simply drop the ACL connection, but then when we needed to connect with it again, there would be a delay. For a cellular phone headset, it could be a real disadvantage to have to wait a few seconds while the phone paged the headset. This would introduce an unacceptable delay in notifying the user that a call is coming in. So, to allow fast audio connections to be made, we want to keep the ACL link, but to save power, we want to drop the link.

The solution is to use the low-power park mode. In this mode, the Bluetooth-enabled audio device remains frequency-hop synchronized by waking up periodically during beacon slots to resynchronize with the master. The master can use beacon slots to reactivate the device, so that when an incoming call arrives, the terminal can be unparked fast enough to answer the call or can start to listen to the music from the beginning of the play. The spacing of the beacons is a trade-off between response times and power saving. Long beacon intervals give a slow response, but require less activity from both master and parked slave. Short beacon intervals give faster response, but require more activity and hence consume more power. See Chapters 1 and 3 for more details on low-power modes.

Differentiating Your Audio Application

So far, we've looked at the basics of writing a Bluetooth audio application, but if you're making a product to sell, you don't want a basic application, you want something special! This section will look at a few of the ways you can differentiate your audio application, adding value for the user. This is the sort of thing you need to do if you want your product to sell better than the next guy's.

Physical Design

Chapter 1 looked at some of the physical factors that make a Bluetooth product succeed, so we won't go into great detail here. But do be sure not to forget the weight, size, and form factor. All of this may be beyond your control, but if you are involved in the original product design, you can contribute to your devices salability by ensuring that these are thought about.

Bluetooth wireless technology is still young. The people buying Bluetooth audio devices today are the classic early adopters—gadget freaks who are willing to take a risk just to have the latest thing. Something that displays the novelty of the device can be quite an important factor—blue LEDs are much more expensive than red or green, but look around and you'll find plenty of Bluetooth products sporting blue flashing LEDs. The reason is that displaying their new technological gadget is an important factor to the early adopters, and that blue LED says "my product is a Bluetooth product." Thinking about apparently trivial items like the color of an LED can be the difference that makes your design stand out and appeal to your target users.

Designing the User Interface

The user interface is the one aspect of your application that has the power to make or break your market success. The qualification process ensures that you've got the technology right, but nobody will stand over you and make sure your product is actually usable! As you write your application, ask yourself if there are ways to hide the complexity of Bluetooth technology.

The profiles constrain what you can do with an application—this is done with good reason: it helps to ensure that products from different manufacturers will interoperate. You might think that if everybody's application is implementing the same profile, there is no real scope for differentiating products at the user interface level. Don't despair—there are *plenty* of things you can do to make sure your application has an edge over the competition.

Many headsets are using a single function button, which is slid from side to side for volume up/down and pressed for various lengths of time to perform other functions. You should balance the complexity of such an interface with the cost and added size involved in having more buttons. What works best will differ from product to product, so think about what works best for *your* form factor.

One factor that is often overlooked in headset design is the possibility of using the audio channel as part of the user interface. Even systems that do not implement voice recognition can quite simply and easily use the audio path as part of the user interface by generating tones to inform the user of events. For example, if a call is disconnected due to link loss, a continuous tone could be sounded for half a second alerting the user that there is a problem. Similarly, if the device has a low battery, a series of tones could be sounded to warn the user that they are about to lose usage of the device. Because the user interfaces are very limited on small mobile audio devices, it is worth considering whether your application can make use of the device's built-in audio facilities to provide a richer user interface.

Enabling Upgrades

One way to differentiate your product is to provide ongoing support for new features, or for future versions of the Bluetooth specification. More and more devices are now providing upgrade facilities for users. If you choose to do this, then you will have to consider how to avoid the upgrade process being run accidentally. This is important because the first stage of a device upgrade often involves wiping code and leaving the device in an unusable state if there is no upgrade code available.

Once you have an interface to start the upgrade process, you will need to consider the route by which you can download code to upgrade to a new version or to add features. Some part of the system will need to check the code to be sure it is a correct authorized version. A checksum should be implemented to ensure the new code is not corrupt, and you may like to also consider incorporating a security code to avoid unauthorized or accidental modification of your device's application.

Many devices are capable of being upgraded, but with the exception of PC applications, it could be argued that very few users ever choose to take advantage of upgrade facilities. However, just because devices installed in the field may not be upgraded, it does not mean that upgrades are not relevant to your application. Often, devices awaiting shipment require an upgrade before delivery; if this might apply to your products, then it is worthwhile providing some route for upgrades to be downloaded to your device. Manufacturers who upgrade old stock awaiting shipment may choose to enable upgrades using special commands which are not publicized to the end users. In this way, they can hide a complex engineering interface from the user's eyes, and prevent accidental use of the upgrade interface.

Improving the Audio Path

As mobile devices become ever smaller, design problems start to appear, particularly with duplex voice systems. In a wired headset, the microphone typically dangles on a flexible cord and is quite well separated from the earpiece. Bluetooth headsets tend to be designed to clip on the ear with the microphone carried on a small boom, which places it close to the user's mouth. This creates two problems: first, the microphone and earpiece are physically closer together, creating the possibility of an audio feedback loop through free space, and second, their linking by the rigid boom creates the potential for acoustic coupling between the microphone and earpiece through the casing of the headset itself.

There can also be resonance effects within the components of devices—rigid cases and printed circuit boards (PCBs) can resonate at particular frequencies, and it is also possible for the coupling between the audio gateway and the headset to affect the audio. Combine all these effects and there can be noticeable impacts on the audio quality perceived by the user.

The primary solution will always lie in good physical design of the product, but there are other things that can be done. Most mobile phones incorporate echo canceling and other such advanced techniques, which use the digital processing power of the phone to reduce unwanted components in the audio signal. Digital signal processing, of course, uses processing time, adding expense and increasing power consumption, so it should only be used in a headset as a last resort.

Summary

Bluetooth wireless technology has a promising future in the mobile phone and handheld devices' audio markets. We have seen that Bluetooth devices can support up to three full-duplex SCO audio channels, or support up to two voice channels with simultaneous data transfer. Those channels use three coding schemes: CVSD, μ-law PCM, and A-law PCM. CVSD is more robust for errors and can support higher quality over good links. However, PCM is cheap and already available in a lot of commercial devices. For maximum compatibility, we really need both.

There are two routes for audio into the Bluetooth system: straight into the baseband or through HCI. The HCI route can experience latency due to flow control of data between host and lower layers. The Bluetooth SCO links provide toll-quality voice suitable for carrying phone calls. For high-quality audio (such as that required for music), the SCO links do not provide sufficient quality. Currently, there is no standardized way of providing high-quality audio across Bluetooth links, but compressed audio (such as MP3) could be sent across an asymmetric ACL link.

There are three audio profiles: Headset, Intercom, and Cordless Telephony. Further profiles are being defined, including those that provide higher quality audio across Bluetooth links. The steps involved in using an audio service are common to all profiles—discover devices perform service discovery, exchange control information, and configure and set up an audio link.

Audio applications can be differentiated in many ways. We considered physical design, user interface design, enabling upgrades, and improving the audio path.

Solutions Fast Track

Choosing a Codec

☑ Codecs (coder/decoders) convert between analog voice samples and the compressed digital format.

☑ The output of the Codecs must be fed into the Bluetooth baseband as a direct input to the baseband (a technique commonly used in Bluetooth chips), or encapsulated in a Host Controller Interface (HCI) packet and fed across the Host Controller Interface.

☑ Bluetooth technology uses CVSD and PCM Codecs. CVSD is more robust in the presence of errors, which is what makes CVSD attractive for use in Bluetooth systems. PCM is cheap and already available in many commercial devices.

☑ There are two types of compression implemented in PCM Codecs: A-law and μ-law. The different types are used by phones in various geographical regions.

Configuring Voice Links

☑ The Bluetooth system transmits data on ACL links and voice on SCO links. SCO links use periodically reserved slots, while ACL links do not reserve slots.

☑ Live audio needs circuit switched channels to guarantee regular delivery of voice information—the receive Codecs need a regular feed of information to provide a good quality output signal. The circuit switched channels are the Synchronous Connection-Oriented links. They occupy fixed slots that are assigned by the master when the link is first set up.

☑ Always remember that Bluetooth technology maintains a maximum of 3×64 Kbps full-duplex SCO voice packets. The SCO links provide voice quality similar to a mobile phone; if higher audio quality is desired, then compressed audio must be sent across ACL links.

☑ Notice that we don't want to modify the voice packets at the L2CAP layer. SCO packets bypass the L2CAP layer.

☑ If you choose to send data at the same time as voice, you will also lose out on error protection on the voice links.

☑ When a link is to be established, use the following procedure: scan or page for an audio device. Use SDP to identify service. Set up ACL connection first for control, then set up SCO connection. During a voice connection, control messages can be sent such as DTMF signals

Choosing an Audio Interface

☑ There are two routes for audio: either a direct link between the baseband and the application layer, or through the HCI. All packets passing through HCI experience some latency.

☑ If the Universal Asynchronous Receiver Transmitter (UART) HCI transport is used, there is no way to separately flow control voice and data, so when data transport is flow controlled, the flow of voice packets across the HCI will also stop. The USB transport provides a separate channel for voice packets; however, USB requires complex drivers.

☑ Not every chip/chip set supports audio. Of those that do, most provide direct access to the baseband, but some do not support audio across HCI.

Selecting an Audio Profile

☑ Three different profiles cover audio applications: the Headset profile, the Cordless Telephony profile, and the Intercom profile. If your product supports several services, it may be appropriate to implement more than one profile. If your application is not covered by one of the profiles, you will have to design a complete proprietary application yourself.

☑ The Headset profile allows the audio signal from a telephone call to be transferred between an audio gateway (AG) and a headset. If you just want to transfer the audio part of a call without control information, then the Headset profile is small, simple, and definitely the one to use.

☑ The Cordless Telephony profile allows incoming calls to be transferred from a base-station to a telephone handset. If you are implementing a base station to pass voice calls to and from a telephone network, then you should use the Cordless Telephony profile.

☑ The Intercom profile allows telephone calls to be transferred across a Bluetooth link without involving a telephone network at all. If you need to initiate voice calls to other Bluetooth devices in the area, but are not passing them on to a network, then you should use the intercom profile.

☑ The Cordless Telephony and Intercom profiles both use Telephony Control Protocol (TCS) commands for control and share the same disconnection procedure. The Headset profile controls the link with AT commands, and does not provide any commands for the headset to terminate the connection.

Writing Audio Applications

☑ In this section, we looked in detail at how a particular profile could be implemented at application level. All inquiry, paging, scanning, and service discovery are the same no matter which profile you implement. Similarly, the audio must be routed into the Bluetooth subsystem somehow, regardless of the audio profile chosen.

☑ The first step will be finding suitable devices in your neighborhood using the Bluetooth Device Discovery procedures.

☑ Once the audio gateway application has found a device that belongs to the audio/headset class of devices, it needs to find out how to connect to the headset service. To do this, it uses Service Discovery Protocol (SDP) and performs a service search for the headset service.

☑ Once the service discovery phase is complete, you can connect to an audio service. The first step is to set up an ACL link. This connection is used to create an L2CAP link using the PSM value for RFCOMM. Next, an RFCOMM channel is set up to control the headset. Once the audio gateway knows that the headset is willing to accept the call, it establishes an audio (SCO) link. The headset must be able to accept all Codecs and all packet types on the link.

Differentiating Your Audio Application

☑ Be sure to consider the weight, size, and form factor in your product design.

☑ The user interface is the most crucial aspect of your application. Ask yourself if there are ways to hide the complexity of Bluetooth technology. Button functions and headset designs offer opportunities for improvement and differentiation.

☑ Another way to differentiate your product is to provide ongoing support for new features or for future versions of the Bluetooth specification.

☑ Improving design and engineering to better the audio path can have a noticeable impact for the user, helping to avoid audio feedback, acoustic coupling, and resonance effects.

Frequently Asked Questions

The following Frequently Asked Questions, answered by the authors of this book, are designed to both measure your understanding of the concepts presented in this chapter and to assist you with real-life implementation of these concepts. To have your questions about this chapter answered by the author, browse to **www.syngress.com/solutions** and click on the **"Ask the Author"** form.

Q: The input to the CVSD encoder is 64 K samples/s linear PCM. How can you create the 64 Kbps encoder output using just using an 8 K samples/s input?

A: It is 64 Kbps but 8 K samples/s. If there are 8 quantization levels per sample, this is the same as saying 64 Kbps. It all depends on the number of distinct levels the sample can represent.

Q: If a Bluetooth SCO link can't carry CD-quality sound, how could you develop a Bluetooth-enabled MP3 player?

A: It is possible, but we have to use ACL channel (maximum asymmetric data rate 732.2 Kbps) audio sent in compressed format, and buffering must be done to allow a constant flow of data to the MP3 decoder, despite delays caused by retransmissions on the ACL link.

Q: Why is CVSD more robust for errors than PCM?

A: First of all, CVSD requires a 1-bit sample length compared to the 8-bits used in PCM, so more samples can be sent in the same bandwidth. Second, since CVSD is a differential scheme and depends on the slope between the symbols (unlike PCM), when the data is corrupted, the effect is less marked, as the signal only has a small difference from the correct signal. Third, the CVSD algorithm incorporates a decay factor, which means that upon receipt of correct data, the output signal will tend towards the correct value.

Chapter 10

Personal Information Base Case Study

Solutions in this chapter:

- **Why Choose Bluetooth Technology?**

- **Using Bluetooth Protocols to Implement a Personal Information Base**

- **Considering the User's View**

☑ **Summary**

☑ **Solutions Fast Track**

☑ **Frequently Asked Questions**

Introduction

The word "personal" keeps coming up whenever people talk about Bluetooth technology. Personal Area Networks, Personal Devices—it's all about bringing communications down to the local personal level. So the next logical step is to use Bluetooth technology to maintain a personal information base! The example we will be working with in this chapter looks at a hospital environment as a case study for implementing Bluetooth technology.

In the past, medical records were limited to a few salient observations. Today, reams of data can be gathered by complex monitoring systems. A lot of that data is lost because it is difficult to move around and store. By creatively using communication applications, we can send the data with the patient so it's easily accessible when needed. By making the database personal and transportable, we guarantee its instant availability. By using Bluetooth wireless technology, we provide an open standard for accessing the data, meaning that if a patient moves from one area or clinic to another, all the data required can accompany them and should be instantaneously accessible—anywhere and anytime.

What would such a Personal Information Base (PIB) device for a medical environment be able to do? It could store all the patient information, such as contact details, digital photographs, calendar of appointments with the doctor and hospital, as well as all the information gathered from tests, be they electronic or manual. These are just the basic details that can be stored. The potential to store more data in any form is infinite.

The advantages of a PIB device for both patient and hospital are security, instant access to almost up-to-date information, electronic and efficient transfer, and safe and compact storage over time.

Figure 10.1 shows what a PIB card could look like and how data can be exchanged by a Data Access Terminal (DAT). Both the PIB and DAT can exchange information with the local server. The local server keeps a copy of the data on the PIB and can synchronize data from other departments and DATs. The synchronised data is backed up to the Central Control, which can then distribute the data to all the hospitals and local servers. The aim of the system is to ensure that the patient's information is stored in at least two places at any one time. Duplicate storage means that data can be recovered if there is a loss of any element of the system.

Figure 10.1 A Personal Information Base System

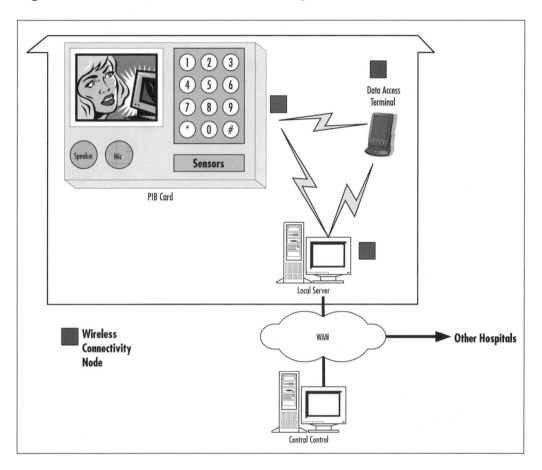

Most of the elements are standard "off-the-shelf" components enabled with either wireless or wired networking. This helps to keep infrastructure costs down, as the elements can be reused for multiple purposes. The software to run the PIB data synchronization and distribution should have an open nonproprietary interface as well as being reliable and robust. This could be either commercially available or public domain.

The only specialized element in the previous figure is the PIB card, as this has very stringent requirements. It has to be mechanically durable, robust, and water-resistant yet at the same time remain low in cost. Since personal information is sacred for the people to whom it belongs, there will be secure communications established which the owner will control, allowing him/her to manage the amount of data that's accessible at any given time.

Why Choose Bluetooth Technology?

There are many communications technologies available, and wireless is not always the best solution for every need. This section looks at the requirements of the PIB device for our sample hospital environment as well as the challenges the system imposes, and considers the factors that influence a choice of communications technology for a PIB.

Requirements for PIB Devices

A hospital environment imposes its own requirements on devices, but many of these overlap with requirements for devices you use in an office or home. The PIB for a hospital environment needs to have all of the following characteristics:

- Low cost
- Easily portable
- Mechanically robust
- Reliable communications
- Hygienic
- Conforms to medical radio restrictions
- User-friendly
- Adequate storage
- Security and access controls

Let us examine each of these requirements in more detail.

It must be a low-cost option. The PIB must be affordably priced, otherwise hospitals or patients will not use them. Exactly what is affordable will depend upon the context. For the UK national health system a target price of $20 to $50 would be acceptable, and for a privately run luxury health clinic, a higher price would be acceptable. In both cases the acceptable price will depend upon the features and capability of the device. The cost of the major components of any such device is likely to come down over time. These major components are re-programmable memory (flash), color liquid crystal displays (LCDs) and robust mechanics. Bluetooth chips are lower cost than other wireless technologies, so they fit well with low-cost requirements.

The PIB must be portable. The PIB should be small in size and comfortable to carry. It needs to be capable of being attached or clipped to the patient, like a name badge. An ideal size is a single slot PC card (100 mm by 50 mm by 4 mm). Bluetooth modules are small in size, so Bluetooth fits well with portable devices. Furthermore, using wireless connections eliminates the need for carrying around bulky cables, and the adapters, which seem to always be needed, to cable different systems together.

It should be mechanically robust. The PIB should be able to take the shock of falling on the floor, being under body weight, and perhaps even being accidentally trodden on! All of the interfaces and the PIB device itself should be durable in everyday uses, and as a target should have a life span of two to three years of constant use. Wires require mechanical connections to be made. Low-cost molded plugs are notoriously unreliable, so wireless technology is ideal for creating a mechanically robust design as all the operating parts can be hidden inside a case.

Communications must be reliable. Transfer of information has to be guaranteed. Radio environments by their very nature are subject to interference, thus making them unreliable. Therefore, it is desirable to have alternative interfaces such as Infrared, which could be used if a radio connection cannot be established for some reason. Such an alternative interface could also be useful for areas where radio operation is not allowed.

The PIB must be hygienic. If the PIB is to be carried around with a patient, it should be easy to clean. By eliminating sockets for wired connectors, crevices which could harbor dirt and germs are removed. This means that wireless technology is ideal for creating a hygienic and easily-cleaned device.

It must conform to medical restrictions. The United States have allowed ISM band within hospitals, for telemetry purposes. However, there may be areas of a hospital where ISM band equipment cannot be used, as it would interfere with sensitive monitoring equipment. Therefore, any devices fitted with Bluetooth wireless technology would have to have an easy way to disable the radio. (This is a requirement for all portable devices, not just medical devices, as airlines do not permit the use of ISM radios on aircraft. In the same way that cellular phones are switched off on aircraft, Bluetooth devices must also have their radios disabled on airplanes.)

It must be user-friendly. Both the PIB device and the Data Access Terminal it connects with must be easy to use. It must be simple to exchange information, add appointments, and enable reminders. Again, this is a requirement that applies to all devices, whether for hospital, home, or workplace.

The PIB device doesn't really need many interfaces except for wireless connectivity, a button, some indication like an LED, to show connectivity and activity, and perhaps a speaker for audio output. This means that an interfacing device is required for extracting and viewing the data.

Adequate storage must be accounted for. A target might be to store information for 5 years, including: personal information and photographs, visits to a GP, hospital and associated notes and information gathered from any tests. X-rays and CAT scans require extremely high-resolution images, so it would probably not be practical to store these within the PIB. Nevertheless, a considerable amount of storage is required, for example anything from 8 to 32MB (Table 10.1 shows the size of this type of information over five years). This is assuming that a simple compression technique is used. The size of the memory should be extensible, either by using a top of the range PIB or by utilizing the wireless connectivity to access old information that may be required, which may be stored on mass storage in another device.

Table 10.1 Typical Example of Personal Information over Five Years

Type of Stored Information	Size of Information over Five Years (KB)
Personal Information	10
Personal Photographs	1,000
Calendar information, including appointments and tasks	1,000
Notes	1,000
Blood tests results	10
Ultrasound scans	4,000
Total	**7,020**

Security and access controls must be adequate. The PIB device is likely to carry confidential information, so the device and the system it connects to must provide adequate protection for that information. This implies that there must be different levels of access to the information in order to maintain confidentially, and whenever data is transferred, it must be protected from eavesdropping. Examples of different access levels could be:

- Access to all information—general doctor and patient (provided the patient is not a minor)

- Restricted access to information—specialist consultant

- Access to information related to current treatment—nurses

The reason for multiple access levels is that not all information is required by all medical staff. For instance, the patient may not wish the chiropodist to know that he/she has visited a sexual disease clinic, as it is not pertinent to the chiropodist's treatment. Bluetooth provides 128-bit security, which can protect data when it is being transferred to and from the PIB. Limiting access to different categories of stored information could be done through a security information based on the PIB which defines those items a particular device can view, as well as those that require authorization. However, security features should be used with caution since the more different the access level required, the more complex the device will be to use.

Implementing Optional Extra Features

There are many more features that would be nice to have but that are not essential to implement a PIB. It would be possible to have basic models available for all patients, and higher cost variants for specialist uses.

An ideal PIB device has many interfaces, some of which would not be necessary when creating a low-cost device. Figure 10.2 shows interfaces that might be used in a high-end system:

- Visual devices like LCD and LEDs
- Input devices like a keypad or possibly buttons
- Microphone/Speaker
- Alternative communication interfaces, namely: Bluetooth, IrDA, and PC Card
- Sensors for motion, pulse rate, and temperature

There are a few internal features to the PIB that are very important:

- Large nonvolatile memory storage
- A small battery that is rechargeable and efficient

These extra interfaces can provide valuable functionality for high-end devices. This section examines the improved functionality that could be offered.

Not everyone will have a Data Access Terminal, so a low-resolution color LCD could be added to provide an instant means of accessing the information. Such an LCD could also be used to display a photo identifying the patient. It could be used for security purposes to show the patient's photograph for confirmation of identity. There are other uses—for example, it can be used for quick language phrase translation, to communicate to non-native speaking patients. The PIB device can be used as a medicine reminder: it could describe the look and feel of the medicine, how many tablets should be taken and even show a picture of the medicine. However, a color LCD adds greatly to the cost of the device, so for the lowest cost, this may not be practical.

Figure 10.2 An Ideal PIB Device

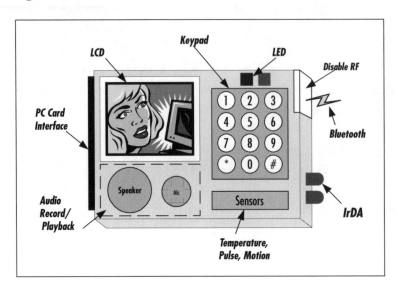

A keypad is useful to answer any questions or enter PIN codes to authorize access to the device, and to control more complex functions on the device. If the PIB does not have a keypad, it would have to use a pre-programmed fixed PIN code, which prevents the user from easily changing the code if they want to bar somebody who was previously granted access to data.

A speaker enables many multimedia options. A microphone could be used to provide Dictaphone capabilities, enabling doctors to record notes directly into the PIB, or allowing patients to record their own memos. This would require a complete audio input system, and could be quite expensive.

LEDs are useful to provide low battery indication. LEDs can also be used to indicate an active communications link; this could be a very useful indication, acting as a reminder that the device is on when entering areas that do not permit use of radio links.

The possibility of having sensors could make the PIB device more acceptable to nurses and other hospital staff. These sensors could be used to detect movement at low or high sensitivity, and would allow hospital staff to be alerted in case the patient has decided to go on a walkabout. It could also be used to establish if the patient is wearing the device, or if it has fallen off. Temperature sensors could be used to monitor the average temperature of the room, or the environment the patient is in. This could be employed to alert the hospital staff of anything abnormal. The PIB device could also have a pulse rate monitor. The monitoring capability will also ease the need to write down the measurements as they could automatically be transferred to the Data Access Terminal.

Alternative communications interfaces might be provided, to cater for circumstances where the wireless radio cannot be used—for instance, near highly sensitive equipment. However, alternative communications interfaces would raise the cost of the PIB. A backup infrared link could add wire-free communications capabilities in areas where radio cannot be used; here, the cost increment isn't great since infrared systems are very cheap, but development costs for dual software systems could be high. It would even be possible to add a PCMCIA PC-card interface for high-speed data exchange, although this would greatly add to the cost and would also negate the advantages of hygiene and reliability, which a wire-free design has.

Specialist monitors or interfaces to monitoring equipment could be added. You could view this as adding monitors to the PIB, (although for more complex and expensive monitors it might be better to think in terms of adding PIB functions to the monitor). A PIB device enabled with monitoring capabilities such as temperature or pulse could continuously monitor and record any abnormal variations. Audible alarms could be triggered if the sensor exceeds either upper or lower programmed thresholds. However, caution should be employed when using a PIB for safety-critical purposes. Wireless links are subject to interference, which makes them unreliable.

Choosing a Wireless Technology for the PIB Device

There are various technologies that could be used to achieve the PIB system. See the brief summary in Table 10.2.

The reasons for choosing Bluetooth as the wireless connectivity for the PIB system are:

- Its physical size is small, and there are many chip vendors to choose from.

- The range is adequate—the lowest power version offers up to a 10 m range, which is sufficient.

- The available choice of chip vendors leads to a competitive market, which means the cost will reach less than $5 over the next two to three years.

- There is a worldwide acceptance of the ISM band used by Bluetooth, which means that the product design can be sold in markets all over the world.

- Products are expected to interoperate if they have been qualified and received a Bluetooth logo. This means that the data terminal side of the Bluetooth link can be implemented with readily available, cost-effective, commercial products.

From Table 10.2, we can see that IrDA is also a good match for the requirements of a Personal Information Base. The advantage of Bluetooth wireless technology is that it is not directional—with infrared technology, the ports on two devices must be lined up, but a Bluetooth device can be accessed while still in the patient's pocket, for example, greatly increasing convenience of use.

Table 10.2 Wireless Communication Alternatives

Technology	Physical layer	Size	Range	Power Consumption	Security	Standards	Software
Infrared	Optical	1 cm by 1 cm, including processor supporting IrDA protocol	Line of sight – 5 m	Very low	Application layer	Worldwide	Complete protocol stack defined
418MHz	Radio	3 cm by 3 cm, including processor	100 m	Medium	Application Layer	Proprietary	Proprietary, however can use whatever is required
Whitetooth	Radio	Not enough information	Range to be determined	Very low	To be defined	Worldwide	To be defined
Bluetooth	Radio	2 cm by 2 cm	From up to 10 m to 100 m, depending on power	Low Application	Part of protocol and at layer	Worldwide defined	Complete protocol stack

Considering the Cost of the PIB

Once the wireless technology is chosen, it is possible to set some cost targets. Our example PIB device is a specialized design to be used in a hospital environment, and as a result, it could be expensive to produce as a product. A target low-end price would be $20 to $40. At these cost levels it is not going to be practical to support all possible optional features, though different subsets of the possible options could be fitted to create various levels of device.

One way to reduce component cost is to produce a single processor system. This means that the processor must not only be able to handle the whole Bluetooth stack for this application, but also the application including the user interface. It also means that the processor must support additional peripheral interfaces, which will mean that hardly any external support devices will be required.

The rest of the infrastructure is robust: networked and Bluetooth-enabled PDAs or desktop computers and a server for local and central control. The cost of these items (including the software) can be targeted at:

- **PDA** $200, per doctor and shared per department
- **Desktop computer** $1500, per department
- **Server** $2500, per major section and per central control

Exploring the Safety and Security Concerns of a Personal Information Base

Access to accurate medical information can be a matter of life and death, so it is important components of a medical information system can't introduce falsified or corrupt data into the system. It's also important to ensure data cannot be lost from the system. In addition, patient confidentiality is an important consideration, and one that should be taken seriously in wireless systems, as communications can potentially be intercepted even by somebody outside the room where data is being exchanged. Finally, medical requirements regarding hygiene and regulations concerning radios in hospitals must be kept in mind when considering any device for hospital use.

Enabling Data Duplication

The aim of data duplication is that data for a patient is stored in two places at any given time. This means that after synchronization, the central database will ensure that any loss of patient information, be it PIB device or a doctor's PDA, can be completely recovered. The reason why this is possible is that no data can be entered in the PIB device on it's own, except for personal notes using the limited local interface. This means data is added to the PIB device by a Data Access Terminal or a desktop computer (local server) pushing new records to the PIB. The Data Access Terminal has a duplicate copy of the new patient data, and can be synchronized to the local and central server.

Wherever data is stored on small mobile devices there is always more risk of data loss than with desktop systems, so data loss is a general problem where mobile data storage is used. The Bluetooth synchronization profile provides a means to ensure data stored on a mobile device is backed up automatically. The synchronization profile could be used to ensure any data entered directly onto the PIB is backed up.

Synchronization software is sometimes very rigid in the way it behaves, as it expects one part of the system, normally the desktop computer or main server, to be the master of the data while the mobile device is a slave to the information. For example, different appointments made by the secretary for one patient at the same time on the server may overwrite a new appointment made on a mobile device. Another area to be careful about is in the use of Universal Time, as different devices may refer to different time zones.

Figure 10.3 shows how a synchronization system could work. The Data Access Terminal pushes data to the PIB, but keeps its own copy of the data. Both the PIB and the Data Access Terminal can synchronize with a local host, which is connected to a local area network. Once data has reached the network, it is backed up across the network. Should network failures occur, backup modem links can be used.

In addition to providing data security, the central control facility also allows patient mobility. If a patient is moved to another hospital, their records can be retrieved from the central backup facility, and a new PIB can easily be set up with all the patient's information.

Ensuring Data Integrity

It is very important that data integrity be maintained on the patient record as decisions cannot be made on data that is in error. A well-known technique for doing this is adding an overall checksum to the end of the patient record.

The overall checksum for the data is a number derived from applying some algorithm to each data element (typically at byte level) of the patient's record. This ensures that if any part of the data is corrupted then the data cannot be trusted and a new copy should be obtained.

Wireless links are prone to errors caused by interference, or by fading of the signal as mobile devices reach the limits of their radio ranges. The Bluetooth baseband implements error checks on data, but these checks will not catch every single error. Therefore, it is a good idea to implement extra error checking on data to be sure any errors that aren't caught by the Bluetooth protocol stack are flagged at the application level.

Figure 10.3 Synchronizing Data with a PIB System

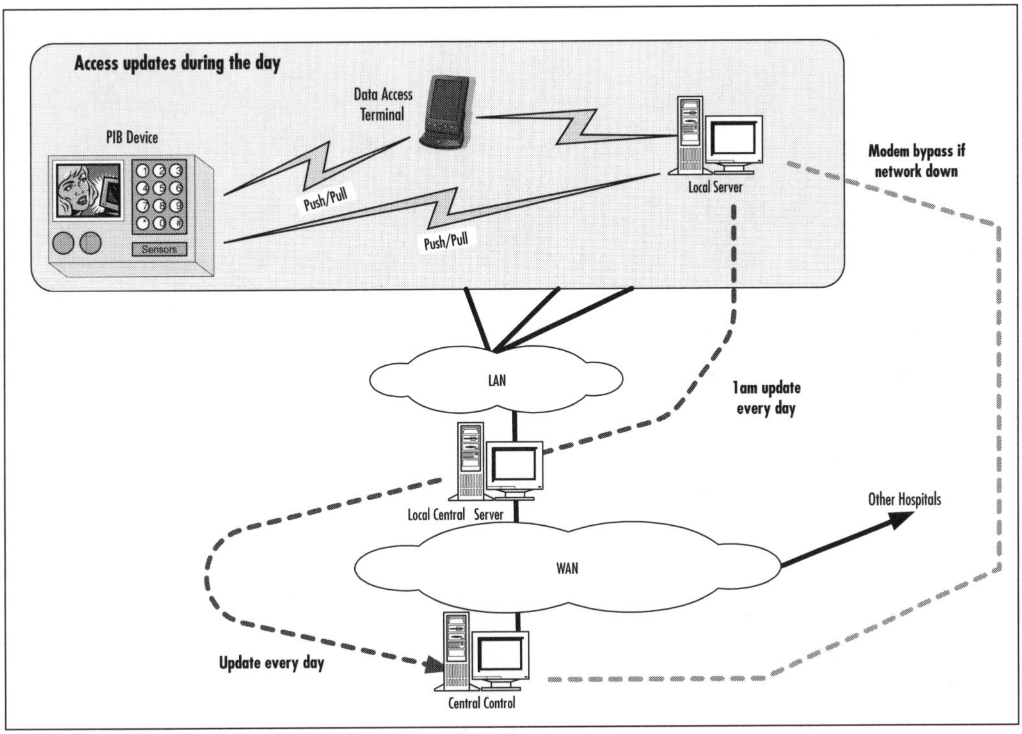

Providing Security

A simple LCD on the PIB device could display a photograph for security confirmation that this device belongs to the correct person. Access to data that normally would be on bedside charts is available using the PIB device; only medical information of a current visit is readable, no other data is viewable, without using PIN code access. Detailed information is only accessible with the use of the Data Access Terminal; this allows the PIN code and other levels of access to be enforced, depending upon the patient or the seriousness of the medical condition. The different levels of security can be provided by Object Exchange or by using password-protected files.

Patient confidentiality is very important. One way of protecting confidentiality is to use a reference code to identify the patient in place of their name. Indeed, in the UK (according to the Data Protection Act) the patient's National Health Service number is used as an indexing method for medical records in order to keep them confidential. Even then, a photograph and other information, such as date of birth, can be used to verify the correct patient. This means that the Data Access Terminal must be able to access a table cross-referencing index numbers to names, so the patient's information can be obtained.

Whenever dealing with protected information, it is important to retain a sense of proportion. In paper-based systems, folders containing medical information can be picked up and read by anybody. The way this information is protected is by keeping it out of sight of patients and staff. While it is good to have extra security, it is all too easy to implement so much security in a system that it becomes virtually unusable. If data is too difficult to access, doctors and patients will undoubtedly resort to using paper notes, thus bypassing all the useful backup features offered by the PIB system. Therefore, user interfaces should be designed with care so that the entry of PINs does not become an onerous task that effectively bars authorized users from the system.

By deploying Bluetooth sensors near the exit of a hospital, any accidental removal of the PIB device can be detected and reported. This is only possible if the device is Bluetooth-functioning, however, so it would still be possible to deliberately remove a PIB by disabling its Bluetooth transmitter.

Meeting Medical Requirements

Mobile phones would be an ideal PIB device since they have all or most of the capabilities described in previous sections. Unfortunately, they cannot be used in hospitals. However, the use of 2.4GHz within US hospitals *has* been cleared. The main example used to demonstrate this was the use of wireless telemetry using 802.11 Wireless LAN. This range also covers Bluetooth operation, although it is not explicitly mentioned, in the ruling. Some medical equipment companies have used this to start producing Bluetooth-enabled products.

As noted earlier, hygiene is a very important requirement for hospitals. This means the PIB device should be made of material that can be easily cleaned and must not have crevices where bacteria can accumulate.

Using Bluetooth Protocols to Implement a PIB

So far, we have seen that Bluetooth wireless technology can fulfill the communication requirements of a PIB. In this section, we will look at some of the details of how the communications protocol stack could work. This section briefly explains the hierarchy of different protocols needed to exchange data, and how those protocols are derived from many different specifications. It also provides an overview of Bluetooth packet layering.

> ## Developing & Deploying...
>
> ### Radio Regulations and the ISM Band
>
> The following reference is from the US Federal Register amended in 2000 to harmonize the use of wireless technologies within hospitals.
>
> *Page 43999 of Federal Register / Vol. 65, No. 137 / Monday, July 17, 2000 / Rules and Regulations*
>
> *47 CFR Part 15 - Changes*:
>
> 15.247 Operation within the bands 902 to 928MHz, 2400 to 2483.5MHz, and 5725 to 5850MHz.
>
> Comment: No change was made to §15.247. As noted in ¶35 of the Final Rule: "... we will continue to allow medical telemetry equipment to operate in the ISM bands under Part 15. While such operation will be permissible, manufacturers and users are cautioned that equipment operating in these bands has no protection from interference from ISM equipment operating under Part 18 of the rules or other low power transmitters operating under Part 15 of the rules."

After this overview, we will go on to explain the details of how the PIM device exchanges information.

Understanding the Bluetooth Specification Hierarchy

The Bluetooth SIG has done a very good job of reusing existing standards and adapting them. This specification reuse means it is possible for protocol stack and applications developers to reuse code. This saves time and improves the robustness and quality of the final system as reused layers have already been tested on other communications systems.

However, there is a drawback to reusing specifications. Reuse means that anyone trying to understand the whole system has many different documents to read: this can become a challenge to understand! To help you find a path through the maze of specifications, this section will summarize all the standards used by the PIB device. Later sections will explain how the standards interact, allowing us to exchange data.

The main aim is to convert the layered (horizontal) approach into a vertical slice so the interaction between the various layers can be easily understood.

The following specifications are used in the PIB device:

- Bluetooth Special Interest Group (SIG)

- Infrared Data Association (IrDA)

- European Telecommunications (ETSI)

- Internet Mail Consortium (IMC)

- Internet Engineering Task Force (IETF)

- Internet Assigned Numbers Authority (IANA)

Figure 10.4 shows an overview of the number of packet layers involved in sending an Object Get Response Packet. Please note that this is a summary—in later sections, we will go into packet details and explain every field with reference to the relevant specification.

When writing applications to run across Bluetooth, you are likely to be using a high-level interface at the top of the Bluetooth protocol stack. However, it is often useful to understand what is happening in the rest of the system. The full data exchanges involved in a PIB system are extremely complex, but it is possible to get a good understanding of how the different stack layers interact using the simplest information exchange: a virtual business card or vCard (see Figure 10.5).

Suppose a Data Access Terminal is gathering information on devices in the area, and it wants to get a vCard object from every device that supports vCards. It must go through a three-step process:

1. The Data Access Terminal inquires to find Bluetooth devices in the area. Each device, which is listening for inquiries, will respond with an FHS packet giving information needed to establish a data connection.

2. For each device found, the Data Access Terminal connects and creates a Service Discovery L2CAP channel and performs Service Discovery on that channel. The Service Discovery Protocol tells the Data Access Terminal whether the device supports vCard transfer, and what parameters are needed to transfer cards (for example, the RFCOMM channel number to be used for this service).

3. The Data Access Terminal shuts down the L2CAP channel and establishes a separate L2CAP channel to RFCOMM. An RFCOMM channel to the OBEX layer is then established. Afterward, an OBEX session is started, enabling the Data Access Terminal to act as a client and pull a vCard from the PIB Device, which acts as a server.

Figure 10.4 Overview of Communications Used in the Personal Information Device

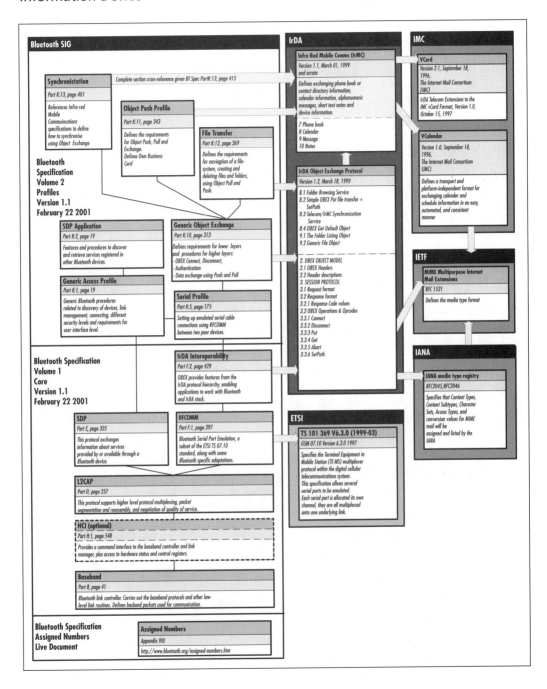

Figure 10.5 Packets Used During vCard Exchange

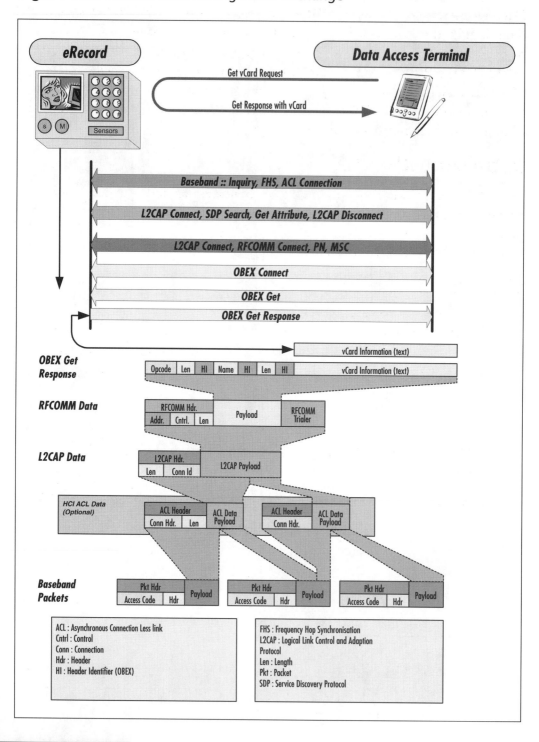

The upper part of Figure 10.5 shows the details of the OBEX session. The Data Access Terminal sends an OBEX Connect across this RFCOMM channel, then the PIB device responds with an OBEX OK, which means that objects can be exchanged. The Data Access Terminal requests an OBEX Get of the local vCard and the PIB device responds with a Get Response, which includes the vCard. The Get Response is shown in Figure 10.5 as it traverses the different layers from vCard to OBEX Response, RFCOMM, L2CAP, Optional HCI ACL Data, and finally, on-air data packets.

Initializing the PIB

In the following section, we will spend more time explaining how Bluetooth operates than how the overall PIB system works. Before we explaining how the Bluetooth PIB device is initialized, enabled, and verified for operation, let's take a look at how the user interacts with it.

Understanding User Interactions

Imagine the following situation. A patient called Mary Clarkson has a check-up scheduled at the hospital. She arrives at the hospital and goes to the receptionist to register herself. The receptionist accesses Mary's patient records and makes sure that Mary has an appointment. Mary doesn't have a PIB device of her own, so the receptionist programs one with Mary's details and gives it to her. If Mary has an out-of-date picture on her records, the receptionist may even take a new photograph and update Mary's records. The following sequence of events check if the PIB device is operating correctly:

1. Mary checks in for her appointment.

2. The receptionist asks Mary for personal details to program into a new PIB.

3. Mary hands over her appointment letter.

4. The receptionist enters the details into her local Data Access Terminal.

5. The Data Access Terminal sends the records to a central server.

6. The central server accesses appointment records and medical history and returns the information to the receptionist.

7. The records do not include a current photo of Mary, so the receptionist takes a photo of her; this could be transferred across a Bluetooth link to the Data Access Terminal.

8. The receptionist programs up a PIB for Mary.

9. Mary is given the PIB. Since it is the first time the record has been accessed over Bluetooth, Mary is asked to enter a password and verify it. The receptionist informs Mary that she has to remember this password since she may be asked to enter it during her stay.

10. Mary can now go off to the wards carrying her records with her in the PIB.

The steps to access both public and private data look very easy, but there is a considerable amount of initialization and protocol that has to be done in order to achieve this level of transfer.

Without going into too much detail, entering the same password for both sides of the link (in this case, the receptionist and patient) translates to the Bluetooth Personal Identification Number. These have to be the same on both devices, otherwise a link will not be established.

If the PIB has a keypad on it, then the password can be entered simply by using the password. If the PIB does not contain a keypad, then it would come with a default password built in. The matching password would be entered on the Data Access Terminal to establish a secure link; an application running across the secure link could then be used to change the password in the PIB.

Obviously, there is a potential problem in regards to patients forgetting their password. Since the information on the PIB is duplicated elsewhere, one solution would be to have a method of resetting the PIB to remove all information, then it could be reinitialized with information from the central server.

Sending and Receiving Information

The previous section referred to receiving data from the PIB device in order to test if the device was functional and if the information was programmed correctly. This section uncovers exactly what goes on when data is exchanged between the PIB device and the communicating device.

Imagine the following situation, where the PIB device replaces the chart at the end of the bed. A doctor (Dr. Merick), who is doing a daily check to diagnose the next course of action for his patients, visits Mary. Each step is illustrated in Figure 10.6.

1. Dr. Merick asks Mary to activate the PIB device by pressing the red button.

2. Mary presses the red button.

Figure 10.6 Exchanging Data

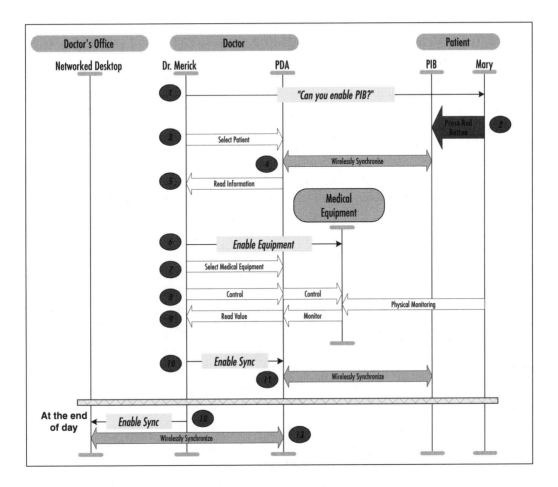

3. The doctor uses his PDA, finds Mary's PIB device, and selects it. On selection, he and Mary may have to enter the password (for simplicity, the password entry has not been shown).

4. The doctor synchronizes with Mary's PIB. This is a two-way synchronization that exchanges any new data between the two devices.

5. The doctor reads any new information, and after a conversation with Mary, adds new notes.

6. The doctor enables the medical equipment to take a measurement of Mary's condition.

7. The doctor uses his PDA and finds the equipment he wants to use. A unique password is entered to use the equipment. This will allow only authorized staff to use the equipment.

8. The doctor gets the control interface on his PDA and remotely controls the device to take the measurements.

9. Blood pressure, temperature measurements, and the doctor's comments and recommendations are recorded on the PIB device.

10. Before the doctor leaves, he synchronizes with Mary's PIB, duplicating the data in the overall system.

11. Later on in the day when the doctor goes to his office, the PDA is synchronized with the local server so that data can be backed up and future appointments can be scheduled.

Now that we understand how Mary and her doctor use the PIB, let's consider what happens at the Bluetooth protocol level.

When Mary presses the button and the Doctor retrieves first Mary's public information, then her medical records, both the doctor and patient begin by exchanging public information. The doctor uses the information to verify that the correct patient is being treated and the PIB can keep a record of who accessed the information. The public information is transferred using the Object Push Profile (BT Profile Spec Part K:11, page 339) and is known as Business Card Exchange (Section 4.4, page 346) using vCards (IMC vCard – The Electronic Business Card Exchange Format, Version 2.1, Sept. 1996).

The role taken by the Doctors PDA is the "Push Client" that wants to initiate the exchange, while the role taken by the patient's PIB device is the "Push Server."

The patient wants this exchange to be as simple as possible, so the patient's PIB will automatically accept the Doctor's information and exchange the public patient information. This means Mary does not have to interact with her PIB beyond enabling it.

Figure 10.7 shows people, devices, and actions involved in the Business Card Exchange.

The doctor is the user of the PDA and asks the patient to press the red button to enable the PIB device.

The patient is the owner of the PIB device and allows the doctor to exchange information without any interaction.

Both the PDA and PIB devices are Bluetooth qualified products and cooperate to allow the exchange of information to happen wirelessly and seamlessly. The high-level steps can be summarized as follows:

1. The doctor asks the patient to press the red button on the PIB device.

2. By pressing the red button, the PIB device is enabled.

3. The PIB device goes through Bluetooth and application initialization.

4. The doctor selects "Get patients?" on his PDA.

5. This initializes the PDA.

6. The PDA does a search for discoverable PIB devices.

7. Discovered PIB devices are displayed in the PDA "Get patients?" window.

8. The doctor uses the remote Bluetooth name to decide which patient is being treated, as this has been programmed with <date_of_entry>, <patient_name>, and <patient_hospital_identification>.

9. The patient is selected and the public information is exchanged. This is the vCard.

10. If the public information is correct, the treatment continues. Otherwise, another patient is chosen.

Figure 10.7 The Business Card Exchange

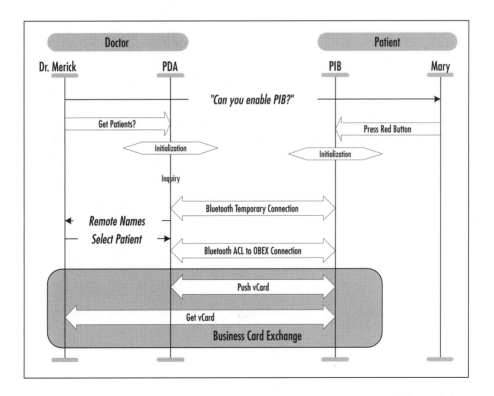

Initialization – PIB Device

When the patient presses the red button, the PIB device initializes the Bluetooth hardware and software. This only happens if there is no active connection present.

We will explain the initialization by using the Host Controller Interface specification (Bluetooth Core Spec Part H:1, page 535), despite the fact that this interface may be collapsed in the final solution.

The most important commands are described in Table 10.3 with reasons for why they are used.

Table 10.3 PIB Initialization Commands

Command	Parameters	Reason
Reset	None	To get the Bluetooth hardware to a known default state.
Set Event Mask	All events enabled	Leave *all events enabled* as default.
Read Buffer Size	None	This allows dimensioning of host data transmitter. Maximum length (bytes) size of data portion of HCI ACL data packet. Total number of HCI ACL data packets that can be stored in the Host Controller. Similar values for SCO data are returned as well.
Set Event Filter	Set auto accept connection from specific Class of Device (in other words, Computer)	This command can be used to control which devices respond to the inquiry process at the HCI level. • All • Specific Class of Device • Specific Bluetooth address It also controls how and which devices connect.
Write Authentication Enable	Disable	
Write Encryption Mode	Disable	

Continued

Table 10.3 (*continued*)

Command	Parameters	Reason
Write Connection Accept Timeout	10 seconds	The time allowed for accepting a connection.
Write Page Scan Activity	Page Scan Interval — Page Scan Window	When a connecting device wants to connect, it "pages" and the connectable device scans for pages (in other words, "page scan").
Write Page Scan Mode	Inquiry and Page	
Write Inquiry Scan Activity	Inquiry Scan Interval— Inquiry Scan Window	When an inquiring device wants to discover, it "inquires" and the slave device scans for inquiries (in other words, "inquiry scan").
Read Bluetooth Address	None	Read Bluetooth address for application use.
Change Local Name	<data> <patient_name> <Hospital Identification>	This name is read by the remote device to establish some sense of description.
Write Class Of Device	Limited Discovery Major Service Class:: Object Transfer Major Device Class:: Computer Minor Class:: Palm-sized PC/PDA	This allows a device wanting to connect to receive a first level description of this device.
Write Link Supervision Timeout	20 seconds	The amount of time allowed to declare a link loss.
Write Scan Enable	Inquiry and Page Scan enabled	

Initialization – Doctor's PDA

Initializing the doctor's PDA employs the same steps for initializing the PIB device, except for the following items:

- Set Event Filter to filter all classes of devices except for Palm devices with OBEX Transfer.

- Disable Page and Inquiry Scans, so scan activity does not need setting.

- The Name reflects <doctors_full_name> <doctors_id>.

- The Class of Device reflects the PDA or small laptop.

Using the Generic Access Profile

The purpose of the Generic Access Profile is to select a suitable connecting device based upon the Inquiry procedure and to get the remote name. The business card exchange doesn't require any security, so this will not occur until critical information has to be exchanged.

For the purposes of the Generic Access Profile (Bluetooth Profile Specification Part K:1, page 23, section 2.2) the doctor's PDA is known as the A-party (the paging or initiator device) and the patient's PIB device is known as the B-Party (the paged or acceptor device).

When the doctor asks the patient to press the red button, the initialization of the PIB places the device into the following mode:

- Limited Discoverable mode for a period of three minutes. This makes sure the device can only be discoverable during that period.

- Connectable mode. The PIB is always in connectable mode when it is powered. This allows other devices that know about the PIB device to connect without going through an inquiry phase.

Afterward, the doctor's PDA is initialized, which places the device into the following mode:

- Non-Discoverable mode. This means that no one can inquire for the device.

- Non-Connectable mode. This means that no one can connect to the device, unless the doctor allows it. This makes sure there are no interruptions when the doctor is dealing with the patient.

Device Discovery

Once both devices are initialized, the doctor's PDA can initiate a one-time inquiry (Bluetooth Core Specification, Appendix IX, page 1041, section 2.2). The inquiry would be initiated by the doctor interacting with a user interface: for instance, by clicking a **Select Patients** icon on the PDA. See Figure 10.8 for an illustration of the device discovery procedure.

Figure 10.8 Detail of Device Discovery Procedure

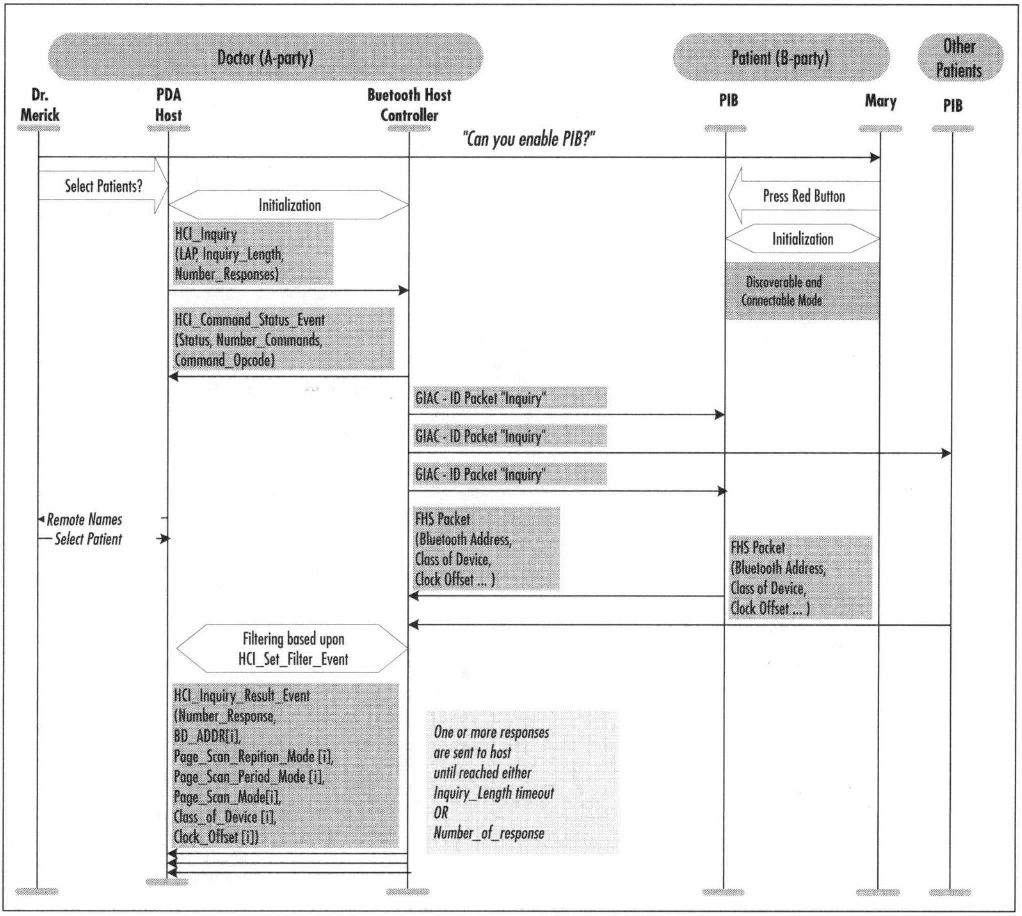

The PDA sends an *HCI_Inquiry* command to its Bluetooth Host Controller; the Host Controller responds with an *HCI_Command_Status_Event*, which acknowledges it has received the command. Then the Host Controller sends out a series of Inquiry packets (ID packets containing the General Inquiry Access Code).

Every device within range (which is in discoverable mode) should hear these packets and respond with an FHS (Frequency Hopping Synchronization) packet. These packets contain all the information the PDA needs to connect with the responding PIBs.

The Host Controller sends the inquiry response information up to the PDA in one or more inquiry result events.

Developing & Deploying…

HCI Implementation Guidelines

There are many possible architectures which can be used to implement a robust PIB system. We have already noted that for the PIB itself, a single processor architecture could provide the cheapest option, but for the rest of the system, it is likely that applications will run on a separate host processor. Let's consider the two processor architectures as defined in Bluetooth Specification Version 1.1 (Part H:1 Introduction, page 584). The communication occurs using HCI (Host Controller Interface) packets. The host is the processor controlling the Bluetooth Host Controller.

Figure 10.9 shows command and dataflows between a host and Host Controller. The dotted line connecting commands with *command complete* events shows how the *command completes* correspond with commands. For every command packet sent, there is a command complete event packet. The *command complete* events may not come back in the same order that the commands were sent. Some commands, such as the *inquiry* command, may take many seconds to implement, so it is likely that sometimes the host will want to send more commands while waiting for a *command complete* event. This means the host must be able to send commands and handle the *command complete* events synchronously.

If a Bluetooth link is established and data is being exchanged, then data from the host can cause flow control events to come back from the Host Controller indicating how empty the data buffers are. This needs to be processed at a higher priority to avoid the Host Controller's buffers overflowing with a consequent loss of data.

Continued

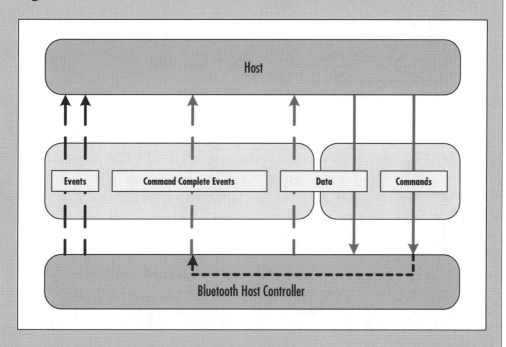

Figure 10.9 Command and Dataflows between a Host and Host Controller

Because events are sent to the host at the same time as the host is sending data and commands to the Host Controller, an asynchronous communications architecture is needed.

The reason why HCI transport also has to be robust is that HCI packets carry a length field, used to calculate where the end of the packet is. If at any moment in time the counting of bytes is lost due to loss of a byte(s), then the synchronization has to be reestablished, at the expense of losing a complete HCI packet.

Version 1.1 of the Bluetooth specification was published with three HCI transports defined: UART, RS232, and USB. RS232 has not been widely implemented, with most Bluetooth adopters seeming to view it as over-complicated. UART was defined for communication between chips on a PCB and does not perform well over links which are subject to errors (as the cabled serial port links to many PCs are). USB is tolerant of errors, but many Bluetooth host controller devices do not implement USB as it is quite a complex protocol. There is currently an HCI working group that is defining a new HCI transport, which, amongst other improvements, provides error detection and correction across serial links.

The *HCI_Inquiry_Result_Event* illustrates one aspect of Bluetooth which is likely to provide a challenge for applications developers. Some Host Controller devices will gather all inquiry responses together in the Host Controller, and just send one *HCI_Inquiry_Result_Event* to the host. Other Host Controller devices will send the host an *HCI_Inquiry_Result_Event* for every inquiry response received. While still other Host Controllers may even send duplicate events if they receive multiple responses from the same device.

If you are able to specify a complete system including hardware and software, you could write an application which was tailored to the behavior of one Host Controller. However, this makes for a system which can be limiting and difficult to upgrade. In the PIB system, one of the requirements is the ability to use a variety of legacy equipment, so there is a requirement to support whatever Host Controller devices fit onto existing equipment.

Whenever writing Bluetooth applications you should be aware that the Bluetooth specifications often include optional parts, and thus behavior is likely to vary subtly between different manufacturer's Bluetooth components. If you want your applications to be robust and useful across a wide range of platforms, you must cater for optional parts of the specification.

Selecting a Device

Once the host has received information that the inquiry is complete, the host can examine the responses and use this information to select a device for a connection. The host gets the Bluetooth device address of each device responding, along with what type of device it is. The response also contains information on how each device scans for paging, which the protocol stack can use during paging to establish a connection.

The central database could provide the doctor's PDA with a lookup facility allowing Bluetooth device addresses to be cross-referenced with patient's names. This only works if the doctor is currently connected to the database, however. If this is the case, then it would be possible to download all the information anyway. The very fact that the doctor is connecting with the PIB to get records means his PDA is not currently networked!

Since there is currently no network connection, the doctor can connect to each PIB in turn, and retrieve their friendly names. These are human-readable names. At it's most basic, the name could be:

```
Mary Smith's PIB
```

The Bluetooth specification allows user-friendly names to be up to 248 bytes long, so the name could be used to convey a limited amount of information, such as a hospital index number, date of admission, date of birth, or a reason for admission. Therefore, the name could be:

```
Mary Smith POMI564 5 November 2001, 9 October 1943, Hip replacement
```

This is certainly very convenient, but care should be taken when employing the user-friendly name in this fashion since the information can be seen by anyone. It is possible that Mary Smith doesn't want the whole world to know her date of birth, or that she is in need of a new hip. Index numbers are often used to protect patient's privacy, so having a device publish name and index numbers immediately provides a way around existing privacy mechanisms.

The issue arises here because the friendly name can be exchanged before authentication and encryption procedures have been performed. When writing Bluetooth applications, you should think about how much information is available unencrypted, and take care to make sure that information sent before encryption is switched on does not compromise a system's privacy or security requirements.

Using the Service Discovery Application Profile

Once Dr. Merick has found Mary's device, the next stage is to use the Service Discovery Protocol. First, a data connection must be established, this could be the same ACL link used to get the friendly name from Mary's PIB.

An L2CAP link is set up on top of the ACL link. The L2CAP link allows multiple services to use the ACL link (in this case, it is set up to the Service Discovery Server). Mary's PIB contains a Service Discovery Server which can tell Dr. Merick 's PDA how to connect with other services running on her PIB.

Dr. Merick 's PDA gets information about OBEX services running on Mary's PIB, including the RFCOM DLCI address which is needed to connect with the services.

The Service Discovery Application Profile provides guidance on how a service discovery session should be set up, how the service discovery protocol should be used, and what parameter values should be used.

Using the Serial Port Profile

Once Dr. Merick 's PDA has all the service discovery information it needs, the L2CAP connection can be torn down, and another L2CAP connection set up to RFCOMM. RFCOMM provides a serial port emulation service which is used by many profiles for communicating with higher layer applications and services.

The usage of RFCOMM is covered by the Serial Port Profile.

Using the Generic Object Exchange Profile

The next stage is for Dr. Merick's PDA to establish an OBEX connection. The messages used are essentially the same as would be used with OBEX across an infrared link. The Generic Object Exchange Profile gives guidance on how to use OBEX across Bluetooth connections.

Using the Object Push Profile

Dr. Merick begins by just getting public information about Mary in the form of a virtual business card or vCard. To do this, his PDA and her PIB use the Object Push Profile. This profile defines how objects with predefined formats are exchanged between Bluetooth devices. Using the Object Push Profile, it is possible to:

- Get public information using the vCard format.
- Get private information using the vCal, and vNotes formats.

This profile uses the facilities of the Generic Object Exchange Profile to exchange data.

Using the File Transfer Profile

Once Dr. Merick has retrieved Mary's card he will want to go on to retrieve medical records with more complex formats. Medical records are not covered by the Object Push Profile, so to retrieve them Dr Merick 's PDA will need to retrieve the data as files using the File Transfer Profile.

Like the Object Push Profile, the File Transfer Profile uses the facilities of the Generic Object Exchange Profile to exchange data. Using the File Transfer Profile it is possible to retrieve files from a remote device. It is also possible to create, delete, and move files on a remote device.

Obviously, you would not want just anybody to be able to come in and alter your medical records. With this in mind, it's possible to set up security access so different users get different levels of access to the file system on a device. A vital part of the design of a PIB system would be making sure that file access was limited, so unauthorized access to files was not permitted. This is necessary to ensure medical records could not be tampered with across the Bluetooth link, either accidentally or maliciously.

The Object Exchange Profile provides OBEX authentication, which can take place independently of Bluetooth authentication. While Bluetooth authentication is extremely secure, it might be desirable to use OBEX facilities to maintain compatibility with existing infrared-based systems.

Figure 10.10 shows how each of the Bluetooth protocols is used in turn to set up layer after layer of connection, culminating in information exchange through OBEX.

Figure 10.10 Information Exchange through the Bluetooth Protocols

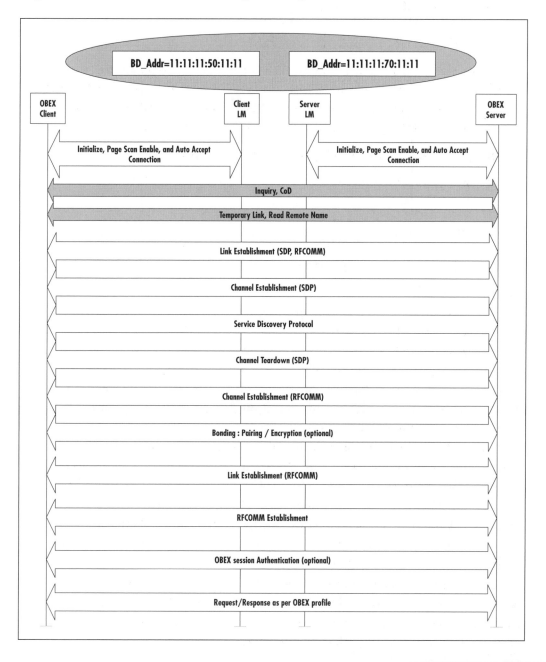

In this section, we will look in more detail at the exchange of OBEX data which actually gets the medical records from Mary's PIB to Dr. Merick 's PDA. To begin with, it is necessary to explain a couple of terms which are fundamental to OBEX operation: *client* and *server* (see Figure 10.11).

Figure 10.11 Using OBEX Clients and Servers

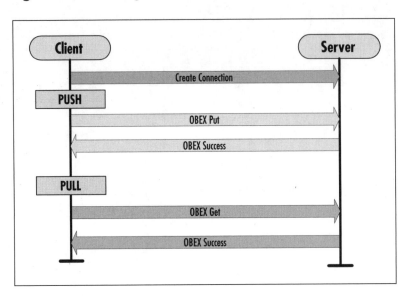

A server is any device that offers a service. That service could be providing data, or storing data. A client, on the other hand, is any entity that wants to take something from a server, or give something to a server. A client usually initiates the connection, and can either push data, (put data onto the server) or pull data (get data from the server).

A device can be both a client and server at the same time. ACL and L2CAP connections made by the client can be reused by the client on the other side. However, the client on the other side needs to create a new RFCOMM channel. Each RFCOMM channel is identified by a DLCI (Data Link Connection Identifier). The DLCI value space is divided between the two communicating devices using an RFCOMM server channel and a direction bit.

Figure 10.12 shows how the RFCOMM address byte can be used to distinguish between server and client direction. The figure summarizes the Part F:1 5.4 DLCI Allocation with RFCOMM Server Channels section in the Bluetooth Core Specification and 5.2.1.2 Address Field section in TS 7.10.

Figure 10.12 Format of OBEX Messages between Client and Server

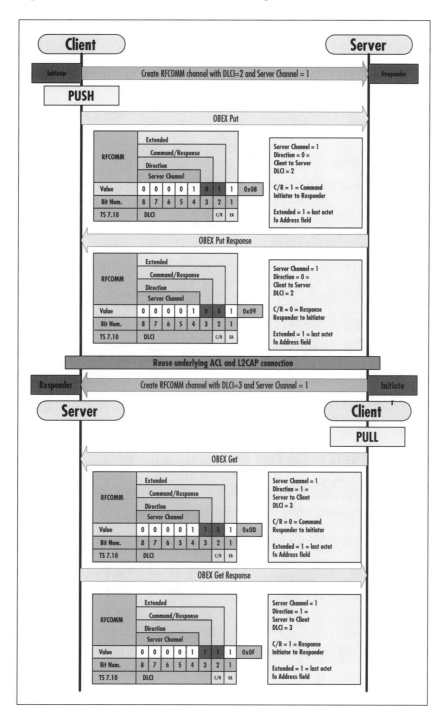

Server applications on initiating devices are reachable on *odd* DLCIs, and server applications on noninitiating devices are reachable on *even* DLCIs.

Depending on whom the initiator or responder device is, the Command/Response bit indicates if the data is a command or a response to a command.

Note that the byte has been shown as it would appear in a packet. This means it is bit-reversed from all the definitions in the specifications. This is clarified by using the appropriate bit numbering.

By using OBEX put and get messages, it is possible for Dr. Merick 's PDA and Mary's PIB to exchange data in any format whatsoever. Only the application that is interpreting the data limits the formats.

However, because of the constraints of size and price it is likely that some types of data would not be stored on PIBs. For instance, as noted earlier, medical images such as X-rays usually require very high resolution, which leads to extremely large files. It is unlikely to be economical to store such files on a PIB. Furthermore, the monitors required to display medical imaging data at a useful resolution currently cost around $20,000 each, so even if Dr. Merick could retrieve an X-ray from Mary's PIB, his PDA would not have the resolution to display it.

Practical issues of what data can be usefully absorbed via the limited user interfaces typically provided by mobile devices should always be considered when designing Bluetooth systems. There is little point to designing a communication system which can push a high quality image to a device if there is no way for that device to display it, or if the image uses up all the device's storage, preventing it from being used for other purposes.

Considering the User's View

A crucial part of any application is the user's view. So, we have to ask ourselves how a PIB will compare with the existing system as far as its users are concerned.

Identifying the System's Users

The immediate users of the system are obvious: the patient and medical staff who directly access the information. However, the system will also have an impact on the staff members who maintain records. Just as the paper-shuffling activities of a hospital are replaced by the automated distribution of information, the staff who maintain the hospital's information systems will also be affected by the PIB system.

In designing applications, you should be aware of all users who will be affected by the system. For large applications, this extends to those who will configure and maintain the system in addition to the direct users.

Identifying System Use Cases

In this case study, we have gone into detail of the most obvious use case for a medical Personal Information Base: carrying records around and communicating them to medical staff. However, there are many more future possibilities for the PIB device.

A PIB device could audibly announce which medicine has to be taken at preprogrammed times, and act as a medicine reminder. Medical compliance, ensuring that patients comply with their program of treatment, is a major obstacle to many treatment programs. In most cases where there is a failure to comply, the patient simply forgot to take their medicine. A portable device, which helped to ensure compliance, offers tangible medical benefits.

With the use of Bluetooth ads, patients passing by a Bluetooth-enabled billboard might download information on events happening in the hospital or any other services that are being offered such as taxis, counseling, and so on. This presupposes that the patient has some way of later viewing the information.

Identifying Barriers to Adoption

With new technology, there are often barriers that prevent adoption of systems. These barriers can mean the difference between the success or failure of an application in the market place. In the case of a medical system, cost, safety, user confidence and usability are all potential barriers to adoption. Issues of cost and safety were considered in our earlier discussions, but in this section, we'll look at user confidence and usability.

For user confidence, one of the biggest challenges for the PIB system is synchronizing the data so that losing the PIB device does not involve a loss of data. It is important for the PIB system to make sure that an authorized person is connected to the correct device, so that the correct information is exchanged with the correct patient, and without any worry of malicious eavesdropping.

Prevention of data loss is very important for user confidence. Data on paper can be seen and felt. Data in electronic format is intangible, and although back-ups may make it safer than paper, there are still issues of confidence which lead many users to feel more secure with paper storage. The system keeps data in two places at any one time so that a single failure in the system will not result in any loss of data; however, it is difficult to protect against double failures in the system. Data will only be synchronized at the central base and then distributed to update any remote changes. For the patient, the role of the PIB in data loss could easily be intimidating. What if you are carrying a device with all your medical information on it and you lose it? What if it should fall into the hands of somebody who would use the information maliciously? To reassure users, security and backup features should be easy to use and unobtrusive, but they also need to be explained well enough to reassure.

For busy medical staff, a system that is both complex to learn and use will not prove welcome. Therefore, to ensure a good user experience, existing interfaces and applications should be reused wherever possible. A new underlying communication system does not necessarily mean that completely new applications must be developed. The Bluetooth protocol stack has been designed to enable a Bluetooth system to fit in with legacy applications, and this should be done wherever possible. Not only does this make it easier for users, it may also make it easier for applications programmers!

For patients, usability translates to doing as little as possible. The device is set up by staff, and most interactions with the PIB are controlled by staff. A patient in a medical environment is already likely to be under stress: it is not the ideal time to start learning a complex user interface! We have shown how the interaction required from the patient can be kept to a minimum.

In designing any Bluetooth application, usability is a potential barrier to adoption that should be considered. Ideally, your application will work straight out of the box, with controls that are obvious to the uninitiated. It can be argued that if the user has to read a manual before using basic features, then your application has failed the usability test!

If you are replacing a legacy system (in this case paper records), you should consider what sort of system your application is replacing, and consider whether your application is as convenient and easy to use as what it replaces. If you are designing a completely new product, your system arguably has greater barriers to overcome, as the user must be convinced they want or need your product. If it is difficult to use, they may never find out how useful your product could have been!

Managing Personal Information Base Performance

The PIB device has many interfaces for communication and for interacting with it, but at the same time it must be extremely power-efficient. This means that the interfaces must only be active when they need to be. Ideally, a PIB device should be able to last for one week (with four hours use a day) before the battery needs to be replaced.

Battery life is very important if uninterrupted access to patient records is required. Each device could be cycled daily, meaning that the only requirement is that it has to run on batteries for a day. This is not a very stringent requirement for a battery-operated device. In comparison, the Bluetooth Human Interface Device profile suggests that a Bluetooth mouse should run for three months!

Bluetooth provides various low-power modes. These modes are most useful when devices wish to remain connected for long periods, but do not have much data to transfer. The PIB system usually establishes connections for short periods of time, exchanges data, then drops the connection. For this sort of usage model, low-power modes are irrelevant. However, if a PIB were used to collect data from a monitor, then it would be expected to remain connected for long periods of time. In this sort of usage scenario, using park or sniff mode would make sense. The PIB could then wake periodically, collect a data update from the monitor, and return to a low power sleep mode for the majority of the time. When collecting data in this fashion, it should be kept in mind that the PIB can have slightly stale data as there are gaps when its radio link is asleep, so data is not being updated.

The PIB must also maintain information from the central system—for example, collecting appointments, or details of test results which have been processed. The PIB could be set to wake every 30 minutes to connect with the nearest networked server and collect any information. This ensures that data is automatically transferred throughout the system.

The user could also request an update, perhaps by pressing a button on the PIB. In this case, it can take up to ten seconds to inquire and find the nearest networked server, and up to another ten seconds to connect with it, going through link and channel setup (this is a worst-case scenario; normally, a link can be established in two to three seconds). This may not sound like a long time, but it can *seem* like an extremely long delay, so it's likely that to convey appointment information quickly it will still end up getting scribbled down on paper and handed to the patient. The strength of the PIB system is not in its speed but in its automated backup facilities, and in the automated distribution and storage of information.

Summary

This case study has looked at a device that does not exist today, but that *can* be created with current technology. Already we are seeing PDAs being used to manage personal appointments as well as information on the move. It is a logical step for large institutions, such as hospitals, to begin to use similar technology to manage their information systems.

Bluetooth wireless technology suits the requirements of a Personal Information Base (PIB) for many reasons:

- The chips/chip sets and associated components are low cost.

- Bluetooth modules typically have a small form factor making them suitable for incorporation in handheld/mobile devices.

- Bluetooth wireless technology is low power, making it suitable for devices which need to run on batteries.

- The technology is available in a wide range of devices (PDAs, phones, laptops) providing a variety of candidates for Data Access Terminals.

- The ISM band used for Bluetooth radio links is available license-free worldwide.

While the PIB system is not safety-critical in itself, it does handle data that may be critical to medical treatment. The integrity and security of that data is paramount. Bluetooth links may introduce errors, but the application can easily compensate by backing up data, and by implementing application level error checks on records. Security of the radio link is also important. This is provided by authenticating communicating devices, and encrypting medical records on air. Finally, password access can protect the PIBs contents should the device itself fall into the wrong hands.

The Bluetooth specifications provide a variety of profiles that lay out rules for using the Bluetooth protocol stack for particular end-user applications. For a Personal Information Base, the Object Push Profile can be used to exchange virtual business cards (vCards), which publicly identify a PIB's owner. The File Transfer Profile can be used to exchange medical records.

The Object Push and File Transfer Profiles both rest on the Generic Object Exchange Profile, which uses the Infrared Data Association's OBEX protocol to exchange data objects. This, in turn, relies on the Serial Port Profile, which uses a modified version of the ETSI TS07.10 specification to emulate serial ports over a radio link (TS07.10 is also used by GSM cellular systems to emulate serial ports). Finally, the Generic Access Profile provides generic procedures related to discovering Bluetooth devices, security levels, and parameters accessible at the user interface.

By using Bluetooth profiles, the PIB application can use standard protocol stacks and features; this enables applications to be easily integrated with existing Bluetooth protocol stacks.

We have looked at a Personal Information Base in a medical context, but many of the elements of this case study are equally applicable to other data exchange applications. As input/output devices come down in price, we are likely to see devices such as the Personal Information Base described in this chapter appearing in more and more contexts.

Solutions Fast Track

Why Choose Bluetooth Technology

☑ The chip's physical size is small, and there are many chip vendors to choose from.

☑ The range is adequate—the lowest power version offers up to a 10 meter range, which is sufficient.

☑ The available choice of chip vendors leads to a competitive market.

☑ There is worldwide acceptance of the ISM band used by Bluetooth.

☑ A Bluetooth-enabled Personal Information Base (PIB) system in our hospital case study would store all patient information and information about visits, prescriptions, x-rays, and test information. It would be encrypted for both doctors and patients, have a user-friendly interface with low resolution screen; and would have a wireless connection to a main computer or Data Access Terminal.

☑ Data loss is avoided using automated backups. Automated backups are enabled by wireless communications.

☑ Encryption and passwords may be used to prevent unauthorized access to data.

☑ Use of radio devices may be restricted in some areas, so it should be possible to easily disable the Bluetooth transmitter.

Using Bluetooth Protocols to Implement a PIB

☑ For a Personal Information Base, the Object Push Profile can be used to exchange virtual business cards (vCards), which publicly identify a PIB's owner. The File Transfer Profile can be used to exchange medical records.

☑ The Object Push and File Transfer Profiles both rest on the Generic Object Exchange Profile, which uses the Infrared Data Association's OBEX protocol to exchange data objects. This, in turn, relies on the Serial Port Profile.

☑ By using Bluetooth profiles, the PIB application can employ standard protocol stacks and features. This enables applications to be easily integrated with existing Bluetooth protocol stacks.

Considering the User's View

☑ In designing any Bluetooth application, usability is a potential barrier to adoption that should be considered. Ideally your application will work straight out of the box, with controls that are obvious to the uninitiated.

☑ Do not redesign existing system interfaces if it is not necessary. Using legacy applications wherever possible can help to ease adoption of new technology.

☑ The PIB device has many interfaces for communication and for interacting with it, but at the same time it must be extremely power-efficient. This means that the interfaces must only be active when they need to be. Ideally, a PIB device should be able to last one week before the battery needs to be replaced.

Frequently Asked Questions

The following Frequently Asked Questions, answered by the authors of this book, are designed to both measure your understanding of the concepts presented in this chapter and to assist you with real-life implementation of these concepts. To have your questions about this chapter answered by the author, browse to **www.syngress.com/solutions** and click on the **"Ask the Author"** form.

Q: How do I know what profiles are appropriate for my application?

A: Each profile provides a profile overview which includes user scenarios. You need to read through the scenarios which the existing profiles offer and pick one which best matches your requirements.

Q: What do I do if there isn't a suitable profile?

A: The Bluetooth SIG will consider applications for new profiles. Contact the SIG via the Bluetooth Web site at www.Bluetooth.com for nonmembers, or www.Bluetooth.org for members.

Q: The PIB used a lot of profiles. Do I have to use profiles if I don't want to?

A: Yes. To get Bluetooth qualification, you must implement profiles which are relevant to the main function of your device. So, if you intend to emulate a serial port, you must use the serial port profile. Of course, there is nothing to stop you from adding extra functionality on top of what the profiles already provide.

Q: What extra considerations are there for medical devices?

A: In the case of the PIB: medical confidentiality and potential life-endangerment (if the medical data is corrupt). There may also be restrictions on using the ISM band in some hospitals, and in some areas of hospitals.

Q: Are there compatibility problems if you have different options on high-end and low-end devices?

A: No, as long as all devices implement a common basic set of functions.

Q: The PIB used Bluetooth PINs and Bluetooth security—how do I know if this will be enough for my application?

A: Bluetooth implements 128-bit security, which is the best currently available on wireless systems. Only you can decide if this is enough for your application. If you feel it isn't, then you are free to add extra security at the application level. For instance, many packages are available for encrypting data on Internet links. These could be reused to provide application level security on Bluetooth links.

Appendix

Bluetooth Application Developer's Guide Fast Track

This Appendix will provide you with a quick, yet comprehensive, review of the most important concepts covered in this book.

❖ Chapter 1 Introducing Bluetooth Applications

Why Throw Away Wires?

☑ You know Bluetooth technology is a good idea if your product satisfies the following six criteria:

1. Adds usability, convenience, or ease-of-use—the Bluetooth Dream!

2. Interference or latency will not affect its primary function.

3. Is tolerant to the connection time overhead.

4. Can afford the limited Bluetooth bandwidth.

5. Battery life or power supply requirements are compatible.

6. The range is adequate.

Considering Product Design

☑ Think about the following items:

- Are you adding end-user value by using Bluetooth technology?

- Does your product's development cycle allow you to add Bluetooth technology to it?

Investigating Product Performance

☑ To know whether Bluetooth technology is right for your product, you must consider:

- Connection times—it can take up to ten seconds to find a device and ten more seconds to connect

- The quality of service—throughput and latency; this will be lower than wired links

- Interference can badly slow down your links, or even cause them to fail

Assessing Required Features

☑ Question whether or not you need to support all the following features:

- Security—you must support it, but will you enable it by default?

- Low power modes—if your product doesn't need them, will it connect with one that does?

- Channel Quality Driven Data Rate—is maximum throughout in noisy conditions important?

Deciding How to Implement

☑ Should your stack be hosted, embedded with application on host, or fully embedded?

☑ Should you design your own PCB (cheap in volume), or buy in a module (faster and easier)?

☑ Battery—if your product is not mains-powered, consider the impact of time spent in different modes on the battery life. Constantly running in scan modes might give you fast connection time, but it will also rapidly drain your batteries. Setting short windows of activity can give almost equivalent performance, and greatly extend your battery life.

❖ Chapter 2 Exploring the Foundations of Bluetooth

Reviewing the Protocol Stack

☑ The protocol stack hides the complexity of the wireless interface and presents, at its highest level, a software interface that resembles that of a wired connection.

☑ Not all the differences between a wired and a wireless interface can be hidden. In particular, the steps required to find and connect to other devices are peculiar to wireless.

☑ Bluetooth devices can contain various combinations of upper stack layers to support various profiles. The Bluetooth specification details a service discovery layer so that devices can find out what services are available and how to connect to them.

Why Unconnected Devices Need to Talk

☑ With Bluetooth devices, the user may not initially know that there are other Bluetooth devices nearby, so a method is required to find them. The Bluetooth equivalent of plugging in a cable is the forming of a connection.

The checks on communications protocols and applications compatibility are actually done once a basic Bluetooth link is established, and are called service discovery.

☑   The procedure used to find devices is called *inquiry*, and the procedure used to connect to devices is called *paging*. In both cases, one device transmits and receives on special sequences of frequencies that are known to all devices. The other device needs to be listening for the transmissions—if a transmission is received correctly, it sends out a reply. Since it knows the sequences used for inquiry and paging, it can work out the correct frequency on which to send the reply.

Discovering Neighboring Devices

☑   Only devices in Inquiry Scan can be discovered.

☑   An inquiry is normally a periodic or user-initiated event.

☑   An inquiry response contains all the information required to connect to a device by paging.

Connecting to a Device

☑   Only devices in Page Scan can accept connections, although they may choose to reject incoming connection requests.

☑   If a page and connection request is successful, then the paging device becomes the master of the piconet and the paged device becomes the slave. An Asynchronous ConnectionLess (ACL) connection now exists between the two.

☑   A master can have connections to several slaves, but a slave can only have a connection to a master. For the upper stack layers, this is the only difference between the two.

Finding Information on Services a Device Offers

☑   The application is responsible for maintaining accurate records of the services it offers in a service database.

☑   An ACL and a Logical Link Control and Adaptation Protocol (L2CAP) connection must exist to a remote device before it can browse the service database using the Service Discovery Protocol (SDP).

☑ The service database contains all the information required for a remote device to identify and connect to local Bluetooth services.

Connecting to and Using Bluetooth Services

☑ A remote device must conduct an SDP query before connecting to a local Bluetooth service, and must support a complementary profile.

☑ Connecting to a service involves first opening L2CAP, then higher layer connections in turn, using the information from the SDP query.

☑ The procedure for using a service is detailed in the appropriate Bluetooth profile.

❖ Chapter 3 Power Management

Using Power Management: When and Why Is It Necessary?

☑ Consider whether your application is suitable for power-managed operation.

☑ Consider the constraints imposed by the application (e.g., maximum response times, characteristics of the data traffic, and so on).

Investigating Bluetooth Power Modes

☑ **Hold mode** One-off event, allowing a device to be placed into hold mode for a negotiated period of time. Hold interval must be negotiated each time this mode is entered.

☑ **Sniff mode** Slave periodically listens to the master and can power save for the remainder of the time. Important to note that data can be transferred while devices are in this mode and a SCO link may be active. Sniff intervals are negotiated once, before sniff is entered, and remain valid until sniff mode is exited.

☑ **Park mode** Parked slave periodically synchronizes with the master and for the remainder of the time can power save. Data packets cannot be sent on a parked connection and the devices must be unparked before a SCO connection can be established. Furthermore, there cannot be an active SCO when its associated ACL is parked.

Evaluating Consumption Levels

☑ All other things being equal, the power consumption of a Bluetooth low power mode depends on the parameters negotiated before that mode is entered.

☑ Page and inquiry scan also have a power consumption cost, so these should be entered only when necessary.

❖ Chapter 4 Security Management

Deciding When to Secure

☑ Secure for protection of data from eavesdroppers.

☑ Create exclusive links between devices.

Outfitting Your Security Toolbox

☑ Authentication verifies that the other Bluetooth device is the device you believe it is, using a link key as the secret password.

☑ Authorization grants permission to a device making a request to use a particular service.

☑ Encryption encodes data being passed between two devices; it requires successful authentication.

Understanding Security Architecture

☑ The Security Manager, which resides in the protocol stack, manages Mode 2 security transparently to the application.

☑ The Host Controller manages Mode 3 security if configured to do so by the application software.

☑ The Security database is configured by the application and specifies when to trigger Mode 2 security procedures as well as which security measures are to be taken.

☑ The device database offers persistent storage for parameters created during the successful completion of security and makes these available for future sessions to reduce security procedures required.

Working with Protocols and Security Interfaces

☑ Mode 2 security is invoked when a client application attempts to establish a connection with the server application and can use authentication, authorization, and/or encryption.

☑ Mode 3 security is triggered by the Host Controller when either an incoming or outgoing request for a radio connection is made. Authentication and/or encryption can be specified.

☑ Application Programming Interfaces support the configuration of the type of security to use and offer a way to insert user input (PIN entry) when required.

Exploring Other Routes to Extra Security

☑ Security measures are to be supported in many profiles, such that if another device wants to invoke a component of the security troika, it will be met with an appropriate response.

☑ In many instances, implementing security is not made mandatory since this is left up to the discretion of the system designer. What is made mandatory in many instances is supporting security as mentioned previously.

☑ Non-discoverable mode as configured into the Host Controller can prevent device detection during the Inquiry process.

☑ Non-accessibility can prevent any device from establishing a radio connection, thereby preventing access.

☑ Applications often have associated with them User IDs and passwords as further measures toward protecting information resident on a server. Authorization, the act of granting permission to a service, is another application-based security measure used by the OBEX transport layer.

❖ Chapter 5 Service Discovery

Introduction to Service Discovery

☑ The term *service discovery* is used to describe the way a networked device (or client) discovers available services on the network. Service discovery makes zero configuration networks possible—the user doesn't have to manually configure the network.

☑ Key features of a discovery protocol are: spontaneous discovery and configuration of network services, low (preferably zero) administrative requirements, automatic adaptation to the changing nature of the network (addition or removal of nodes or services), and interoperability across platforms.

☑ Bluetooth Service Discovery is protocol-dependent; it mandates the use of the underlying Bluetooth communication protocol as the basis for service discovery. However, Bluetooth SDP could indeed be implemented using other underlying transport mechanisms, and higher-level protocols (such as TCP/IP) may be run over Bluetooth.

Architecture of Bluetooth Service Discovery

☑ For a particular service (and there may be many services on one device) a *service record* contains a description of that service. The description takes the form of a sequence of *service attributes*, each one describing a piece of information about the service.

☑ Within the SDP server, each service record is uniquely identified by a *service record handle*. A *service class* defines the set of service attributes that a particular service record may have. In other words, a service record is a particular instance of a class of services.

☑ A service attribute is a name-value pair that includes an *attribute ID* and an *attribute value*. The attribute ID uniquely identifies the attribute within the scope of the service record.

☑ An attribute value can contain data of arbitrary complexity, rather than just simple types. This is accomplished using *data elements*. A data element is made up of a header and a data field.

☑ The Service Discovery Protocol includes a set of Protocol Data Units (PDUs) that contain the basic requests and responses needed to implement the functionality of Bluetooth Service Discovery. An SDP PDU contains a PDU ID, a transaction ID, and a parameter length in its header. Its body contains some number of additional parameters, depending on which type of transaction the PDU contains.

Discovering Services

☑ Every Bluetooth device can contain a Service Discovery Server (SDS) that advertises the services available on that particular device, be it a mobile phone, PDA, or something else. It can do this by making available the service records that describe those services.

☑ The Bluetooth-defined Class of Device (CoD) value can tell a discovering device if a connection should be opened to the discovered device—it doesn't have to open a connection to the SDS and check the Service Discovery Database (SDDB) of the discovered device, "short-circuiting" service discovery.

☑ The Bluetooth Service Discovery Protocol allows for services to be discovered on the basis of a series of attributes with values of type UUID. In reality, when talking about discovering specific services, one of the most important attributes of a service, if not *the* most important, is the *ServiceClassIDList*.

Service Discovery Application Profile

☑ The SDAP is a usage scenario describing the functionality a Service Discovery Application (SrvDscApp) should provide to an end user on a local device (LocDev) so that user can discover services on a Remote Device (RemDev). The SDAP doesn't specify an API that will provide this functionality, but suggests primitives that can be mapped to an API.

☑ Most profiles detailed in the Bluetooth specification have a service discovery component that specifies the structure and content of the service record that accompanies the service (or application) and which realizes the profile. The SDAP (in addition to dealing with application functionality for service discovery) specifies the procedures that an application realizing a profile must use to perform service discovery. If these procedures are upheld, interoperability is ensured.

Java, C, and SDP

☑ As part of Java Community Process (JCP), a set of standard Java APIs for Bluetooth is being developed and is due for publication at the end of 2001. Implementations of this standard will allow programmers to implement Bluetooth applications within the J2ME environment in a standard and portable way.

☑ A key element of the J2ME specification is the Generic Connection Framework (GCF), a mechanism that allows a programmer to create different types of networking connections through a standard Connector interface. This would allow programmers to quickly produce Java Bluetooth applications by applying existing techniques and design patterns.

Other Service Discovery Protocols

☑ The Bluetooth SDP may be integrated with a number of the other service discovery protocols, including Salutation, UPnP, Service Location Protocol (SLP), and Jini.

☑ The Salutation architecture defines a uniform way of labeling devices (fax machines, printers, copiers, and also phones, PDAs, and general electronic equipment) with descriptions of their capabilities and with a single, common method of sharing that information.

☑ Salutation is "transport independent," that is, a separate Transport Manager may be written for each underlying transport required, and the Salutation Manager, which provides the core functionality of the system, remains transport neutral.

☑ SLP is a language-independent protocol for automatic resource discovery on IP-based networks. Like some of the other service discovery protocols, it makes use of UDP/IP multicast functionality in TCP/IP. This makes it particularly useful for networks where there is some form of centralized administrative control, such as corporate and campus networks.

☑ Jini is a distributed service-oriented architecture, considered an extension of the Java language and platform. Services communicate with each other using a service protocol, which is defined as a set of interfaces in Java. The standard itself provides a base set of interfaces to facilitate core interaction between services. A key component of Jini is the *lookup* service.

☑ Communication between services in Jini occurs using Java Remote Method Invocation (RMI). RMI is a Java-based extension to traditional remote procedure call (RPC) mechanisms. One important extension is that it enables actual code, not just data, to be exchanged between services.

☑ Universal Plug and Play (UPnP) defines a set of lightweight, open, IP-based discovery protocols that allow appliances to exchange and replicate relevant data between themselves and the PCs on the network. UPnP is a "wire-only" protocol—it defines the format and meaning of what is transmitted between members of the network and says nothing about how the standard is actually implemented. It requires TCP/IP and HTTP to be present to operate.

☑ UPnP uses the Simple Service Discovery Protocol (SSDP) to discover services on IP-based networks. SSDP can be operated with or without a lookup or directory service in the network. SSDP operates on the top of the existing open standard protocols, using the HTTP over both Unicast UDP and Multicast UDP.

The Future of SDP

☑ SDP is one of many protocols that deal with the concept of service discovery. One of the key issues is interoperability of the various protocols.

☑ In the immediate future of SDP, the Bluetooth SIG is defining the Extended Service Discovery Protocol. This "new" protocol is expressed as a profile (dependent on the Generic Access Profile) and allows the Universal Plug and Play (UPnP) protocol suite to run over a Bluetooth stack. Though not proposed at present, a similar profile could be developed for the Jini service discovery protocol.

❖ Chapter 6 Linux Bluetooth Development

Assessing Linux Bluetooth Protocol Stacks

☑ The standard kernel source tree only recently accepted the Bluez Bluetooth stack, but it may not yet possess all the features some application developers require. It requires Linux 2.4.4 or greater.

☑ IBM's BlueDrekar is a nice-looking implementation distributed in binary form for x86 platforms running 2.2.x. Source is not freely available to the general public.

☑ The OpenBT project is a not-as-nice open source project that works for most things an embedded developer would want. Source is available and has been used on x86, ARM9, ARM7, MIPS, and PowerPCs.

Understanding the Linux Bluetooth Driver

☑ The OpenBT stack implements TTY drivers for RFCOMM, SDP, and stack control.

☑ The Bluetooth driver must be stacked over a lower-layer hardware driver that implements a TTY.

☑ Any legacy application that uses a TTY can use RFCOMM once another application sets up the underlying RFCOMM connection.

☑ SDP, connection setup, and stack control are accomplished with *ioctl* calls.

☑ No interface exists for SCO, or L2CAP, although *ioctls* are available to support most HCI commands.

Using Open Source Development Applications

☑ The OpenBT source tree comes with some applications: btd/btduser, sdp_server, and BluetoothPN.

☑ The difference between btd and btduser is that btd is meant to work with the kernel mode Bluetooth driver while btduser works with the user mode Bluetooth driver. Many people prefer btduser since it is less prone to lock up your system if things go badly. However, the OpenBT developers do not support it as well as btd.

☑ The sdp_server application provides you with an SDP database server daemon. Once you've installed the Bluetooth driver, you can start this daemon and it will automatically receive and respond to SDP queries from remote devices.

☑ This application provides a GUI that displays the SDP database on a remote device. It provides some examples of how to make SDP requests and process their results.

☑ The quickest, most useful way to establish and exploit a Bluetooth connection from Linux is to use the standard GNU network applications over PPP. And the easiest way to do that is with the btd application.

Connecting to a Bluetooth Device

☑ An application manager must set up the driver stack over the hardware TTY and initialize the Bluetooth driver. This can be any application; the OpenBT source tree does not provide a general stack manager.

☑ Client applications must obtain the Bluetooth Device address of the remote device and—for RFCOMM connections—the channel number of the remote service in order to establish a connection.

☑ Once a connection is established, any application can use the TTY associated with the connection for data transfer.

☑ The driver indicates a disconnection event with a hang-up of the associated TTY.

Controlling a Bluetooth Device

☑ Use *ioctl* calls to control the device and get information about device status.

☑ Use /proc/bt_status to get information about device status.

☑ A stack manager must be able to deal with link loss and system shutdown requests. It should provide an interface for users as well as other processes like power management to signal shutdown requests.

❖ Chapter 7 Embedding Bluetooth Applications

Understanding Embedded Systems

☑ Embedded systems commonly have many tasks running simultaneously. Since the processor can only run one line of code at a time, a scheduler swaps between tasks running a few instructions from each in turn.

☑ On BlueCore, your application task is called through an interpreter referred to as the Virtual Machine, which interprets a few of your instructions each time it is called. This interpreter means that even if you write code in an endless loop, the other tasks in the system will still get to run. The Virtual Machine's interpreter also stops you from accessing areas of memory which are needed for other tasks.

☑ Tasks communicate by sending messages to one another, using areas of memory which are set up as queues. The first message in the queue is the first out, so these are sometimes called FIFOs (First In First Out).

☑ Application software can interact with hardware using interrupts. There are two pins on BlueCore which will generate an interrupt when they change state. An application can register to be notified when these interrupts happen.

☑ When you close a switch, the contacts usually bounce off one another. This bouncing causes the switch to oscillate, making and breaking a connection. This means that if a switch (such as a pushbutton, or keypad) is connected to an interrupt line, you will get many interrupts as the switch closes. BlueLab provides debounce routines.

Getting Started

☑ To create embedded applications to run on CSR's BlueCore chip, you need BlueLab and a Casira. The Casira must be configured to run BCSP.

Running an Application under the Debugger

☑ The PC is connected to the Casira with a serial cable and an SPI cable.

☑ The Casira must be loaded with a null image containing an empty version of the Virtual Machine.

☑ Applications running under the debugger on the PC can then use facilities on the Casira, so they can access PIO pins and the BlueCore chip's radio while still having full PC debugging facilities.

Running an Application on BlueCore

☑ You must make a special firmware build linking your application with a Virtual Machine build to run your application on the Casira.

☑ Your application should be fully debugged before you build it for BlueCore, since on-chip debugging facilities are very limited.

☑ You can communicate with the Virtual Machine on BCSP Channel 13 using VM Spy.

Using the BlueLab Libraries

☑ A selection of libraries provide ANSII C support as well as access to the Bluetooth protocol stack, PIO pins, and various operating system facilities such as scheduling, timers, messaging, and so on.

Deploying Applications

☑ If you do not have RFCOMM in your build, you can upgrade devices in the field using the Device Firmware Upgrade (DFU) tools. Otherwise, you must program the flash using an interface similar to the SPI interface.

❖ Chapter 8 Using the Palm OS for Bluetooth Applications

What You Need to Get Started

☑ In order to begin using Bluetooth technology, you will need to have a Palm OS device with at least 4MB of memory that is running Palm OS version 4.0 or greater. Alternatively, you may wish to develop using the Palm OS Emulator, often the easiest and fastest way to create new application.

☑ In addition to a Palm 4.0 device, you will need to have the Bluetooth Support Package installed. The Bluetooth Support Package consists of several .prc files that work together. The latest version of the Bluetooth support .prc files, along with the Bluetooth header files and several pieces of example code, can be found in the Bluetooth area of the Palm Resource Pavilion at www.palmos.com/dev/tech/bluetooth.

☑ In addition, you will also want to have a copy of the Palm OS 4.0 SDK documentation, also available on the Palm, Inc. Web site.

Understanding Palm OS Profiles

☑ The Palm OS currently supports five Bluetooth profiles defined in the Bluetooth 1.1 Specification: the Generic Access Profile, the Serial Port Profile, the Dial-up Networking Profile, the LAN Access Profile, and the Object Push Profile.

☑ Generic Access Profile (GAP) is a general look at the overall process of carrying out a Bluetooth transaction without regard to the nature of that transaction, and is background for all the other profiles.

☑ The new virtual serial driver (VDRV) in the Bluetooth Support Package provides support for the Serial Port Profile.

☑ The Network Library (NetLib) supports the Data Terminal role of both the *Dial-up Networking* and *LAN Access Profiles*.

☑ The new Bluetooth Exchange Library implements the Object Push Profile, much in the same way that the Exchange Manager supports IR-based Object Exchange Protocol (OBEX) push.

☑ If none of the profiles cover what you are trying to do, don't despair—the Palm OS also provides a robust API that allows you direct access to the SDP, RFCOMM, and Logical Link and Control Adaptation Protocol (L2CAP) layers of the Bluetooth stack, along with calls to allow you to manage the Bluetooth-specific concerns like discovery and piconet creation.

Updating Palm OS Applications Using the Bluetooth Virtual Serial Driver

☑ Using the Bluetooth Virtual Serial Driver allows existing serial-based applications to quickly be updated to take advantage of Bluetooth technology. The VDRV itself is "glue code" that allows Bluetooth functionality to be accessed though a more traditional API. Using the VDRV also gives you an advantage in writing multi-transport applications.

☑ Virtual Serial Drivers in the Palm OS are individual .prc files of type *vdrv* and are used throughout the new Serial Manager interface, much the same way as traditional physical serial ports are used.

☑ Since most Bluetooth radios are not capable of simultaneously listening for an inbound connection and trying to create an outbound connection, an instance of the Bluetooth VDRV also needs to know whether it is initiating or accepting the connection. Since a traditional serial API does not present a mechanism for passing all of this extra information, Palm OS 4.0 has added a new call, *SrmExtOpen()* (found in SerialMgr.h), to the new Serial Manager API.

☑ A VDRV client-only application might be useful when you know that the Palm device will always be playing a client-based role, and therefore never need to accept a connection.

☑ Applications and the VDRV use the Bluetooth Library in different modes. Because of this difference, the VDRV will not be able to open while the application is holding the Bluetooth stack open.

☑ Setting up the serial port as a server does not cause the driver to go out and create an ACL or RFCOMM connection, it merely sets up the port as a listener. Like a normal serial port, the VDRV will not alert the application when an incoming connection is established, the application will simply begin to receive data from the port.

Using Bluetooth Technology with Exchange Manager

☑ You can make an Exchange Manager-based application Bluetooth-aware with just a few lines of code. The Bluetooth Exchange Library registers itself for the *exgSendScheme*, so if you've already updated your application to take advantage of the *exgSendScheme*, it should work with Bluetooth technology as soon as you have installed the Bluetooth .prc files.

☑ The Exchange Library allows applications to send data blocks without having to worry too much about the underlying transport.

☑ The VDRV and Exchange Manager simplify using Bluetooth technology by encapsulating it inside familiar and easy to use interfaces, but the simplification also hides functionality and increases overhead.

Creating Bluetooth-Aware Palm OS Applications

☑ If your application requires direct access to Bluetooth protocol layers or management functions, then you will need to make use of the Bluetooth Library (BtLib) API.

☑ Even when using the Bluetooth Library directly, a Palm OS application cannot put the Palm device or the remote device into park, hold, or sniff modes. Also, while an application can request that a given link be authenticated or encrypted, for security reasons the application is not allowed to specify the authentication passkey or insist that a device be added to a list of trusted (or bonded) devices.

☑ The Bluetooth Library API is fairly large, and can generally be divided into six sections: Common Library calls, management calls, socket calls, SDP calls, services calls, and security calls.

☑ If your application is going to receive *inbound* connections, you should check to make sure the radio's accessibility mode has been set to allow connection and (if desired) discovery. The accessible state of the device is determined by the user's settings in the Bluetooth Preferences Panel.

☑ If you plan to have your application create *outbound* Bluetooth connections, you will probably want to perform a device discovery in order to allow the user to select the remote device(s) with which she wished to create a connection. The Bluetooth Library offers two similar calls that handle the entire discovery experience, including inquiry, name retrieval, and user selection, *BtLibDiscoverSingleDevice()* and *BtLibDiscoverMultipleDevices()*.

☑ Bluetooth piconets have a star formation: one master connected to up to seven active slaves. Once a successful call *BtLibPiconetCreate()* call has been made, up to seven simultaneous ACL connections can be established. Depending upon the usage model for your application, you may wish to have the piconet master actively create outbound connections, wait for inbound connections from remote devices, or both.

☑ The L2CAP and RFCOMM protocol layers are exposed in the Bluetooth API through a sockets-based interface. The ability to create and receive RFCOMM and L2CAP connections is entirely independent of the device's role in a piconet.

☑ Applications or protocols that run on top of L2CAP must be able to handle the flow control themselves, while applications that run on top of RFCOMM can make use of its built-in flow control. Also, an RFCOMM listener is only capable of supporting one connection at a time, while a L2CAP listener can receive an unlimited number of connections. If your application involves functionality covered by a Bluetooth profile, you will not have to make a choice of which layer to use, as the profiles provide guidance on how to use the Bluetooth protocol stack.

☑ L2CAP identifies available listeners by a Protocol Service Multiplexor (PSM), which can be thought of as similar to an IP port. The RFCOMM protocol uses a simple enumeration called a Server ID to distinguish its listeners. You can let remote applications know which PSM and Server ID to connect to by advertising them with SDP.

☑ The Bluetooth Library offers an extensive set of APIs for working with SDP.

Writing Persistent Bluetooth Services for Palm OS

☑ The Palm OS allows services to run on an as-needed basis by implementing the OBEX service in the IR implementation. While the client side of OBEX starts up in response to a user action (the "beam" command), the service side of OBEX is brought up by the OS when an inbound IR connection is detected. Palm OS's IR service implementation is able to avoid the overhead of the OBEX service and IR stack when they are not in use.

☑ Although multiple services can be registered, once a given service begins a session, the other services become unavailable until it completes its session.

☑ Services are simply pieces of code that register for and respond to Bluetooth service notifications, normal Service Manager notifications of type *BtLibServiceNotifyType* (btsv). When the application is launched in the normal manner, it displays controls that allow the user to enable and disable the service, which can correspond to registering and unregistering for the Bluetooth service notification.

The Future of Palm OS Bluetooth Support

☑ In the near future, Bluetooth technology will address the issues of Layer 3 (Network level) support in the Bluetooth communication protocol stack. New specifications will define a network layer for communications between all the members of a piconet (not just master to slave), as well as inter-piconet communication issues.

☑ Roaming and scatternets will also be addressed.

☑ The eventual goal is the creation of true ad-hoc networks, self-configuring network groupings that grow and change as the user's environment changes.

☑ New editions of the Palm OS Bluetooth Library will expand the Palm OS's Bluetooth capabilities without compromising existing applications.

❖ Chapter 9 Designing an Audio Application

Choosing a Codec

☑ Codecs (coder/decoders) convert between analog voice samples and the compressed digital format.

☑ The output of the Codecs must be fed into the Bluetooth baseband as a direct input to the baseband (a technique commonly used in Bluetooth chips), or encapsulated in a Host Controller Interface (HCI) packet and fed across the Host Controller Interface.

☑ Bluetooth technology uses CVSD and PCM Codecs. CVSD is more robust in the presence of errors, which is what makes CVSD attractive for use in Blue-tooth systems. PCM is cheap and already available in many commercial devices.

☑ There are two types of compression implemented in PCM Codecs: A-law and μ-law. The different types are used by phones in various geographical regions.

Configuring Voice Links

☑ The Bluetooth system transmits data on ACL links and voice on SCO links. SCO links use periodically reserved slots, while ACL links do not reserve slots.

☑ Live audio needs circuit switched channels to guarantee regular delivery of voice information—the receive Codecs need a regular feed of information to provide a good quality output signal. The circuit switched channels are the Synchronous Connection-Oriented links. They occupy fixed slots that are assigned by the master when the link is first set up.

☑ Always remember that Bluetooth technology maintains a maximum of 3 ★ 64 Kbps full-duplex SCO voice packets. The SCO links provide voice quality similar to a mobile phone; if higher audio quality is desired, then compressed audio must be sent across ACL links.

☑ Notice that we don't want to modify the voice packets at the L2CAP layer. SCO packets bypass the L2CAP layer.

☑ If you choose to send data at the same time as voice, you will also lose out on error protection on the voice links.

☑ When a link is to be established, use the following procedure: scan or page for an audio device. Use SDP to identify service. Set up ACL connection first for control, then set up SCO connection. During a voice connection, control messages can be sent such as DTMF signals

Choosing an Audio Interface

☑ There are two routes for audio: either a direct link between the baseband and the application layer, or through the HCI. All packets passing through HCI experience some latency.

☑ If the Universal Asynchronous Receiver Transmitter (UART) HCI transport is used, there is no way to separately flow control voice and data, so when data transport is flow controlled, the flow of voice packets across the HCI will also stop. The USB transport provides a separate channel for voice packets; however, USB requires complex drivers.

☑ Not every chip/chip set supports audio. Of those that do, most provide direct access to the baseband, but some do not support audio across HCI.

Selecting an Audio Profile

☑ Three different profiles cover audio applications: the Headset profile, the Cordless Telephony profile, and the Intercom profile. If your product supports several services, it may be appropriate to implement more than one profile. If your application is not covered by one of the profiles, you will have to design a complete proprietary application yourself.

☑ The Headset profile allows the audio signal from a telephone call to be transferred between an audio gateway (AG) and a headset. If you just want to transfer the audio part of a call without control information, then the Headset profile is small, simple, and definitely the one to use.

☑ The Cordless Telephony profile allows incoming calls to be transferred from a base-station to a telephone handset. If you are implementing a base station to pass voice calls to and from a telephone network, then you should use the Cordless Telephony profile.

☑ The Intercom profile allows telephone calls to be transferred across a Bluetooth link without involving a telephone network at all. If you need to initiate voice calls to other Bluetooth devices in the area, but are not passing them on to a network, then you should use the intercom profile.

☑ The Cordless Telephony and Intercom profiles both use Telephony Control Protocol (TCS) commands for control and share the same disconnection procedure. The Headset profile controls the link with AT commands, and does not provide any commands for the headset to terminate the connection.

Writing Audio Applications

☑ In this section, we looked in detail at how a particular profile could be implemented at application level. All inquiry, paging, scanning, and service discovery are the same no matter which profile you implement. Similarly, the audio must be routed into the Bluetooth subsystem somehow, regardless of the audio profile chosen.

☑ The first step will be finding suitable devices in your neighborhood using the Bluetooth Device Discovery procedures.

☑ Once the audio gateway application has found a device that belongs to the audio/headset class of devices, it needs to find out how to connect to the headset service. To do this, it uses Service Discovery Protocol (SDP) and performs a service search for the headset service.

☑ Once the service discovery phase is complete, you can connect to an audio service. The first step is to set up an ACL link. This connection is used to create an L2CAP link using the PSM value for RFCOMM. Next, an RFCOMM channel is set up to control the headset. Once the audio gateway knows that the headset is willing to accept the call, it establishes an audio (SCO) link. The headset must be able to accept all Codecs and all packet types on the link.

Differentiating Your Audio Application

☑ Be sure to consider the weight, size, and form factor in your product design.

☑ The user interface is the most crucial aspect of your application. Ask yourself if there are ways to hide the complexity of Bluetooth technology. Button functions and headset designs offer opportunities for improvement and differentiation.

☑ Another way to differentiate your product is to provide ongoing support for new features or for future versions of the Bluetooth specification.

☑ Improving design and engineering to better the audio path can have a noticeable impact for the user, helping to avoid audio feedback, acoustic coupling, and resonance effects.

❖ Chapter 10 Personal Information Base Case Study

Why Choose Bluetooth Technology

☑ The chip's physical size is small, and there are many chip vendors to choose from.

☑ The range is adequate—the lowest power version offers up to a 10 meter range, which is sufficient.

☑ The available choice of chip vendors leads to a competitive market.

☑ There is worldwide acceptance of the ISM band used by Bluetooth.

☑ A Bluetooth-enabled Personal Information Base (PIB) system in our hospital case study would store all patient information and information about visits, prescriptions, x-rays, and test information. It would be encrypted for both doctors and patients, have a user-friendly interface with low resolution screen; and would have a wireless connection to a main computer or Data Access Terminal.

☑ Data loss is avoided using automated backups. Automated backups are enabled by wireless communications.

☑ Encryption and passwords may be used to prevent unauthorized access to data.

☑ Use of radio devices may be restricted in some areas, so it should be possible to easily disable the Bluetooth transmitter.

Using Bluetooth Protocols to Implement a PIB

☑ For a Personal Information Base, the Object Push Profile can be used to exchange virtual business cards (vCards), which publicly identify a PIB's owner. The File Transfer Profile can be used to exchange medical records.

☑ The Object Push and File Transfer Profiles both rest on the Generic Object Exchange Profile, which uses the Infrared Data Association's OBEX protocol to exchange data objects. This, in turn, relies on the Serial Port Profile.

☑ By using Bluetooth profiles, the PIB application can employ standard protocol stacks and features. This enables applications to be easily integrated with existing Bluetooth protocol stacks.

Considering the User's View

☑ In designing any Bluetooth application, usability is a potential barrier to adoption that should be considered. Ideally your application will work straight out of the box, with controls that are obvious to the uninitiated.

☑ Do not redesign existing system interfaces if it is not necessary. Using legacy applications wherever possible can help to ease adoption of new technology.

☑ The PIB device has many interfaces for communication and for interacting with it, but at the same time it must be extremely power-efficient. This means that the interfaces must only be active when they need to be. Ideally, a PIB device should be able to last one week before the battery needs to be replaced.

Glossary

Term/Acronym	Expanded Acronym	Definition
ACL	Asynchronous ConnectionLess	A low-level Bluetooth data connection.
ADC	Analog to Digital Converter	Hardware used to convert analog signals (such as voice) into a digital format.
AG	Audio Gateway	A device that takes audio (for instance from a telephone call), and sends it across a Bluetooth link. For example, when a cellular phone is connected to a Bluetooth headset, the cellular phone is acting as an audio gateway.
API	Application Programmers Interface	A software interface designed specifically for application programmers. APIs aim to present features in easy-to-use ways.
ARM	Advanced RISC Machines	A Cambridge, UK-based company that manufactures a powerful range of processors. These have proved popular for embedding in Bluetooth chips.
Authentication		A procedure whereby one Bluetooth device checks that the link keys on another Bluetooth device match its own link keys.
BD_ADDR	Bluetooth Device Address	A unique address allocated to every Bluetooth device on manufacture.
Bonding		A process where two Bluetooth devices which share a secret PIN code connect, generate a link key (which can later be used for authentication and encryption), then disconnect.
CID	Channel Identifier	A number used by L2CAP to identify a logical channel.
Codec	Coder-Decoder	A hardware subsystem that converts audio samples into a compressed data stream.
CVSD	Continuous Variable Slope Delta Modulation	An error tolerant Codec used in Bluetooth audio systems.
DA	Directory Agent	An agent that accumulates service information and forms a repository of that information.
DAC	Digital to Analog Converter	Hardware used to convert digital signals into an analog format (such as voice).
DLCI	Data Link Connection Identifier	A number identifying one of the emulated serial ports carried on an RFCOMM connection.
Encryption		A procedure whereby link keys are used to generate encryption keys, and the encryption keys are then used to encode data so it cannot be read by anyone who does not know the keys.
GAP	Generic Access Profile	A profile that provides the basic operation rules for all Bluetooth devices. For instance, it defines the timing rules for inquiry and paging.

Term/Acronym	Expanded Acronym	Definition
GPL	Gnu General Public License	A free license attached to much open source code.
HCI	Host Controller Interface	An interface that allows a Bluetooth host to communicate with a Bluetooth device. Various transport layers are possible: UART, USB, and RS232.
HID	Human Interface Device	A device used to interface between a human and a computer (for instance, a mouse, keyboard, joystick, or tracker ball).
Hold mode		A device in hold mode is temporarily inactive until a hold timer expires. A master might use hold mode to allow slaves to save power if it knows it will not communicate with them for a while—for example, when it is connecting to a new slave.
IETF	Internet Engineering Task Force	The body that defines Internet specifications.
Ioctl	Input output control	An interface for controlling data transfer. There are a set of standard *ioctl* control calls used in UNIX and Linux.
IP	Intellectual Property	Designs, patents, and so forth, which are intangible but have an owner.
IP	Internet Protocol	The higher layer networking protocol that runs on Internet connections. Layered on TCP (Transmission Control Protocol) for reliable communications, or UDP (User Datagram Protocol) for unreliable communications.
IPC	InterProcess Communications	Usually First In First Out (FIFO) queues carrying messages between processes.
IR	Infrared	Infrared light is used for optical communications as another alternative to cabled connections.
IrDA	Infrared Data Association	An association which defines specifications for OBEX, vCal, vCard, and so on.
L2CAP	Logical Link Control and Adaptation Protocol	The part of a Bluetooth Protocol Stack that multiplexes several higher layer logical links onto one underlying physical link. L2CAP also provides segmentation and reassembly to adapt large higher layer packets onto the smaller packet sizes handled by HCI and the lower layers.
LAP	LAN Access Point	Bluetooth-enabled device for accessing a LAN, which supports the LAN Access Profile.

Term/Acronym	Expanded Acronym	Definition
LAP	Lower Address Part	Part of a Bluetooth device address, or other Bluetooth access code (such as an inquiry access code).
Ldisc	Line Discipline	Line Discipline controls the format and rules you use when reading input from a terminal (TTY) line. Examples of line disciplines include: *raw*, *cbreak*, *select()*, *ioctl()*.
Link Key		Numbers up to 128 bits long which are used in Bluetooth security procedures.
Link Manager		The layer that establishes and configures Bluetooth connections. The Link Manager is usually implemented on a Bluetooth chip.
LMP	Link Management Protocol	The protocol that two Link Managers use to communicate when they are setting up and configuring Bluetooth connections.
MMI	Man Machine Interface	Input and output devices used by a human to interface to a machine. For example, a keypad and a display could make up an MMI.
MOS	Mean Opinion Score	A testing method used to assess audio quality—because this is such a subjective quantity, it cannot be measured by instrumentation, so users are surveyed and asked to score the quality of signals. The opinion scores of many users are taken and the mean average is used to provide the MOS.
NetLib	Network Library	The Palm OS library, which supports networking functions.
OBEX	Object Exchange	IrDA protocol which allows exchange of data objects, as well as providing facilities for specifying directories, and creating and deleting objects and folders.
Park mode		A device in park mode has given up the active member address that identifies it as part of a piconet. It is inactive except for occasional beacon slots when it wakes up to listen for unpark messages that can be used to reactivate it. Parked devices are allocated special access window slots in which they can request the master to reactivate them by unparking.
Pairing		A process whereby two Bluetooth devices generate a link key that can be used later for authentication and encryption. For devices to pair successfully, they must have matching PIN codes.
PCM	Pulse Code Modulation	A type of Codec used in Bluetooth, and also used in cellular phones.

Term/Acronym	Expanded Acronym	Definition
PDA	Personal Digital Assistant	A small handheld computing device such as a Palm device.
PDU	Protocol Data Unit	A single package of information carrying a message in a format specific to one protocol layer. PDUs are used for peer-to-peer communication between local and remote protocol entities. For instance, SDP client and server communicate with SDP PDUs.
Piconet		A network of Bluetooth devices consisting of a master device and one to seven slave devices.
PIN	Personal Identification Number	A number used for security procedures to verify that the user is authorized to use a system.
PPP	Point-to-Point Protocol	An Internet protocol used for transporting datagrams across point-to-point links.
PRC	Palm Resource	A file containing a set of resources for a Palm OS software module.
Profile		A set of instructions for how to use a protocol stack to implement an end-user service. For example, the Bluetooth Headset profile describes how to use the Bluetooth protocol stack in a headset.
PSM	Protocol Service Multiplexor	A number used by L2CAP to identify which protocol or service is connected to a channel.
PSTN	Public Switched Telephone Network	The networks provided by telephone service providers to carry subscriber's telephone calls.
RFCOMM	Radio Frequency COMMunications port	The serial port emulation layer of the Bluetooth protocol stack.
SA	Service Agent	An agent that advertises information about a service on behalf of the service.
SAFER+	Secure And Fast Encryption Routine Plus	The algorithm used by Bluetooth devices to generate link keys used for authentication and encryption.
SCO	Synchronous Connection Oriented	A low-level Bluetooth duplex voice connection. To set up a SCO connection, you must first set up an ACL (data) connection.
SDAP	Service Discovery Application Profile	A profile that gives rules for using Service Discovery Protocol in an application.
SDP	Service Discovery Protocol	A peer-to-peer protocol that allows a client Bluetooth device to ask a server device whether it supports a service, or to browse through a list of services. SDP can also be used to retrieve information on how to connect to a service.

Term/Acronym	Expanded Acronym	Definition
SIG	Special Interest Group	A group that shares a common interest, and joins together to pursue goals related to that interest. The Bluetooth SIG shares an interest in Bluetooth.
Sniff mode		A low-power mode where a device only wakes up to listen for data in periodic sniff slots.
TCS-Bin	Telephony Control Protocol Binary	A specification based on ITU-T Q.931, which allows telephone calls to be transferred across Bluetooth links.
TTY	TeleTYpe	The abbreviation that originally referred to tele-types connected to mainframes. It is now used for data terminals. (Linux has a TTY command that prints the filename of the terminal connected to standard input.)
UA	User Agent	An agent that performs service discovery tasks for a client.
UART	Universal Asynchronous Receiver Transmitter	A device that supports transfer of data in a serial bit stream.
UI	User Interface	Also called MMI (Man Machine Interface). Graphical User Interface (GUI) and Command Line Interface (CLI) are both types of user interfaces.
UPnP	Universal Plug and Play	A system that allows wireless devices to find one another, advertise services, and exchange status monitoring and control information.
USB	Universal Serial Bus	A high-speed standard for data connections, which allows many devices to be connected to one hub device. USB is often used on PCs.
UUID	Universally Unique Identifiers	128-bit numbers that are guaranteed to be unique across all space and time (or at least until A.D. 3400).
VDRV	Virtual Serial Driver	A Palm OS driver that provides virtual serial ports.

Index

S